Biocomposites in 3D and 4D Printing

Scrivener Publishing
100 Cummings Center, Suite 541J
Beverly, MA 01915-6106

Publishers at Scrivener
Martin Scrivener (martin@scrivenerpublishing.com)
Phillip Carmical (pcarmical@scrivenerpublishing.com)

Biocomposites in 3D and 4D Printing

Materials, Techniques, and Applications in Regenerative Medicine

Edited by

Rishabha Malviya

Department of Pharmacy, School of Medical and Allied Sciences
Galgotias University, Greater Noida, Uttar Pradesh, India

and

Mukesh Roy

Office of Clinical Research, MCI Baptist Health South Florida, Miami, USA

Scrivener Publishing

WILEY

This edition first published 2026 by John Wiley & Sons, Inc., 111 River Street, Hoboken, NJ 07030, USA and Scrivener Publishing LLC, 100 Cummings Center, Suite 541J, Beverly, MA 01915, USA

For more information about Scrivener publications please visit www.scrivenerpublishing.com.

Wiley Global Headquarters
111 River Street, Hoboken, NJ 07030, USA

For details of our global editorial offices, customer services, and more information about Wiley products visit us at www.wiley.com.

The manufacturer's authorized representative according to the EU General Product Safety Regulation is Wiley-VCH GmbH, Boschstr. 12, 69469 Weinheim, Germany, e-mail: Product_Safety@wiley.com.

Library of Congress Cataloging-in-Publication Data

ISBN 978-1-394-44036-8

Cover image: Generated with AI using Adobe Firefly
Cover design by Russell Richardson

Set in size of 11pt and Minion Pro by Manila Typesetting Company, Makati, Philippines

Printed in the USA

10 9 8 7 6 5 4 3 2 1

Contents

Foreword

Technological developments, such as 3D and 4D printing technologies, contribute to an evolutionary change in biomedical innovation. These technologies provide unparalleled prospects for designing and fabricating complex, patient-specific biomedical composites that are precise, functional, and biocompatible. Biocomposites are transforming tissue engineering, regenerative medicine, and personalised healthcare. Integrating biocomposites into 3D and 4D printing represents a significant step forward in the development of living, adaptable, and functional constructions capable of healing, replacing, or enhancing human tissues. 3D printing enables complex geometries and biomimetic structures, whereas 4D printing develops designs that can respond to environmental stimuli, alter shape, and grow over time.

This book is an updated and informative compilation that captures the core of this rapidly evolving field. Its structure is both logical and insightful, taking readers from fundamental concepts to specialised applications and, finally, to the regulatory and future-oriented issues that will influence the use of these technologies in clinical practice.

The first section introduces the reader to the fundamentals of 3D and 4D printing, as well as the range of materials used in biocomposite manufacturing. This sets the basis for the second segment, which explores specialized applications such as cardiovascular and orthopaedic therapies, as well as the development of dynamic scaffolds. The inclusion of biocompatibility, immunogenicity, and commercialisation ensures that the discussion is grounded in the reality of medical translation. The third section provides a forward-looking view on the field's ethical, regulatory, and personalised medicine dimensions, emphasising the importance of balancing innovation and accountability.

This is not just an educational work; it is an interdisciplinary discussion involving engineering, medicine, and materials science. It will be a significant resource for researchers, clinical experts, and industry experts working to turn the possibilities of biocomposite-based 3D and 4D printing into practical medical solutions. The contributions in this volume represent a shared vision for the future, one in which adaptable implants and altered tissues are not experimental innovations but rather fundamental components of routine medical treatment. It's a world in which medicine is fully personalised, healing is expedited by design, and the lines between biology and engineering are seamlessly merged.

I congratulate Dr. Rishabha Malviya for compiling such an esteemed group of contributors and curating a work that not only educates but inspires. As we approach a new era in regenerative medicine, this book will serve as a guiding light for present practitioners and a spark for future advances.

Happy Reading.

Dr. Dhruv Galgotia
CEO, Galgotias University,
Greater Noida, India

Preface

The emergence of 3D and 4D printing technology has transformed biomedical engineering by making it possible to create dynamic tissue constructs, implants tailored to the individual patient, and functional biomaterials that work with human physiology. The combination of natural and synthetic materials, known as biocomposites, has become a potential approach for constructing structures that are both biologically active and mechanically strong. The goal of this book is to give readers a thorough understanding of biocomposites and their applications. It offers a comprehensive outlook that incorporates biomedical applications, fabrication methods, material science, and regulatory considerations.

The book is divided into three sections: Introduction to 3D and 4D Printing, Specialised Applications, and Regulatory Considerations and Future Perspectives. The first section highlights the significance of choosing the right biocomposite materials and optimising manufacturing parameters while introducing readers to the foundational concepts, fabrication techniques, and material innovations at the foundation of 3D and 4D bioprinting. The second section explores the practical applications of these technologies in regenerative medicine, including biocomposite-based approaches for bone regeneration, cartilage and soft tissue repair, vascular and cardiac tissue engineering, and the fabrication of dynamic, stimuli-responsive tissues via 4D printing. The importance of biocomposites in personalised medicine, commercialisation routes, and ethical and legal frameworks is covered in the third section.

With its useful insights for scientists, material engineers, clinicians, and industry experts, the book stimulates more creativity and multidisciplinary cooperation. It offers a balanced resource for both academic and professional audiences by fusing theoretical underpinnings with discussions that are driven by applications. We extend our gratitude to Martin Scrivener and the team at Scrivener Publishing for their effort and guidance in publishing this work.

The Editors
July 2026

Part 1

INTRODUCTION TO 3D AND 4D PRINTING

1

3D and 4D Printing Technologies: An Overview

Diptee Gupta*, **Swatantra K.S. Kushwaha and Ashwani Kumar Chaturvedi**

Department of Pharmaceutics, Krishna Institute of Pharmacy and Sciences, Kanpur, Uttar Pradesh, India

Abstract

Additive manufacturing, or 3D printing (3DP), has been widely studied over the past decade for its potential to create advanced materials and structures. It gained popularity in industries and academic institutions due to its speed, accurate prediction, and lower cost. 3D printing combines various scientific fields like materials science, engineering, mechanics, and data processing. The newest idea in the pharmaceutical and biomedical industries, 4-dimensional printing (4DP), allows printed objects to change their morphology by using time as the fourth dimension. It has a huge promise for dose customization and medicine creation. 4D printing involves creating 3D objects that can change shape over time or in response to environmental stimuli. This represents a major innovation in additive manufacturing, providing a streamlined transition from concept to reality with functionality built directly into the materials. With 4D printing, it is possible to produce a variety of active, programmable materials that can transform from one shape to another. The biomedical fields of biosensing, immunotherapy, drug delivery, tissue engineering and regeneration, implants, and medical devices all make extensive use of biopolymers. Bio-ink for 3D and 4D printing may be made from a variety of biodegradable and non-biodegradable polymeric components. This article explores the uses of 3DP and 4DP in biomedicine and healthcare, including tissue engineering, artificial organs, medication delivery, and pharmaceuticals. 4DP has several potential uses in medical treatments and pharmaceuticals, but its widespread usage depends on future improvements and study. In several areas, including the medical and healthcare sectors, earlier advancements with smart

**Corresponding author*: dipteegupta4166@gmail.com; ORCID: 0009-0008-0118-9956
Swatantra K.S. Kushwaha: ORCID: 0000-0002-1004-6323
Ashwani Kumar Chaturvedi: ORCID: 0000-0003-2694-2840

Rishabha Malviya and Mukesh Roy (eds.) Biocomposites in 3D and 4D Printing: Materials, Techniques, and Applications in Regenerative Medicine, (3–38) © 2026 Scrivener Publishing LLC

materials are therefore expected to serve as innovators of 4DP, which can provide a more comprehensive viewpoint.

***Keywords*:** 3D printing, biopolymers, additive manufacturing, bio-ink, 4D printing

1.1 Introduction

Research in drug delivery aims to overcome challenges like improving drug effectiveness, ensuring patients take their medication as prescribed, and customizing treatments for individuals. This involves discovering new drugs, creating better formulations, and using new manufacturing methods. Recently, there has been a focus on new manufacturing techniques, such as 3D printing, to explore their potential applications in the pharmaceutical field [1].

Additive manufacturing (AM), often known as 3D printing (3DP), has been thoroughly studied over the past decade for its potential to create advanced materials and structures. It gained popularity in organizations and educational institutions because of its swiftness, accurate prediction, and lower cost.

3D printing combines various scientific fields like materials science, engineering, mechanics, and data processing. Complex structures can be built using 3D printing techniques that is not created by conventional methods. Notable discoveries with 3D printing include remotely operated robots, bio-inspired designs, microbial environments, tissues, and medicine delivery systems [2].

Three-dimensional (3D) printing is a revolutionary innovation that uses the consequent layer-after-layer material adhesion like plastic, metal, ceramics, or even living cells to create 3D physical items from digital models. Complex, adaptable structures can be produced using 3D printing from a 3D computer-aided design file (CAD). Charles W. Hull originally introduced the idea of 3D printing in the 1980s when he used stereolithography to create polymer items. Carl Deckard invented Selective Laser Sintering (SLS), a laser-based technique for hardening powdered material.

According to ISO/ASTM 52921:2013, the various 3D printing or AM techniques have been described in a number of processes. These incorporate primarily material extrusion, material jetting, binder jetting, powder bed fusion, direct energy deposition, vat photo-polymerization, and sheet lamination procedures. Depending on the ink's composition and the printing procedure, substances are constructed layer by layer using an extruder, chemical agents (like binders), or a laser (sintering/melting technology).

The field of biomedical engineering, which includes human health, has been stated to have the most potential uses of 3DP. AM has been effectively used in dental implants and tailored prostheses, and it has been utilized as a technique for bioengineering tissue with a wide range of compositions, from the framework of arteries and organs to bone and teeth. 3DP technology in tissue engineering includes gathering accurate tissue and organ data for model creation, converting the data into an electricity signal to operate the printer, creating a printing procedure, and ensuring cell viability throughout the production process. In the emerging discipline of rejuvenation medicinal care., 3D bioprinting has been used to create 3D cell frameworks [3].

While 3D printing is already an innovative technology in commercial and pharmaceutical fields, 4D printing is still new but possesses significant potential for the future of manufacturing. A key drawback of 3D-printed structures is stationary and stiff in character; they maintain their original shape and typically serve only one function. To address this limitation, scientists have developed four-dimensional (4D) printing technologies, which integrate smart materials into the 3D printing technique to construct flexible elements.

Furthermore, 3D printing has facilitated progress in personalized treatment of patients by permitting the production of treatment plans personalized to each patient through the printing of their anatomical structures. Examining a real-world representation of the patient's anatomy prior to surgery helps medical personnel organize more successfully than depending just on 3D pictures derived from MRI or CT scans [3].

4D printing involves creating 3D materials that might vary configuration gradually or in respond to external stimuli. It represents a major innovation in additive manufacturing, providing a more efficient transition from concept to reality by incorporating characteristics precisely into the materials as in Figure 1.1. With 4D printing, it is acceptable for manufacture a variety of active, programmable materials that can transform from one shape to another.

In comparison to the current 3DP process, 4DP is another significant advancement and is a prime example of additive manufacturing (AM). 4DP is another intelligent innovation that allows for the building of flexible and dynamic constructs while overcoming the limits of 3DP. It is vital

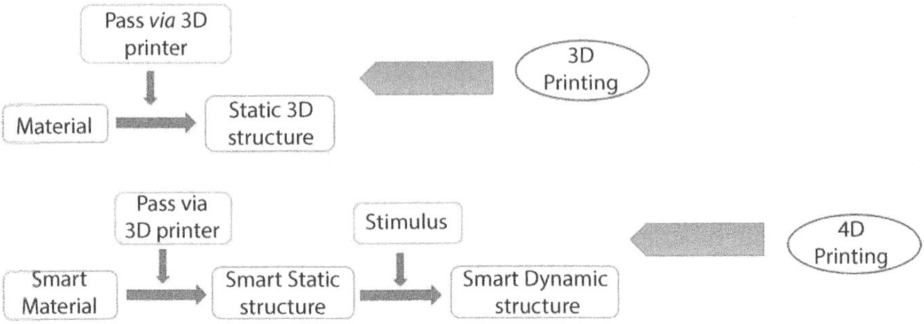

Figure 1.1 3D and 4D printing with their differences.

to know about 3DP, its evolution, and printing procedures since it will help us comprehend more effectively a recent notion of smart 4DP and its discovery.

1.2 Historical Evolution of 3D and 4D Printing

Charles W. Hull first came up with the belief of 3D printing in the 1980s when he employed stereolithography to build polymer products. Carl Deckard came up with selective laser sintering (SLS), which involves the use of a laser to fuse powdered material together. In 1986, Charles "Chuck" Hull designed the stereolithography machine, the first 3D printer, and started 3D Systems. Scott Crump explained fused deposition modeling (FDM) in 1988 and co-founded Stratasys with his wife, Lisa Crump. These firms became pioneers in 3D printing and rapid prototyping. 3D printing was utilized by business designers and engineers for roughly twenty years. This was altered in 2005 with the RepRap project, an open-source project to democratize 3D printing. In 2009, Bre Pettis, Adam Mayer, and Zach "Hoeken" Smith, participants in RepRap, created MakerBot, one of the first consumer 3D printer firms, raising consumer awareness and interest in 3D printing [4].

Skylar Tibbits, the initiator of MIT's Self-Assembly Lab, originally used the phrase "4DP" in a February 2013 TED Talk. These talks are public lectures made available online to share ideas. Tibbits demonstrated preliminary examples of 4D printed objects that had the ability to modify shape over time. One strand of these were printed with inflexible segments and flexible hinges comprised of smart materials that shifted shape when submerged in water [5].

One year after following the TED talk, in 2013, the first 4D printing research paper was published. In this article, the printed portion was altered into an intricate framework using the shape memory effect (SME) depend on the printed active composites (PACs) concept and subsequently 4D printing research has exponentially increased globally [6]. Some of the basic improvements of 4D printing is the fact that it can shift its shape over time as a result of a preprogrammed computer command.

Soon after, other scientists also created 4D printed autonomous systems. For example, Ge and colleagues made active composite material hinges and rigid plates to form devices that could be programmed to modify their shape when the temperature varied. This, in turn, gave rise to 3D printed self-folding origami applied to robotics and more. There were also biomedical applications, such as printed scaffolds that would change and enclose insulin-secreting cells [7].

For the purpose of getting a better knowledge of the advancements of 4D printing in comparison to several other significant forthcoming technologies worldwide, it is necessary to examine the Gartner hype cycle. It evaluates a number of the world's most important modern innovations and how their compatibility with forthcoming commercial developments and services within the period of 5 to 10 years from now. Based on the hype cycle, 4D printing is currently at the phase of sparking technology. During the next 10 years, expectations and progress will increase, and it will become a popular topic of study [8].

1.3 Key Elements of 4D Printing

A variety of manufacturing techniques have been used to create 4D smart structures, which are fabricated to be objects that can change over time and adjust to their surroundings. The 4D printing technique, which builds on the framework of 3D printing, adds one more dimension time along with the x, y, and z axis. The resulting products can undergo alterations in the design throughout time, frequently in reaction to outside stimuli like water or heat [9].

Next-generation materials that react to stimuli and 4D bioprinting processes combine to create dynamic 3D-printed conceptual frameworks. It is important to consider all five of these criteria while using 4D printing as in Figure 1.2 [10].

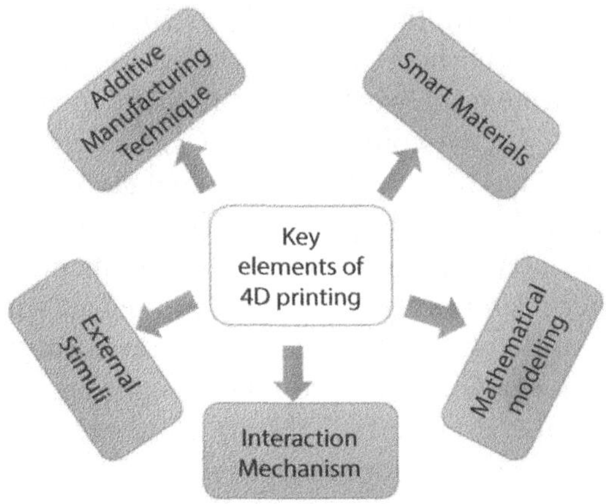

Figure 1.2 Key elements of 4D printing.

1.3.1 Additive Manufacturing Process

The initial step in the procedure is to print a 3D structure using one or more intelligent materials in the right amount. The AM technique produces printing materials directly from software commands, without the need for any tools. Selective laser sintering (SLS), selective laser melting (SLM), stereolithography (SLA), fused deposition modeling (FDM), direct ink writing (DIW), electron beam melting (EBM), and other procedures are examples of AM methods [11]. The majority of procedures may print 4D content as long as the printer and printing are compatible.

1.3.2 Smart Materials

The printing material must be sensitive to stimuli and embedded layer after layer. These types of substances are sometimes referred to as smart materials (SMs) or customizable materials. The nature of stimuli relied on the type of smart material, and the material's capacity for becoming oneself is determined by how it reacts to the stimuli. When exposed to specific stimuli, they may change their form [12].

1.3.3 Stimuli

The third element is the stimulus that will be applied during the 4D printing process. Stimuli may be physical, chemical, or biological in nature. The stimuli used are determined by the application's particular demands.

Light, moisture, electric/magnetic energy, temperature, ultraviolet light, and so on are examples of physical stimuli. The chemicals, pH level, and application of free radicals and reducing agents are exemplars of chemical stimuli. Additionally, glucose and enzymes are examples of biological stimuli [13]. After involving a stimulus, the shape undergoes modifications in substance or composition, such as tension relaxation, molecular mobility, or transitions across phase, that cause the shape to deform.

1.3.4 Mathematical Modeling and the Mechanism of Interaction

A personalized substance has a propensity to change in the desired way if introduced to an external stimulus. The amount of time that the stimulus will work on the personalized substance during mathematical modeling must be planned. Theoretical and mathematical models should be designed to forecast and link the network between the process, material, stimuli, and interaction methods, assuring the desired form or characteristics modification [14].

1.4 The Mechanics of 3D and 4D Printing

Composites of polymer are made using a variety of techniques, each with specific benefits and drawbacks. Various criteria, involving the need of raw materials, production efficiency and accuracy, economic concerns, and the desired functionality of the materials, determine the selection of a particular 3D printing technique [15]. Since every technique has unique advantages and disadvantages, it is essential to equivalent the processes selected with the particular needs of the proposed use.

The conceptual model is frequently divided into thin, horizontal layers, and each layer is printed one after the other by the printer to create the structure. 3D printing is an affordable method that does not require costly molds or other tools for punching, milling, or building structures. Depending on the source, there are several methods for producing 3D-printed goods.

Advanced manufacturing techniques have evolved in large part due to rapid prototype development. The American Society for Testing and Materials (ASTM) has categorized AM into seven processes: Direct Energy Deposition, VAT Photo Polymerization, Material Jetting, Binder Jetting, Material Extrusion, Powder Bed Fusion, and Sheet Lamination.

Irrespective of the AM technique, an exhaustive Computer Aided Design (CAD) model of the physical framework is needed to build a 3D structure [16].

1.4.1 Stereolithography

The photo initiator, specific radiation contacts conditions, and any additional UV absorbers, dyes, or pigments influenced the SLA. The earliest 3D printers to operate were stereolithographic/SL machines, that produced 3D models, prototypes, components, and patterns. Stereolithography is the oldest 3D printing method, which was initially launched in the market.

Charles Hull introduced and patented 3D printing in 1984, despite the fact that a lot of research in this field was done in the 1970s. In order to create extremely precise parts, Photopolymer resins are used in SL's laser-based process, where they react with the laser and solidify as a result [17].

Stereolithographic apparatus (SLA), foremost easily accessible 3D printing technique, was created by Charles W. Hull in the late 1980s. It was initially accepted as an affordable and effective method of producing models and personalized prototypes. These days, SLA is one of the commonly used techniques for solid freeform manufacturing.

SLA's capacity to create high-quality materials of distinctive dimensions is its primary benefits; items ranging in size from submicron's to decimetres have been created by this approach. SLA produced SMPs with precision spanning 0.1 mm to 1 μm, and it can create features as thin as 20 μm. In complex biomedical devices where tiny, precise components are needed for body deployment, this shows that stereolithography is a good fit [18].

1.4.2 Fused Deposition Modeling

Fused-deposition modeling (FDM), denoted to as melt material extrusion (MME) or fused filament fabrication (FFF), is an additive manufacturing technology that relies on the extrusion of strands of thermoplastic. Scott Crump created FDM technology in the early 1990s, which was later introduced by Stratasys INC, USA.

Polylactic acid (PLA), acrylonitrile butadiene styrene copolymer (ABS), polycarbonate (PC), and polyurethane (PU) are among the thermoplastics that have been used to create filaments that are appropriate for FDM. These materials have varying degrees of firmness, elastic expansion, and sturdiness. FDM printers may be used to create a number of health care equipment, such as customized release dosage forms for drug administration methods, and are easy to use, affordable, and dependable [19].

There is a framework for the FDM 3D printers with some degree of mobility and are configured to move vertically. Using a heated extruder, the plastic filament is melted and lay down layer after layer onto a build construction that is consistent with the 3D data that is transmitted to the printer. Every coating that is smeared solidifies and sticks to the one before it [20].

1.4.3 Powder Bed Fusion

A source of heat, like a laser or charged particle, is utilized in the Powder Bed Fusion (PBF) 3D printing process to fuse powdered material layer by layer. Selective Laser Sintering (SLS), Selective Laser Melting (SLM), Direct Metal Laser Sintering (DMLS), Electron Beam Melting (EBM), and Multi Jet Fusion (MJF) are all important techniques. Although PBF can be costly, sluggish, and post-processing-intensive, it enables complicated geometrical structures, minimal material consumption, and individually tailored components.

Materials employed in this technique include, for example, metals, composites, ceramics, polymers, and hybrids. A substance is fused or melted, using warmth as the fundamental idea behind powder bed fusion AM methods [21].

1.4.4 Selective Laser Sintering

Selective laser sintering (SLS) method was created by Carl Deckard in 1987. SLS is a 3D printing technique which can produce results quickly, accurately, and with a variety of surface finishes. Items made of ceramic, plastic, and metal may be produced by selective laser sintering. SLS developed a three-dimensional object by employing an intense laser to sinter polymer particles.

SLS is a powder-based 3D printing technique that creates final parts by fusing layers of material with a laser. Every cross section of a 3D design is traced by the laser onto a powder bed. The building framework drops once a layer is finished, enabling the development of a second layer on top of the first. This technique is repeated until all layers are constructed and the part is finished [22].

1.4.5 Binder Jetting

A revised kind of inkjet technology is used in binder jetting. At the Massachusetts Institute of Technology (MIT), this method was first originated by utilized an inkjet rather than lasers to adhere objects together.

It is the technique of precisely spraying a binder onto a powder bed of component material to combine it one layer at a time, creating or printing the desired part. Once a layer is finished, the powder sheet drops gradually, and a roller or blade smoothed the powder over the surface of the bed prior to the subsequent jet head passing. It is allowing the binder for the next layer to form and fuse with the one before it. This method is similar to other powder bed systems [23].

1.4.6 Direct Energy Deposition

In contrast to other 3D printing techniques, this approach is utilized for maintenance and repair in preference to element manufacturing. A high-energy source is utilized to bind and feed metal material in Directed Energy Deposition (DED) 3D printing process. DED techniques simplify the process of creating materials by melting the component as soon as it's layered. The directed energy deposition technique is conceptually same to material extrusion; But the nozzle isn't locked to just one axis and is able to shift in several dimensions. Moreover, the method works with metals and hybrids based on them, either as powder or wire, but it may also be utilized for ceramics and polymers [24].

1.4.7 Laminated Object Manufacturing

It is a rapid designing technique which fabricate frameworks using laminates of paper, plastic, or metal that are effectively epoxied. A laser cutter is then needed to make the incision or model into the required form. Among the benefits of that technique are its ability to produce full-color prints, its affordability, its ease of handling materials, and its ability to reuse additional components. The complex geometrical components may be produced using laminated object manufacturing (LOM) at a reduced cost and with less operating time [23].

1.5 Materials Utilized in 3D and 4D Printing

1.5.1 3D Printing Materials

Materials for 3D printing must pass tests to produce a plastic filament, be trace-binding during the extrusion process, and be suitable for end-use applications as 3D printed parts or objects. These tests ensure the materials pass through extrusion and meet the required standards.

In order to achieve the intended outcome, certain 3DP research have been performed on a numerous agent, including metal, ceramics, wood, magnetic materials, natural and synthetic polymers and liquids. Materials utilized in the printing process include granules, resin, pellets, and powder. The specific needs and limitations of the desired application should be carefully considered before selecting a polymer 3D printing process.

Throughout the years, 3D printing of polymer combinations has been created, and novel components are being researched and studied. A polymer matrix composite material combined with 3D printing has a great deal of promise for industrial production with remarkable dynamic functionality and durability [25].

1.5.1.1 Polymers

Polymers are the most often utilized materials in 3D printing, particularly in the processes of stereolithography (SLA) and FDM. There are various kinds of polymers used in 3D printing, all with own distinct properties and purposes.

1.5.1.1.1 Polylactic Acid

Due to its affordability and convenience of usage, it is widely utilized material for 3D printing. PLA comes in a distinct range of colors, such as fluorescent, metallic, and translucent. Although it is not as robust or long-lasting as other materials, it is an excellent option for situations where aesthetics is crucial.

The material is perfect for complex patterns since it can be printed at high resolutions. It is not ideal for uses where the item will be exposed to a lot of friction or damage from use, though, because it isn't as resilient as some other materials [26].

1.5.1.1.2 Acrylonitrile Butadiene Styrene

A variety of substances, such as layered silicates, ultrahigh-molecular-weight polyethylene, styrene ethylene butadiene styrene, montmorillonite clay, and other petroleum- and bio-based compounds, are compatible with acrylonitrile butadiene styrene (ABS). Because of its strong mechanical properties, such as its resistance to impact, heat, chemicals, and abrasion, ABS is ideal for many applications, such as electrical enclosures, tool handles, high-wear toys, and car trim parts. Due to its propensity for warping, ABS has to be printed on a heated print bed using bed adhesive [27].

1.5.1.1.3 Polyethylene Terephthalate Glycol

Due to its rigidity, flexibility, and ease of printing, Polyethylene terephthalate glycol (PETG) is appropriate for range of applications, such as printer parts, protective components, and mechanical parts. Nevertheless, PETG is readily damaged and prone to moisture.

Glycol is added to PETG, a modified form of PET (polyethylene terephthalate), and it increases the material's lifespan and utility by preventing crystallization and stiffness [28]. Because of its safety features, it is appropriate for printing components of medical devices that come into touch with food or are utilized in medical applications.

1.5.1.1.4 Nylon

Nylon, commonly known as polyamide (PA), is a popular material employed in powder bed fusion 3D printing. As a result of its exceptional flexibility, rigidity, as well as robustness, it is perfect for uses requiring long-lasting components, including mechanical parts, tools, or functioning models.

Yet, Mohd Radzuan *et al.* clarified; problems for example weak adhesion and delamination can arise when utilizing nylon in 3D printing with higher loadings of fillers. However, the mechanical qualities can be enhanced, and stronger samples can be produced with a post-drying procedure and optimal printing conditions [29].

1.5.1.1.5 Polycarbonate

Polycarbonate (PC) is one of the toughest and among the resistant polymers included in 3D printing. It resists physical harm and heat and can tolerate temperatures greater than 110°C. Many applications use polycarbonate, including electrical, mechanical, or parts for automobiles, where durability and strength are crucial. The performance of the polycarbonate components made using the material extrusion process is significantly influenced by depth of the layers and infill density settings for 3D printing. Although thicker layers required less energy, a higher infill density and a lower layer thickness produced a larger compression strength [30].

1.5.1.2 Other Polymers

1.5.1.2.1 Water-Soluble Polyvinyl Alcohol

A common foundation material for complex 3D printing with extensions is water-soluble polyvinyl alcohol (PVA). Despite being moisture sensitive and potentially difficult to deal with, PVA filament is a great support material. Due to its suitable fluidity, sustainability, and accessibility, PVA is becoming a growing trend in 3D/4D printing [31].

1.5.1.2.2 Acrylonitrile Styrene Acrylate

A substance suitable for practical applications and resistant to surroundings is acrylonitrile styrene acrylate (ASA), especially in the automotive industry. ASA filament is appropriate for a number of applications because of its exceptional ultraviolet (UV) resistance, thermal resilience, and sturdiness. For example, it can withstand harsh environments and sunlight without suffering considerable degradation [32].

1.5.1.3 Metal

Metals, such as iron, nickel, cobalt, chromium, gold, silver, stainless steel 316L, titanium alloys Ti6Al4V and Hastelloy, are widely used in 3D printing of materials. These have high mechanical strengths comparable to bone, making them appropriate for 3D printing constructs. However, there are restricted 3D metal printing processes. Metal rusting, deterioration, and possible toxicity are being considered for long-lasting metal installation in 3DP.

In selective laser melting, gold is a precious metals used as an initial component. Good thermal conductivity is the primary reason for the selection of gold as an initial component. Gold is a fundamental material utilized in the selective laser melting process to produce jewellery and elaborates [33].

The SLM technique uses a variety of metals as raw materials, including silver. It allows one to create jewellery and ornaments with intricate patterns that are otherwise out of the question to make through traditional methods. Additionally, it decreases production costs by not having to use high-cost forms and minimizing the dust, as well as providing leftover powder to be recycled for the next generation of creation [34].

The most frequently used metal in 3D printing is stainless steel. For outstanding strength and hardness requirements, stainless steels can be utilized in a numerous application, mostly in molding and tools. The properties of 3D printing are largely complex for stainless steels. One of the most important raw materials used in the SLM and EBM processes is titanium [35]. This material has considerable applications in the biomedical sector.

1.5.1.4 Ceramics

Both non-metallic and metallic ceramics have been used as materials for frameworks in 3D printing since they are very strong and are biocompatible. Ceramics' strong mechanical, electrical, chemical, and thermal features are utilized by numerous industries in a broad cross-section, which extends to biomedical engineering, electronics, machinery, and aerospace industries. Selective laser melting as well as fused deposition modeling can both be used to design ceramic-based objects.

Ceramics' look, biocompatibility, and physicochemical attributes render them a suitable application in biochemical, biomedical, and diagnostic purposes. One of the principal hurdles in additive manufacturing development is sharp melting point of ceramics as paralleled to metal and polymers [36].

Ceramics, such as pure alumina (Al_2O_3), pure zirconia (ZrO_2), lithium oxide (LiO_2), silicon di oxide (SiO_2), silicon carbide (SiC), silicon nitride (Si_3N_4), and lead zirconate titanate (PZT), are employed as raw materials. They aid in the development of light weight components including turbine rotors, gears, intricate parts, and transducers. Alumina and zirconia are more suitable for implants than other metals due to their corrosion resistance alongside their capacity to sustain extreme temperatures [37].

Ceramics can be 3D printed using inkjet, powder bed fusion, paste extrusion, or stereolithography. The initial technique for creating dense ceramic prototypes that might not require subsequent evaluations is inkjet. The ceramic components were 3D produced using a stereolithography (SLA) printer by combining certain monomers with a UV photo initiator (PI) that are pre-ceramic. When these polymer networks were ignited, a ceramic without pores and with constant distortion was produced. Compared to ceramic foams of a comparable density, the silicon oxycarbide micro lattice and honeycomb substances with cells created using this method are more resilient. The green bodies, or advanced ceramics, were made utilizing a stereolithography-based AM process implementing light susceptible resins filled with 3-, 8-, and 25–60 weight percent alumina.

Selective laser sintering (SLS) is another method for 3D printing ceramic powders. However, cracks may emerge in ceramic components as a result of extreme temperature from convergence heating and cooling to room temperature.

A new method for ceramic-matrix hybrids is called selective laser gelation (SLG), that mix the sol-gel and SLS processes. The green body is sintered at a higher temperature to harden the ceramic dust after the unused dust has been discarded. The laser-activated binder momentarily retains the powder in the proper form. After sintering, the additional binder can be drawn out [38].

Indirect SLS is a technique commonly employed for ceramic-glass and ceramic-polymer hybrids. Superior intricate ceramic networks can be created by 3D printing polymer-derived ceramifiable monomers, such as silicon oxycarbide, polymerizing them with UV light, and then fusing them at elevated temperatures. A natural ceramic found in human teeth and bones is called hydroxyapatite (HAp). For creating substrates with the same durability as real bone, HAp and equivalent ceramics are ideal due to their unique properties.

1.5.1.5 Hydrogels

Hydrogels are excellent objects for 3D printing because they may be used as ink, especially in extrusion-based processes. The hydrogel is made up of a 3D crosslinked network of hydrophilic polymers. The penetration of solvent molecules into a network of polymers, hydrogels expand. They are also utilized as padding among tissue layers or to sustain 3D-printed structures.

Hydrogels are classified as homopolymers, copolymers, or semi-interpenetrating networks (semi-IPN) based on their manner of manufacture. Poly (hydroxyl ethyl methacrylate) (PHEMA) and polyethylene glycol (PEG) are examples of homopolymers, whereas MA and carboxymethyl cellulose combine two kinds of monomers [39].

A linear polymer can enter into a different interconnected system without forming a chemical bond, semi-IPN is created, as demonstrated by the acrylamide/acrylic acid copolymer. For solution polymerization, solvents, such as ethanol, water, benzyl alcohol, and water–ethanol combinations, are frequently used. In the realm of organ and tissue regeneration, 4DP has recently greatly sparked research on self-healing hydrogels [40].

1.6 Smart Materials Used in 4D Printing

In 4D printing, the use of intelligent/smart materials is fundamental to the ability to receive, transfer, and process stimuli. Materials undergo activation, which involves shape morphing or functional alteration, leading to structural transformations. Countless smart polymers, for example, liquid crystal polymers (LCPs)/liquid crystal elastomers (LCEs), smart hydrogels, and SMPs, have been successfully synthesized; however, all do have certain liabilities.

Numerous smart polymers have poor reversibility in changes or exhibit hardly any reaction even after long terms of time. Among the more sophisticated materials are smart materials that are gaining attention due to their ability to alter their properties—like dimensions, color, or shape—based on external stimuli such as moisture, temperature, light, or electricity and magnetism [41].

This type of responsive material adds a new aspect to 3D printing, which has been termed 4D printing. Creating a mixture of intelligent responsive and inactive/rigid components having various thermomechanical attributes can enable joints, curves, twists, or hinges must be created on connections. These parts respond to stimuli by generating differential stresses.

Smart or intelligent materials are those that can change their configuration or geometry and even their physical characteristics, including toughness, resistance, and Young's modulus, in response to stimulus from outside like light, heat, magnetic fields, and radiation [42].

A 4D printing smart material must be able to be printed and have autonomous form memory that reacts to external stimuli. Another important characteristic for biomedical applications is biocompatibility; however, additional features can be more important based on the printing method selected and the ultimate purpose.

The shape memory effect (SME) refers to a semi-plastically and significantly pre-deformed object's capacity to restore its former shape when presented with the relevant stimuli. These substances are sometimes referred to as soft active materials (SAMs). Smart materials use their "smartness" to respond to external stimuli regarding sensors, controls, and shape recovery. The SME is frequently induced by a number of stimuli, including mechanical stress (mechanoresponsive), chemicals (chemo responsive), light (photo responsive), and cooling and heating (thermoresponsive) [43].

1.6.1 Shape Memory Polymers

Shape memory materials (SMM) are substances that may alter their physical form when revealed to a stimulant in a non-instantaneous and unplanned approach. They are extensively used in 4D printing. Shape memory alloy (SMA), shape memory polymer (SMP), shape memory hybrid (SMH), shape memory ceramic (SMC), and shape memory gel (SMG) are the several types of SMMs. A shape memory composite (SMC) may be made by joining these SMMs with an additive. SMA and SMP are the two typically employed substances for 4D printing out of all the ones that are available [44].

Shape-memory polymers (SMPs) have gained greater interest due to their academic and technological relevance. They can change from an improvised programmed shape to a stable structure when exposed to an external stimulus. Due to their elastic properties, these polymers are ideal for the complex geometric changes required for 4D printing [8]. For the creation of interactive 4D objects, SMPs are very helpful since the SME can be designed and regulated.

SMPs may be employed in a numerous of sectors, such as production industry and aerospace, due to their affordable price, compact size, effortless finishing, and great programming mobility. Their biodegradability and biocompatibility make them ideal for fabricating biomedical devices.

Polyesters, including poly (lactic acid) (PLA) and polycaprolactone (PCL), as well as polyether urethane, are the most popular category of biodegradable SMPs, as per Wang *et al.* An oligomer resin that cross-links with a monofunctional monomer resin are the two components of dual-component polymers, which are employed to achieve the SME needed for 4D-printed geometries. The monofunctional monomer serves as the chain of polymers' linear backbone. These SMPs' shape memory behaviour is caused by two main features: switching segments (soft components) and net points (hard components) [45].

The traditional training of SMPs includes six steps: heating, deforming, cooling to set, and unloading, actuating, and final cooling. 4D-printed materials showcase features like shape-memory, self-assembly, self-actuation, self-sensing, and self-healing.

Ge *et al.*'s study, which created a multi-material approach for micro-stereolithography manufacturing of shape memory polymers, is a great example of how SMPs can be used for 4D printing. They used important chemicals like Sudan I, Rhodamine B, and the photo initiator Phenylbis (2,4,6-trimethyl benzoyl) phosphine oxide (BAPO) [12]. They also used methacrylate-based monomers in their smart ink. They could accurately customize the glass transition temperature (Tg) to accomplish rigidity control at different temperatures, such as 43°C and 56°C, by varying the amounts of PEGDMA, BPA, DEGMA, and BMA (Figure 7d). Because of this accuracy, their printed flower's bloom motion was able to develop in two separate stages [46].

Shape-memory polymers are currently used to build stents that react to variations in temperature when implanted into patients. The intricate shapes needed for stents make traditional manufacturing techniques challenging and time-consuming. With the use of 4D printing, however, stents of any size may be produced quickly and affordably. SMPs have been used to produce a cardiovascular stent with a tiny, indefinite framework that can be heated to enable less-invasive implantation. After insertion, the stent cools and reverts to its bigger, more therapeutically useful form, where it stays indefinitely. This method has been used to build "smart" stents in addition to tracheal stents that allow for patient customization. These so-called "smart" structures may be remotely triggered and steered magnetically thanks to iron oxide included in a PLA-based ink that reacts to temperature.

This invention is a major breakthrough in medical technology as it enables doctors to customize care to fulfil the unique necessities of every patient. Additionally, the use of magnetic guidance may improve patient outcomes by increasing the accuracy of stent insertion and operation [47].

Permanent net points are important in remembering the initial shape that exists alongside segments responsible for material distortion in the molecular matrix of a standard SMP. These net points, which are unaffected by the programming stage, might arise from physical interactions or chemical bonds. In the manufacturing of the SMP or during the post-processing/printing phase (e.g., UV curing), crosslinking—the most prevalent kind of chemical net point can be added [48].

When compared to metallic smart materials and inorganic ceramics, SMPs provide a number of benefits, comprising low density, easier manufacturing, chemical resilience, great resistance to stress, and recoverable stresses. In comparison to SMAs, SMPs are comparatively cheap to make and may be made transparent. They are especially well suited for use in biological domains like drug delivery systems (DDS) because of their biodegradability, biocompatibility, and variable rate of disintegration. Their application in AM methods has been boosted by their low melting points (which enhance printability) and low cost of production as compared to other materials. Compared to printed hydrogels, 4D printing SMPs can reach significantly quicker printing rates and higher structural toughness. SMPs show potential as smart materials for biomedical device fabrication, and further study is needed to fully realize their potential [49].

1.6.2 Hydrogels in 4D Printing

A hydrogel is made up of cross-linking polymer chains produced from hydrophilic compounds. The capacity of hydrogels to absorb enormous quantity of water without dispersing is due to the links' three-dimensional network arrangement. Their ability to return to their initial size after drying and their considerable expansion when absorbing water distinguish them from dry-state polymers. As has already been shown with SMPs, a variety of triggering stimuli, including variations in temperature, pH, ion concentration, or the application of electric and magnetic fields, may cause hydrogel-based products to change in structure and/or capability.

Shape morphing, also known as smart hydrogels, is more frequently achieved by utilizing substances that react to basic solvent absorption/desorption processes. Furthermore, depending on the unique features of the area immediately affected, regulated swelling would result from water diffusion into the network to varying degrees. Regardless, a swelling mismatch must be created in the structure to control the changes in the object's size and form. This may be done, for example, by specifying the material's composition and positioning during the printing process. Multi-material structures can be created by combining hydrogels with different hydration capacities.

A single hydrogel type that can exhibit varying degrees of swelling across the framework (anisotropic swelling) is an alternative. It could be achieved by adding anisotropic particles to the hydrogel or by forming areas of the material with varying densities and levels of crosslinking [49].

Hydrogel composites have been effectively used in 4D printing procedures to produce flexible, sustainable constructions in a number of instances. As an illustration, Ding *et al.* demonstrated how to print a thermoresponsive hydrogel material with an integrated SME. This work was the initial attempt to suggest a complementary AM method which incorporates the shape memory propensity into its components during printing. It differs from typical processes that include printing a mixture of both active and inert elements to generate differential stresses or thermomechanic training after printing. It was possible to create high-resolution structures with incorporated adjustable stresses by managing the photopolymerization during printing.

The main methods of 3D bioprinting hydrogels are:

- Extrusion-based bioprinting- Uses Gelatin-methacrylate (GelMA) material with techniques like SLA and DLP.
- Inkjet-based bioprinting-Prints using tiny droplets of material.
- Laser-based techniques- Uses dynamic SLA to create GelMA frameworks.

Huang, along with colleagues, digitally printed a hydrogel composite, attaining a complicated yet realistic construct by adjusting the localized swelling. Instead of altering the sequence of the printing process, it was done by carefully regulating the light exposure period. Mao *et al.* utilized PolyJet technology to fabricate a self-folding hydrogel structure by combining hydrogel with SMP and elastomer layers, enabling it to change shape and revert to its original form. When the material was submerged in water with low and then high temperatures, it folded itself. When submerged in hot water, the structure unfurled and settled on its initial configuration [51].

The usage of hydrogels is restricted in dry conditions since they can only be activated in environments that are dependent on water or moisture. The hydrogels' ability to change form in reaction to moisture in the human body makes them advantageous for biomedical applications, even though it could be a drawback for other uses.

When the printed hydrogels were exposed to water, their volume expanded by 200%, demonstrating their ability to change form. In the study of the concepts of hydrogel 3D manufacturing, M. Jamal *et al.* emphasized the creation of bi-layer structures based on Polyethylene Glycol (PEG) using traditional photolithography. The hydrogel's differential swelling made it possible for the structures to fold in a regular and feasible manner.

Numerous sectors, including biomedicine, which encompasses tissue fabrication, prosthetic organs, and medication delivery devices, and agriculture, utilize them. These hydrogel composites provide a novel approach to 4D printing for a range of uses, including robotics and specially made sensors [52].

A smart hydrogel that can respond to temperature that was created with alginate/poly(N-isopropylacrylamide) (PNIPAAm) inks to create an elegant valve is a noteworthy instance. The smart valve might be sealed to lower the flow rate through this valve. Furthermore, a substantial amount of research has been done on hydrogels that are chemo responsive and pH responsive.

1.6.3 Liquid Crystal Elastomers/Liquid Crystalline Polymers

Covalent bonds are used to join tiny liquid crystal molecules to create liquid crystal polymers, or LCPs. LCPs with strong cross-linking can be used to create liquid crystal networks. A family of unique elastomers known as LCEs is elastotic in both liquid crystal and isotropic phases. They exhibit high molecular interaction, flexibility, and extrinsic sensitivity because they combine the rubbery flexibility of polymer channels with the anisotropy of liquid crystal. By altering the liquid crystal components' arrangement in response to outside inputs, they may create a significant macroscopic form shift [53].

External triggers, such as warmth, light, and electrical fields, can activate structural reactions in LCEs, making them flexible and responsive. At a certain transition temperature, also known as the isotropic transition temperature (Ti or Tm), LCEs go from an anisotropic phase to an isotropic phase. Particularly, LCEs provide unique qualities in the 4D printing industry. To create, fix, and gather carbon nanotubes (CNTs) and xLCEs for soft robotic devices intended at lower-temperature settings, for example, Yang *et al.* used infrared (IR) light (808 nm) [54].

The resultant 4D-printed smart LCE item displays the capacity to transition in between the high-order pattern configured pattern and the low-order structure established by the printing technique. Barnes and colleagues' investigation employed a customized LCE ink that included RM257 mesogens and chain extenders 2,2′-(ethylenedioxy) diethanethio (EDDET) and pentaerythritol tetrakis (3-mercaptopropionate) (PETMP) [55].

When exposed to UV light, azobenzene molecules isomerize from the trans to the cis state, which causes the polymer network to expand. In 4D printing, this process has been used to produce light-responsive activating silicone bilayers. Azobenzene molecules in 4D-printed LCEs cause strain and mesogen misalignment by enhancing light sensitivity at room temperature. A 3D-printed LCE strip containing azobenzene was shown by Ceamanos *et al.* to compress boosting the load in the presence of ultraviolet light, and regain its original dimension under blue light [56].

1.6.4 Self-Healing Polymers

Researchers have incorporated self-healing processes into materials, drawing inspiration from nature, to allow them to restore functioning, prolong their lifespan, and fix internal or exterior damage. They are referred to as self-healing materials. Due to their benefits, self-healing materials are

Table 1.1 Detailed of stimuli responsive materials used in 4D printing.

Smart materials	Description	Materials used	Applications
Thermoresponsive materials	It worked by altering the molecular properties which modify their structural shape	Poly-Nvinylcaprolactam (PNVCL), Poly-Nisopropylacrylamide (PNIPAM), Polyethylene glycol, Collagen,	Utilized mostly in drug release and biomedical engineering. Also used in many soft robotics and smart sensing functioning [61].
pH responsive material	Have functional chemical groups, such as acid, pyridine, carboxyl, or sulfonic phosphate. These chemical groups react to variations in the pH of the environment by delivering protons, which stimulate biomaterials to expand or contract.	Hyaluronic acid (HA), Collagen, Alginate (Alg), Gel, Keratin & Chitosan (CS), Poly histidine Poly acrylic acid (PAA), Poly L-glutamic acid	used in a number of biological uses, including as glucose sensors, drug delivery systems, and gene delivery systems. utilized in Soft robots, valves, biocatalysts, and colloidal stabilization [62].
Magneto-responsive materials	Using magnetic and electric fields to cause particular responses in bioactive materials	para- and ferro-magnetic nanoparticles (MNPs) additives	offer regulated and less invasive therapeutic activity in DDSs. Used in the micro-gripper for printed hydrogels. By applying a magnetic field, It also functions as a remote controller for the magnetic substance [63].
Moisture-responsive smart materials	Generally, expand, fold, twist, or deform in various ways after absorbing or losing moist in the humid atmosphere or cell culture media. When exposed to moisture, printed structures reformed and recovered.	Cellulose, Silk fibroin, Polyurethane, Poly (ethylene glycol) diacrylate	Beneficial for applications involving soft robotics and actuation. Using tunicate cellulose nanocrystals (TCNCs) in polymeric networks to create tendril-inspired hydrogel artificial muscles [64].

(Continued)

Table 1.1 Detailed of stimuli responsive materials used in 4D printing. (*Continued*)

Smart materials	Description	Materials used	Applications
Photo-responsive smart materials	Allow the structures to undergo volume transformations, bending, contractions, and other modifications in size or shape in respond to light stimuli. Developed by incorporating chromophores with polymer resins.	Polylactic acid / Fe_3O_4 Acrylonitrile Butadiene styrene, Polycarbonate	utilized in the fabrication of drug delivery system. Used in fabrication of self-assembled nanoparticles and self-folding polymer sheets [65].
Electro-responsive smart materials	By altering the electric field's intensity and the contraction, expansion, or folding of endophyte bioactive components.	Polyacetylene Polyaniline Polythiophene Polyphenyl acetylene	Explored to create structures for brain and muscular TE applications. These gels can be employed as biomedical and soft materials, as well as artificial muscle, sensors, actuators, and lenses, for medical applications [66].
Biological-responsive materials	Employ biological stimulus such as glucose, biomolecules, and bioenzymes. In this case, bioenzymes alter the polymers, causing them to expand, and disrupt peptide sequences and biomolecules.	Protein and various enzyme	Thrombin and alkaline phosphatase, two bioenzymes, activated bioinspired hydrogels. It causes fibrin accumulation and calcification on the 4D-printed object [67].
Piezo-electric materials	These have unique characteristics of smart materials include sensitivity to mechanical stress and stress in the presence of electric fields, produce electricity or distort	Polyvinylidene fluoride (PVDF) $BaTiO_3$ Barium titanate (BTO)	Used in development of Sensors, energy harvesting, Medical imaging device [68].

considered an emerging category of sophisticated elements that have been used in many different professions [57].

Despite having a minimal environmental effect, self-healing polymers give manufactured goods robustness, dependability, and product longevity. Soft robots, for instance, are prone to breakage and frequently need their broken parts replaced completely. This is expensive, laborious, and wasteful. An option that is more cost-effective and eco-friendlier is the development of soft robots with self-healing polymers [58].

The incorporation of self-healing characteristics into 4D printing materials permit for the inventions of constructions that are resilient to severe conditions continues to operate for prolonged periods, and minimizes the ecological effect of disposing of injured objects. With broad potential across several sectors and applications, this amazing property of self-healing materials marks a substantial development in 4D printing technology. When flaws such as cracks or scratches appear, self-healing materials work by starting the healing method on their own [59].

1.7 Stimuli Responsive Smart Materials Utilized in 4D Printing

Materials that offer a way to achieve a responsive and smart reaction in a product and that can generate a variety of additional capabilities are referred to as smart or intelligent materials. The ultimate structure of the material is accurately restrained in 4D bioprinting by initiating an external stimulus. Stimulus-responsive materials are those that alter shape in response to environmental stimuli.

A wide range of external stimuli can be used, such as chemical (pH, dampness, and biological molecules.), biological (enzymes, glucose, and cell adhesion forces), and physical (temperature, ultrasound imaging, light, magnetism, and electric field) [60]. The stimuli-responsive smart materials utilized in 4D printing are briefly explained in Table 1.1.

1.8 4D Printing Applications in Biomedical Field

1.8.1 Bone Reconstruction/Tissue Engineering

Bone has the ability to heal itself after a fracture or when little flaws need to be repaired. 4D bioprinting improves bone healing by transforming the shape of 3D-printed scaffolds in reaction to environmental stimuli. This is particularly useful for uneven bone defects. There is revolutionary potential for orthopaedic surgery in the developing domain of 4D printing for bone tissue applications, which offers opportunities for improved integration of bones, decreased surgical invasiveness, and decreased chances of infection. Recent years have seen reports of the use of 3D printing to create complex scaffolds for bone alternatives and reconstructions. The robust use of FDM of ceramic and/or metal-reinforced polymers to replicate native tissue with comprehensive mechanical properties has been demonstrated.

Recent progress can be roughly divided into four groups, with the goal of closely replicating the biomechanical properties of natural bone. These are:

a. Injectable stimuli-responsive hydrogels;
b. Shape memory scaffolds;
c. Effective transformation processes; and
d. Neovascularization and neural development to promote bone development and mineral accumulation [69].

The characteristics of 4D bioprinting materials to alter their function over time makes it an amazing method for creating bone grafting scaffolds. Several major benefits have been achieved with SMPs like polylactide (PLA), polycaprolactone (PCL), and polyurethane (PU), along with various copolymers. Even though there have been many 3D printing accomplishments, the use of time-sensitive intelligent designs in bone regeneration is still a fairly intriguing and unexplored idea.

Wang *et al.* developed photothermally sensitive structures out of β-tricalcium phosphate (TCP) and poly (lactic acid-co-trimethylene carbonate) (PLA-TMC). They added osteogenic peptide and black phosphorus nanosheets (BPNSs) to make them stronger. The results showed that the β-TCP/PLA-TMC-based scaffolds changed form when near-infrared radiation (NIR) was used. It was possible because the scaffolds contained light-responsive BPNSs [70].

Miao *et al.* demonstrated the to print four-dimensional bone-regeneration structures that adopt to the body's temperature. Using SLA, a sustainable soy bean oil epoxidized acrylate was printed, and it demonstrated the ability to create permeable biodegradable scaffolds. It assists in the development of multipurpose human bone marrow mesenchymal [71].

Chawla *et al.* created cartilage callus, which may then be turned into bone tissue by resorption and reshaping. They used a two-step process with a bioprinted scaffold made of silk and gelatin, loaded with special stem cells. First, they turned these cells into cartilage over three weeks under chondrogenic modification. After that cartilage converted into bone tissue within two weeks [72].

The ability of 4D bioprinting to respond *in vivo* is vital for meeting international standards. Liu *et al.* used HA/PCL-based SMP to create translucent frameworks for insertion in mandibular bone reconstruction for BTE. At appropriate temperatures, SMP-based scaffolds shrink into transitory deformed-pore forms and rebound to spongy scaffolds. Furthermore, these intelligent 4D-printed scaffolds show great promise for implantation under adaptive *in vivo* environments.

A number of polymers, such as nanocellulose, alginate, and PU, have been used across different AM techniques, such as SLS and inkjet bioprinting, for the production of cartilage tissues. Due to the cartilage scaffold's high level of shape transformation, a study employing its 4D bioprinting showed encouraging cartilage formation outcomes. The scaffolds were made using dual-layer hydrogels, where one layer had a reduced expanding ratio than the other, to enable a shape-morphing response.

1.8.2 Smart Stents

A medical device called a stent is placed into arterial systems to lower the risk of thrombosis and stop them from collapsing. Stents are small, short tubes that are inserted into hollow tubes like veins, arteries, or other tissues like the ureter. Examples of these are coronary artery stents, carotid artery stents, and airway stents. Stents are typically used to keep these hollow tubes open as well as also assist in the release of antiplatelet and antiproliferative medicines. More recently, experts used 4D printing technology to create excellent quality SMP-based thermoresponsive cardiovascular stents to tackle standard manufacturing problems [73].

Ge *et al.* developed the prominent -quality stents having different diameters, heights, different ligament dimensions, inter-ligament angles, and joint counts using an excellent quality PμSL AM technique. They are designed to expand once inserted into the body and to be tiny enough to allow for minimally painless implantation. The majority of stents are used to enlarge blood arteries or the trachea [74].

A sophisticated 4D biofabrication technique for hollow self-folding tubes with an optimal diameter of 20 m was created by Ionov *et al.* utilizing shape-morphing biopolymer hydrogels. When the concentration of Ca^{2+} ions changes, the polymer undergoes reversible shape changes. The self-folded hydrogel-based tubes maintain cell survival for a minimum of 7 days refrain from reducing cell lifespan, and this procedure has no detrimental effects on the vitality of the created cells [75].

SMPs' capacity to regain their previous forms will be useful for a variety of uses, particularly stents. The two issues that 4D bioprinted stents still need to resolve are biological compatibility and conforming to the physiological attributes of humans.

A hollow tube known as a tracheal-bronchial stent is used to widen the human airways and facilitate breathing. When exposed to constant wear stress from breathing, conventional trachea stents show an unresponsive character. As a result, personalized stents which are free of anxiety and stimuli-responsive have been created using 4D bioprinting technique. 4D-printed transforming stents detect ambient conditions and react to constant strain caused by the body's temperature.

A tracheal stent having shape memory functionality has been created by printing an SMP bioink derived from a methacrylated polycaprolactone source. When the stent is in the proper place, it can be minimized in size, placed into the body, then eventually reverted to its previous shape when the ambient temperature rises. A Fe3O4/PLA-based tracheal stent that was 4D printed by Zhang *et al.* displayed outstanding shape memory behaviour in response to UV light and magnetic stimulation [76]. Furthermore, this stent indicated low volume and unwanted damage. According to research, 4D bioprinting technology can be utilized to create synthetic frameworks for tracheal and intravascular stents.

1.8.3 Self-Healing Implants

Self-healing implants, which are made of 4D-printed components with the ability to self-healing in reaction to outside stimuli like temperature, mechanical stress, or chemical signals. It is among the most exciting applications of 4D printing within the pharmaceutical field. There are several benefits to using self-healing materials in medical devices, especially when it comes to implanted devices, wound healing, and bone scaffolds. In implanted medical devices, self-healing materials can successfully lower the possibilities of mechanical deterioration or injury over time [77].

In the fields of bone tissue engineering, 4D-printed bone scaffolds composed of composite materials can help in the early process of healing. it also heals themselves minor fractures that could damage the scaffold's structural stability. In situations where the body continuously applies mechanical pressures to the implant site, such in bone fractures or replacements of joints, this feature greatly extends the lifetime and dependability of implants [78].

Cellular tissue regeneration can be aided by 4D-printed materials that could be engineered to liberate bioactive substances, such as growth factors or antibacterial chemicals, in reaction to surrounding alterations. Furthermore, depending on the stage of wound healing, 4D-printed bandages or wound dressings can modify their characteristics (such as porosity or stiffness) to facilitate a quicker recovery. Because there may be less need for ongoing monitoring and follow-up therapies, this flexibility can also improve patient comfort and save healthcare expenses [79].

1.8.4 4D Printing in Drug Delivery System

The drug delivery system can provide a stable and uninterrupted resource of drugs by controlling the rate of medication release. With very high accuracy, 4D bioprinting can be used to make several different DDSs, comprising tablets, transdermal systems, orodispersible films, dermal skin patches, nanosuspensions, and vaginal and rectal delivery systems. In addition to customizing mechanical durability or structure, 4D bioprinting aids in the delivery of biomolecules by reacting to outside stimuli. We are currently designing smart automated bandages with pH-sensitive sensors. These sensors may release drugs in reaction to external stimuli, particular signals, and improper clinical conditions. Due to their potential to incorporate medications, antibodies, as well as biological substances, hydrogels have been used extensively as careers for drug delivery.

Vehse *et al.* synthesized poly (ethylene glycol) diacrylate (PEGDA) scaffolds using micro-stereolithography that may be employed as drug release vehicles. According to Gioumouxouzis *et al.*, polyacrylics (Eudragit®), polylactic acid (PLA), and polyvinyl alcohol (PVA) are ideal biodegradable SMPs that have an ability to form therapeutically effective structures for drug delivery systems [80]. This work created a controlled release drug delivery system by printing a filament blend of PVA, mannitol, and hydrochlorothiazide (HCTZ) using FDM. Either hot-melt extrusion or submersion of the filament in drug solutions might be used to integrate the medication into the polymeric filament.

The ingestible tablets developed for ulcer therapy include a pH-responsive shell. The tablet releases the specified dosage of medicine when the stomach environment becomes acidic. This novel strategy was further expanded in a confirmation of concept research, where an SMP-based capsule was created to control release of drug using internal expansion of cracks processes. Xin *et al.* created dynamically adaptable shape-morphing microrobots (SMMRs) based on pH-responsive hydrogel. Different expansion rates were programmed encoded into the manufactured SMMRs [81].

The shape-morphing micro fish (SMMF) may incorporate doxorubicin (DOX) by shutting and expanding its mouth in PBS (pH ~ 7.4) and a slightly acidic solution (pH < 7). Gupta *et al.* created a 4D-printed capsule with a core-shell hydrogel structure which allowed the medications to be released at predetermined times and places. The capsule's shell comprised poly (lactic-co-glycolic acid) (PLGA) and gold nanorods (AuNRs), while the inner section contained biomolecules, ethylene glycol, and poly (vinyl alcohol). When exposed to a laser stimulation, these photo-responsive AuNRs exploded and discharged drugs [82].

AI disclosure statement: The author acknowledges the use of AI-based tools for minor language editing or grammatical correction. However, all intellectual contributions, analyses, and conclusions presented in this chapter are entirely the author's own.

References

1. Shokrani, H., Shokrani, A., Seidi, F., Mashayekhi, M., Kar, S., Nedeljkovic, D., Kuang, T., Saeb, M.R., Mozafari, M., Polysaccharide-based biomaterials in a journey from 3D to 4D printing. *Bioeng. Transl. Med.*, 8, 4, e10503, 2023 Jul.
2. Grira, S., Khalifeh, H.A., Alkhedher, M., Ramadan, M., 3D printing algae-based materials: pathway towards 4D bioprinting. *Bioprinting*, e00291, 2023.
3. Rana, A., Mittal, A., Vashist, C., Rajput, S., Sridhar, S.B., Malviya, R., Exploration of 4D Printing and its Applications in the Biomedical Sciences. *Curr. Pharm. Des.*, 2025.
4. Matias, E. and Rao, B., 3D printing: On its historical evolution and the implications for business, in: *2015 Portland International Conference on Management of Engineering and Technology (PICMET)*, 2015 Aug 2, IEEE, pp. 551–558.
5. Gazzaniga, A., Foppoli, A., Cerea, M., Palugan, L., Cirilli, M., Moutaharrik, S., Melocchi, A., Maroni, A., Towards 4D printing in pharmaceutics. *Int. J. Pharm.: X*, 5, 100171, 2023 Dec 1.
6. Ge, Q., Qi, H.J., Dunn, M.L., Active materials by four-dimension printing. *Appl. Phys. Lett.*, 103, 13, 2013 Sep 23.
7. Shiblee, M.N., Ahmed, K., Kawakami, M., Furukawa, H., 4D printing of shape-memory hydrogels for soft-robotic functions. *Adv. Mater. Technol.*, 4, 8, 1900071, 2019 Aug.
8. Kuang, X., Roach, D.J., Wu, J., Hamel, C.M., Ding, Z., Wang, T., Dunn, M.L., Qi, H.J., Advances in 4D printing: materials and applications. *Adv. Funct. Mater.*, 29, 2, 1805290, 2019 Jan.
9. Arif, Z.U., Khalid, M.Y., Zolfagharian, A., Bodaghi, M., 4D bioprinting of smart polymers for biomedical applications: Recent progress, challenges, and future perspectives. *React. Funct. Polym.*, 179, 105374, 2022 Oct 1.

10. Wu, J.-J., Huang, L.-M., Zhao, Q., Xie, T., 4D printing: history and recent progress. *Chin. J. Polym. Sci.*, 36, 563–575, 2018, doi:10.1007/s10118-018-2089-8.
11. Zhang, Z., Demir, K.G., Gu, G.X., Developments in 4D-printing: a review on current smart materials, technologies, and applications. *Int. J. Smart Nano Mater.*, 10, 205–224, 2019b, doi:10.1080/19475411.2019.1591541.
12. Invernizzi, M., Turri, S., Levi, M., Suriano, R., 4D printed thermally activated self-healing and shape memory polycaprolactone-based polymers. *Eur. Polym. J.*, 101, 169–176, 2018.
13. Yadav, V.K., Amari, A., Gacem, A., Elboughdiri, N., Eltayeb, L.B., Fulekar, M.H., Treatment of fly-ash-contaminated wastewater loaded with heavy metals by using fly-ash-synthesized iron oxide nanoparticles. *Water (Basel)*, 15, 908, 2023, doi:10.3390/w15050908.
14. Liu, K., Tebyetekerwa, M., Ji, D., Ramakrishna, S., Intelligent materials. *Matter*, 3, 590–593, 2020, doi:10.1016/j.matt.2020.07.003.
15. Tan, L., Zhu, W., Zhou, K., Recent progress on polymer materials for additive manufacturing. *Adv. Funct. Mater.*, 30, 43, 2003062, 2020, doi: 10.1002/adfm.202003062.
16. Boydston, A.J., Cao, B., Nelson, A., Ono, R.J., Saha, A., Schwartz, J.J., Thrasher, C.J., Additive manufacturing with stimuli-responsive materials. *J. Mater. Chem. A*, 6, 20621–20645, 2018.
17. Ng, W.L., Lee, J.M., Zhou, M., Chen, Y.-W., Lee, K.-X.A., Yeong, W.Y., Shen, Y.-F., Vat polymerization-based bioprinting—Process, materials, applications and regulatory challenges. *Biofabrication*, 12, 022001, 2020.
18. Bagaria, V., Bhansali, R., Pawar, P., 3D printing- creating a blueprint for the future of orthopedics: current concept review and the road ahead! *J. Clin. Orthopaed. Trauma*, 9, 3, 207–212, 2018, doi: 10.1016/j.jcot.2018.07.007.
19. Dumpa, N.R., Bandari, S., Repka, M.A., Novel Gastroretentive Floating Pulsatile Drug Delivery System Produced *via* Hot-Melt Extrusion and Fused Deposition Modeling 3D Printing. *Pharmaceutics*, 12, 52, 2020.
20. Haq, M.I., Raina, A., Ghazali, M.J., Javaid, M., Haleem, A., Potential of 3D Printing Technologies in Developing Applications of Polymeric Nanocomposites, in: *Tribology of Polymer and Polymer Composites for Industry 4.0*, pp. 193-210, Singapore: Springer Singapore, 2021, doi: 10.1007/978-981-16-3903-6_10.
21. Gorji, N.E., Saxena, P., Corfield, M., Clare, A., Rueff, J.P., Bogan, J., González, P.G., Snelgrove, M., Hughes, G., O'Connor, R., Raghavendra, R., A new method for assessing the recyclability of powders within Powder Bed Fusion process. *Mater. Charact.*, 161, 110167, 2020 Mar 1.
22. Shishkovsky, I. and Scherbakov, V., 4D manufacturing of intermetallic SMA fabricated by SLM process. *Laser 3D Manuf. V*, 10523, 1052311, 2018.
23. Ngo, T.D., Kashani, A., Imbalzano, G., Nguyen, K.T., Hui, D., Additive manufacturing (3D printing): A review of materials, methods, applications and challenges. *Composites, Part B*, 143, 172–96, 2018 Jun 15.

24. Sing, S., Tey, C., Tan, J., Huang, S., Yeong, W., 3D printing of metals in rapid prototyping of biomaterials: techniques in additive manufacturing. *Rapid Prototyp. Biomater.*, 17–40, 2020, doi: 10.1016/b978-0-08-102663-2.00002-2.
25. Alaimo, G., Marconi, S., Costato, L., Auricchio, F., Influence of meso-structure and chemical composition on FDM 3D-printed parts. *Compos. B Eng.*, 113, 371–80, 2017.
26. Liu, Z., Wang, Y., Wu, B., Cui, C., Guo, Y., Yan, C., A critical review of fused deposition modeling 3D printing technology in manufacturing polylactic acid parts. *Int. J. Adv. Manuf. Technol.*, 102, 2877–2889, 2019.
27. Moradi, M., Beygi, R., Yusof, N.M., Amiri, A., da Silva, L.F.M., Sharif, S., 3D Printing of Acrylonitrile Butadiene Styrene by Fused Deposition Modeling: Artificial Neural Network and Response Surface Method Analyses. *J. Mater. Eng. Perform.*, 32, 2016–2028, 2023.
28. Kotomin, S.V., Kramarev, D.V., Obidin, I.M., Polunin, S.V., Influence of 3D Printing Conditions of Polyethylene Terephthalate Glycol on the Mechanical Properties of Products Based on It. *Polym. Sci. Ser. A*, 64, 617–623, 2022.
29. Radzuan, N.A.M., Khalid, N.N., Foudzi, F.M., Royan, N.R.R., Sulong, A.B., Mechanical Analysis of 3D Printed Polyamide Composites under Different Filler Loadings. *Polymers*, 15, 1846, 2023.
30. Vidakis, N., Petousis, M., David, C.N., Sagris, D., Mountakis, N., Karapidakis, E., Mechanical Performance over Energy Expenditure in MEX 3D Printing of Polycarbonate: A Multiparametric Optimization with the Aid of Robust Experimental Design. *J. Manuf. Mater. Process.*, 7, 38, 2023.
31. Mallakpour, S., Tabesh, F., Hussain, C.M., A new trend of using poly (vinyl alcohol) in 3D and 4D printing technologies: Process and applications. *Adv. Colloid Interface Sci.*, 301, 102605, 2022 Mar 1.
32. El Magri, A., Ouassil, S.E., Vaudreuil, S., Effects of printing parameters on the tensile behavior of 3D-printed acrylonitrile styrene acrylate (ASA) material in Z direction. *Polym. Eng. Sci.*, 62, 3, 848–60, 2022 Mar.
33. Olakanmi, E.O., Cochrane, R.F., Dalgarno, K.W., A review on selective laser sintering/melting (SLS/SLM) of aluminium alloy powders: Processing, microstructure, and properties. *Prog. Mater Sci.*, 74, 401–77, 2015 Oct 1.
34. Ni, J., Ling, H., Zhang, S., Wang, Z., Peng, Z., Benyshek, C., Zan, R., Miri, A.K., Li, Z., Zhang, X., Lee, J., Three-dimensional printing of metals for biomedical applications. *Mater. Today Bio*, 3, 100024, 2019 Jun 1.
35. Do, A.-V., Khorsand, B., Geary, S.M., Salem, A.K., 3D printing of scaffolds for tissue regeneration applications. *Adv. Healthcare Mater.*, 4, 1742–1762, 2015.
36. Varghese, G., Moral, M., Castro-García, M., López-López, J.J., Marín-Rueda, J.R., Yagüe-Alcaraz, V., Hernández-Afonso, L., Ruiz-Morales, J.C., Canales-Vázquez, J., Fabrication and characterisation of ceramics *via* low-cost DLP 3D printing. *Bol. Soc. Esp. Cerám. Vidr.*, 57, 1, 9–18, 2018 Jan 1.

37. Chen, Z., Li, Z., Li, J., Liu, C., Lao, C., Fu, Y., Liu, C., Li, Y., Wang, P., He, Y., 3D printing of ceramics: A review. *J. Eur. Ceram. Soc.*, 39, 4, 661–87, 2019 Apr 1.
38. Henriques, B., Pinto, P., Silva, F.S., Fredel, M.C., Fabris, D., Souza, J.C.M., Carvalho, O., On the mechanical properties of monolithic and laminated nano-ceramic resin structures obtained by laser printing. *Compos. B Eng.*, 141, 76–83, 2018.
39. Miao, S., Castro, N., Nowicki, M., Xia, L., Cui, H., Zhou, X., Zhu, W., Lee, S.-J., Sarkar, K., Vozzi, G., *et al.*, 4D printing of polymeric materials for tissue and organ regeneration. *Mater. Today*, 20, 577–591, 2017.
40. Li, J., Wu, C., Chu, P.K., Gelinsky, M., 3D Printing of Hydrogels: Rational Design Strategies and Emerging Biomedical Applications. *Mater. Sci. Eng. R Rep.*, 140, 100543, 2020.
41. Chauhan, A., Rajput, S., Malviya, R., Sridhar, S.B., Production and Multidisciplinary Applications of Bioplastic: A Biodegradable and Eco-Friendly Alternative to Plastic. *Curr. Green Chem.*, 2025.
42. Yuan, C., Roach, D.J., Dunn, C.K., Mu, Q., Kuang, X., Yakacki, C.M., Wang, T.J., Yu, K., Qi, H.J., 3D Printed Reversible Shape Changing Soft Actuators Assisted by Liquid Crystal Elastomers. *Soft Matter*, 13, 55585568, 2017.
43. Shin, D.-G., Kim, T.-H., Kim, D.-E., Review of 4D printing materials and their properties. *Int. J. Precis. Eng. Manuf. Technol.*, 4, 349–357, 2017.
44. Lee, A.Y., An, J., Chua, C.K., Two-Way 4D Printing: A Review on the Reversibility of 3D-Printed Shape Memory Materials. *Engineering*, 3, 663–674, 2017.
45. Xia, Y., He, Y., Zhang, F., Liu, Y., Leng, J.A., Review of Shape Memory Polymers and Composites: Mechanisms, Materials, and Applications. *Adv. Mater.*, *33*, 2000713, 2020.
46. Sydney Gladman, A., Matsumoto, E.A., Nuzzo, R.G., Mahadevan, L., Lewis, J.A., Biomimetic 4D Printing. *Nat. Mater.*, 15, 413–418, 2016.
47. Zarek, M., Mansour, N., Shapira, S., Cohn, D., 4D printing of shape memory-based personalized endoluminal medical devices. *Macromol. Rapid Commun.*, 38, 2, 1600628, 2017.
48. Song, Z., Ren, L., Zhao, C., Liu, H., Yu, Z., Liu, Q., Ren, L., Biomimetic non-uniform, dual-stimuli self-morphing enabled by gradient four-dimensional printing. *ACS Appl. Mater. Interfaces*, 12, 6351–6361, 2020, https://doi.org/10.1021/ acsami.9b17577.
49. Sidler, H.J., Duvenage, J., Anderson, E.J., Ng, J., Hageman, D.J., Tate, M.L.K., Prospective design, rapid prototyping, and testing of smart dressings, drug delivery patches, and replacement body parts using Microscopy Aided Design and ManufacturE (MADAME). *Front. Med.*, 5, 348, 2018.
50. Dong, Y., Wang, S., Ke, Y., Ding, L., Zeng, X., Magdassi, S., Long, Y., 4D printed hydrogels: fabrication, materials, and applications. *Adv. Mater. Technol.*, 5, 6, 2000034, 2020 Jun.

51. Huang, L., Jiang, R., Wu, J., Song, J., Bai, H., Li, B., Zhao, Q., Xie, T., Ultrafast digital printing toward 4D shape changing materials. *Adv. Mater.,* (Deerfield Beach, Fla.), 29, 7, 2016 Dec 9.
52. Kumar, D., Malviya, R., Sundram, S., Sridhar, S.B., Cellulose in 3D Printing for the Development of Biomedical Devices, in: *Sustainable Cellulose Materials for Biomedical Applications*, pp. 679–720, Springer Nature Singapore, Singapore, 2025 Nov 1.
53. Zang, T., Fu, S., Cheng, J., Zhang, C., Lu, X., Hu, J., Xia, H., Zhao, Y., 4D printing of shape-morphing liquid crystal elastomers. *Chem. Bio Eng.*, 1, 6, 488–515, 2024 Jun 3.
54. Yang, Y., Pei, Z., Li, Z., Wei, Y., Ji, Y., Making and Remaking Dynamic 3D Structures by Shining Light on Flat Liquid Crystalline Vitrimer Films without a Mold. *J. Am. Chem. Soc.*, 138, 2118–2121, 2016.
55. Barnes, M., Sajadi, S.M., Parekh, S., Rahman, M.M., Ajayan, P.M., Verduzco, R., Reactive 3D Printing of Shape-Programmable Liquid Crystal Elastomer Actuators. *ACS Appl. Mater. Interfaces*, 12, 28692–28699, 2020.
56. Ceamanos, L., Kahveci, Z., López-Valdeolivas, M., Liu, D., Broer, D.J., Sánchez-Somolinos, C., Four-Dimensional Printed Liquid Crystalline Elastomer Actuators with Fast Photoinduced Mechanical Response toward Light-Driven Robotic Functions. *ACS Appl. Mater. Interfaces*, 12, 44195–44204, 2020.
57. Jiang, Y., Ng, E.L., Han, D.X., Yan, Y., Chan, S.Y., Wang, J., Chan, B.Q., Self-healing polymeric materials and composites for additive manufacturing. *Polymers.*, 15, 21, 4206, 2023 Oct 24.
58. Kuang, X., Chen, K., Dunn, C.K., Wu, J., Li, V.C., Qi, H.J., 3D printing of highly stretchable, shape-memory, and self-healing elastomer toward novel 4D printing. *ACS Appl. Mater. Interfaces*, 10, 8, 7381–8, 2018 Feb 28.
59. Wang, J., Lin, X., Wang, R., Lu, Y., Zhang, L., Self-healing, photothermal-responsive, and shape memory polyurethanes for enhanced mechanical properties of 3D/4D printed objects. *Adv. Funct. Mater.*, 33, 15, 2211579, 2023 Apr.
60. Rastogi, P. and Kandasubramanian, B., Breakthrough in the printing tactics for stimuli-responsive materials: 4D printing. *Chem. Eng. J.*, 366, 264–304, 2019 Jun 15.
61. Krstic, M., Miladinovic, Z.R., Barudzija, T., Mladenovic, A., Suljovrujic, E., Stimuli-responsive copolymeric hydrogels based on oligo (ethylene glycol) dimethacrylate for biomedical applications: An optimisation study of pH and thermoresponsive behaviour. *React. Funct. Polym.*, 170, 105140, 2022 Jan 1.
62. Hussain, K., Aslam, Z., Ullah, S., Shah, M.R., Synthesis of pH responsive, photocrosslinked gelatin-based hydrogel system for control release of ceftriaxone. *Chem. Phys. Lipids*, 238, 105101, Aug. 2021.

63. Nichterwitz, M., Honnali, S., Kutuzau, M., Guo, S., Zehner, J., Nielsch, K., Leistner, K., Advances in magneto-ionic materials and perspectives for their application. *APL Mater.*, 9, 3, 2021 Mar 1.
64. Erol, O., Pantula, A., Liu, W., Gracias, D.H., Transformer hydrogels: A review. *Adv. Mater. Technol.*, 4, 1900043, 2019, doi:10.1002/admt.201900043.
65. Shi, H., Wang, C., Ma, Z., Stimuli-responsive biomaterials for cardiac tissue engineering and dynamic mechanobiology. *APL Bioeng.*, 5, 1, 2021 Mar 1.
66. Ali, I., Xudong, L., Xiaoqing, C., Zhiwei, J., Pervaiz, M., Weimin, Y., Haoyi, L., Sain, M., A review of electro-stimulated gels and their applications: Present state and future perspectives. *Mater. Sci. Eng. C.*, 103, 109852, 2019 Oct 1.
67. Hill, J., Wildman, R., Mata, A., Exploiting the fundamentals of biological organization for the advancement of biofabrication. *Curr. Opin. Biotechnol.*, 74, 42–54, 2022 Apr 1.
68. Grinberg, D., Siddique, S., Le, M.Q., Liang, R., Capsal, J.F., Cottinet, P.J., 4D Printing based piezoelectric composite for medical applications. *J. Polym. Sci. B Polym. Phys.*, 57, 109–115, 2019, doi:10.1002/polb.24763.
69. Mota, C., Camarero-Espinosa, S., Baker, M.B., Wieringa, P., Moroni, L., Bioprinting: From Tissue and Organ Development to in Vitro Models. *Chem. Rev.*, 120, 10547–10607, 2020.
70. Wang, C., *et al.*, Advanced reconfigurable scaffolds fabricated by 4D printing for treating critical-size bone defects of irregular shapes. *Biofabrication*, 12, 4, 45025, 2020.
71. Miao, S., Zhu, W., Castro, N.J., Nowicki, M., Zhou, X., Cui, H., Fisher, J.P., Zhang, L.G., 4D printing smart biomedical scaffolds with novel soybean oil epoxidized acrylate. *Sci. Rep.*, 6, 1, 27226, 2016 Jun 2.
72. Chawla, S., Sharma, A., Bandyopadhyay, A., Ghosh, S., Developmental biology-inspired strategies to engineer 3D bioprinted bone construct. *ACS Biomater. Sci. Eng.*, 4, 10, 3545–60, 2018 Sep 11.
73. Prakash, A., Malviya, R., Singh, D.P., Transformative Potential and Healthcare Applications of 3D Printing. *Curr. Pharm. Des.*, 30, 42, 3311–21, 2024 Dec.
74. Ge, Q., Sakhaei, A.H., Lee, H., Dunn, C.K., Fang, N.X., Dunn, M.L., Multimaterial 4D Printing with Tailorable Shape Memory Polymers. *Sci. Rep.*, 6, 31110, 2016.
75. Zhou, W., Qiao, Z., Nazarzadeh Zare, E., Huang, J., Zheng, X., Sun, X., Shao, M., Wang, H., Wang, X., Chen, D., 4D-printed dynamic materials in biomedical applications: chemistry, challenges, and their future perspectives in the clinical sector. *J. Med. Chem.*, 63, 15, 8003, 2020 Apr 7.
76. Shafranek, R.T., Millik, S.C., Smith, P.T., Lee, C.U., Boydston, A.J., Nelson, A., Stimuli-responsive materials in additive manufacturing. *Prog. Polym. Sci.*, 93, 36–67, 2019 Jun 1.
77. Kantaros, A., Petrescu, F.I., Ganetsos, T., From Stents to Smart Implants Employing Biomimetic Materials: The Impact of 4D Printing on Modern Healthcare. *Biomimetics.*, 10, 2, 125, 2025 Feb 19.

78. Wu, S.D. and Hsu, S.H., 4D bioprintable self-healing hydrogel with shape memory and cryopreserving properties. *Biofabrication.*, 13, 4, 045029, 2021 Sep 27.
79. Zhou, X., Yu, X., You, T., Zhao, B., Dong, L., Huang, C., Zhou, X., Xing, M., Qian, W., Luo, G., 3D Printing-Based Hydrogel Dressings for Wound Healing. *Adv. Sci.*, 11, 47, 2404580, 2024 Dec.
80. Mirani, B., Pagan, E., Currie, B., Siddiqui, M.A., Hosseinzadeh, R., Mostafalu, P., Zhang, Y.S., Ghahary, A., Akbari, M., An Advanced Multifunctional Hydrogel-Based Dressing for Wound Monitoring and Drug Delivery. *Adv. Health Mater.*, 6, 1–15, 2017.
81. Baniasadi, M., Yarali, E., Bodaghi, M., Zolfagharian, A., Baghani, M., Modeling multi-stimuli-responsive shape memory polymers with reduced working temperature. *Int. J. Mech. Sci.*, 106082, 2020, indi.
82. Mu, X., Bertron, T., Dunn, C., Qiao, H., Wu, J., Zhao, Z., Saldana, C., Qi, H.J., Porous polymeric materials by 3D printing of photocurable resin. *Mater. Horiz.*, 4, 442–44, 2017.

2

Materials for 3D and 4D Bioprinting

Sheetal S. Samant[1]*, Vijay A. Jagtap[2], Shruti V. Kolambkar[3] and Faizan A. Beerwala[4]

[1]*Department of Pharmacognosy, Yashwantrao Bhonsale College of Pharmacy, University of Mumbai, Sawantwadi, Maharashtra, India*
[2]*Department of Pharmaceutical Chemistry, Yashwantrao Bhonsale College of Pharmacy, University of Mumbai, Sawantwadi, Maharashtra, India*
[3]*Department of Pharmacognosy, KLE College of Pharmacy, KLE Academy of Higher Education and Research, Belagavi, Karnataka, India*
[4]*Department of Pharmacology and Toxicology, KLE College of Pharmacy, KLE Academy of Higher Education and Research, Belagavi, Karnataka, India*

Abstract

3D and 4D bioprinting is important in tissue engineering and the regenerative medicines. The biomaterials used in this technique mimic the original tissue, and these biomaterials play the important role in the success of this bioprinting technology. This chapter provides detailed information on materials used in 3D and 4D bioprinting. It also discusses the characteristics, merits and demerits of these materials. Collagen, alginate, and gelatin are some of the most widely used natural polymers. These natural polymers are biocompatible and they are also biodegradable, that is why they have become very popular and are widely used in bioprinting. Synthetic polymers are also used in bioprinting because they offer great mechanical strength. Polycaprolactone, polylactic acid, and polyglycolic acid are some of the most commonly used synthetic polymers in bioprinting. Complex tissues structures can be generated using hydrogels that are derived from natural and synthetic polymers. Tissues that can change shape in response to stimuli can be created using 4D bioprinting technology. Such stimuli-responsive tissues can be created by using hydrogels and shape-changing polymers. This chapter provides thorough information on materials used in 3D and 4D bioprinting. It also

**Corresponding author*: sheetal.samant99@gmail.com; ORCID: 0000-0002-6681-9292
Vijay A. Jagtap: ORCID: 0000-0002-6386-4290
Shruti V. Kolambkar: ORCID: https://orcid.org/0009-0002-9613-1621
Faizan A. Beerwala: ORCID: https://orcid.org/0000-0001-9464-3899

Rishabha Malviya and Mukesh Roy (eds.) Biocomposites in 3D and 4D Printing: Materials, Techniques, and Applications in Regenerative Medicine, (39–62) © 2026 Scrivener Publishing LLC

provides insights into the applications, future research and trends in 3D and 4D bioprinting technology.

***Keywords*:** Bioprinting, tissue engineering, 3D printing, biomaterials, hydrogels, regenerative medicine, 4D printing

2.1 Introduction

2.1.1 Overview of Bioprinting Technologies

Intricate tissue structures are developed due to rapid advancement that is done recently in the field of bioprinting technology. The automated methods used in bioprinting provides the stability and accuracy to this technology. The most crucial part of the bioprinting technology is the bioink, which is comprised of the active cells, hydrogels and biomaterials. Personalized tissues can be developed using this dynamic composition. These generated tissues have same property and functions as original tissues. The tissue structures of preferred form can be created in bioprinting when bioink undergoes cross-linking or stabilization. Bioinks can be created using the natural, synthetic, and hybrid biomaterials. Bioink can also be formulated using only the cell clusters, this avoids the use of different biomaterials. The biological, rheological, and mechanical property of the desired tissue and bioink should be similar, so that the bioprinted structure can function properly and amalgamate well with the body [1]. 3D structures made from biomaterials using conventional techniques are having less accuracy and flexibility. Biologically functional products can be designed due to advancements that is carried in the biofabrication techniques. Living cells, bioactive molecules and cell aggregates are properly assembled by using this biofabrication techniques. Functional structures are then obtained by subjecting this composition to tissue maturation process [2, 3]. It is now possible to construct complex biological structures by using this innovative techniques, that is bringing a very positive impact in the field medical science. Damaged tissues and organs can be regenerated using this technology. Disease modeling and tissue repair is also possible using this technology. Miniature organs can be developed using this technology. Bone, blood vessels, skin, and cartilages are some of the tissues that can be bioprinted [4]. 4D bioprinting has made a huge progress by fabricating tissue structures that are capable of responding to stimuli. 4D bioprinting technology is advanced version of 3D bioprinting. 4D bioprinting helps to create tissue structures that change shape on stimuli. The organs and tissue structures created using 4D bioprinting mimics the original organs and it can readily integrate with the human body by using less invasive surgery. The healthcare sector can be immensely developed by using this 4D bioprinting technology. Artificial organ can be created using this 4D bioprinting technology. Such organs are capable of mimicking the original organ and can easily integrate with the human body without any hassle [5]. Hydrogels and shape memory polymers can be used in 4D bioprinting to create tissue structures that can change property to stimuli. Treatment can be tailored to individual's healthcare needs of the patient by using the technique of 4D bioprinting, this promotes the concept of personalized healthcare solutions [6].

2.1.2 Evolution from 3D to 4D Bioprinting

Dynamic and responsive structures can be created using 4D bioprinting technology. 4D bioprinting uses advanced biofabrication techniques. Static biological structures are created by 3D bioprinting. Stimuli-responsive material is used to create biological structures. These stimuli-responsive material are capable of changing their property, shape, and functionality on application of stimuli. 4D bioprinting has caused a rapid advancement in drug-delivery, tissue engineering, and regenerative medicines. The tissues generated through 4D bioprinting can mimic natural morphogenic processes. They have ability to morph over time. They can change on application of stimuli like temperature or light. Bioinks and smart biomaterials are used to create such types of tissues. The evolution from 3D to 4D led to development of dynamic shape-morphing tissues from static tissue structure. Precise control over morphing pattern, enhancement in tissue alignment, and improved structural and functional property can be achieved by utilizing granular support hydrogels. These hydrogels have good viscoelastic property [7]. Complex tissues structures of cartilage, bones and vascular networks can be created using 4D bioprinting. These tissue structure have good structural and functional properties [8]. Programmable, shape-changing tissues of osteoid, chondroid, and vascular system can be produced by 4D bioprinting by using intelligent biomaterials [8]. Innovative drug delivery systems can be developed by using the technique 4D bioprinting, which will help in fulfilling the specific requirements of the patients. The drug delivery system developed has many beneficial properties. They are made up of bio-erodible, and they degrade after delivering the required quantity of the drug. They have multi-chamber system. This promotes the concept of personalized medicines, where patients receive treatment based on their requirements [6]. Regenerative medicines is greatly transformed by 4D bioprinting as it has developed intelligent and biomimetic structures. They also promote regeneration of tissues. The external stimuli lead to changing of the shape of these structures, which makes these tissue structures flexible. These structures have resembling characteristics of the native tissue. 4D bioprinting leads to development of sophisticated and flexible structures. Personalized adaptive response and improved biological interfaces can be easily achieved by the integration of multi-material and biocomposite approach [9]. 3D bioprinting helps in creation of static structures, which have fixed shape and properties, while 4D bioprinting helps in creation of stimuli-responsive structures. 4D bioprinting uses smart and stimuli-responsive composites. This has led to development of bioengineered organs that can not only mimic the original tissues but are also compatible with the body [5]. The tissue structures created from 4D bioprinting are influenced by living cells, but the structures developed in 3D bioprinting were static. 4D bioprinting can develop structures that have the capacity to undergo self-repair and cellular remodeling. This allows the human cells to change the tissue architecture over time [10].

2.1.3 Importance of Material Selection in Bioprinting

The survival and function of the constructed tissue is greatly dependent on the choice of materials used to create the tissue structure. Cell compatibility, printing capability, and therapeutic potential is greatly influenced by the choice of the material. When the material is selected properly, it will help in development of tissue construct that has good performance.

The material should promote the growth and proliferation of cells. Polysaccharide-based hydrogels are most commonly used material for bioprinting because they are biocompatible. They also have good properties, which make them suitable candidate for bioprinting [11]. Biocompatibility with living tissues and immune response must be also taken into consideration while selecting the material. When the material is biocompatible, it has many advantages. Such material promotes proper interaction with tissues and also support growth of cells which will ultimately help in development of construct which will function properly. Cytotoxicity and immunogenicity are the factors that require careful consideration, while selection of material [12]. Printability is greatly affected by flow behavior and mechanical property of the bioink. Proteins and polysaccharides are also explored to check whether they are suitable material for bioprinting. These materials can be used for bioprinting. They can maintain structural stability during and after bioprinting [13]. Materials are selected based on their applications. Natural polymers like silk fibroin and collagen are also explored to check their compatibility with biological systems [14]. Mechanical property and biocompatibility of tissue construct is completely dependent on selection of material. The effectiveness of bioprinting process can be enhanced by selection of proper materials which will properly integrate with the body [15]. The activity of the cells, mechanical properties, and printing process is completely dependent on bioprinting materials [16].

2.2 Fundamentals of Bioprinting Materials

2.2.1 Key Properties of Bioprinting Materials

The success of bioprinting is completely dependent on materials that is used for bioprinting. These materials must be such that they must maintain structural integrity by supporting cell viability, proliferation and differentiation. Cell growth and tissue formation are promoted by the use of hydrogels that provide gel-like environment. The bioprinting materials should be biocompatible. They should have good mechanical property. They should undergo cross-linking to form stable structures. All these characteristics helps in forming functional tissue constructs that can be used for biomedical applications. Compatibility of bioink is the important factor, as it should not produce any undesired immune response in the body after implantation. Cell viability, proliferation, and differentiation is achieved by using bioprinting material that is biocompatible, non-toxic, and it should produce minimal immune response in the body. Natural and synthetic polymers can be used in bioprinting. Each material has its own merits, demerits, and applications. It is very important to select proper material for bioprinting as the structural integrity and biological function of that tissue construct is dependent on the bioprinting material [12]. Good cellular compatibility is one of the most important property of bioinks. It should have good rheological property as it will enhance the process of bioprinting. The bioinks have cross-linking property that helps them to change from sol to gel [17]. The shape and structure of bioprinted tissue is dependent on the mechanical property of bioinks. The biomaterial must be selected by considering its chemical stability, mechanical strength, and biocompatibility. Some hydrogels have weak mechanical property, so they need to modified to improve their property. When a hydrogel-based bioprinted structure is implanted, it should be stable at that site and should exhibit good mechanical property and performance. There is a need to focus on strategies that can help in enhancing the mechanical property of biomaterials and the bioprinted constructs [18]. Some hydrogels have self-healing property that can return to their original shape after deforming to other shape. This enhances the

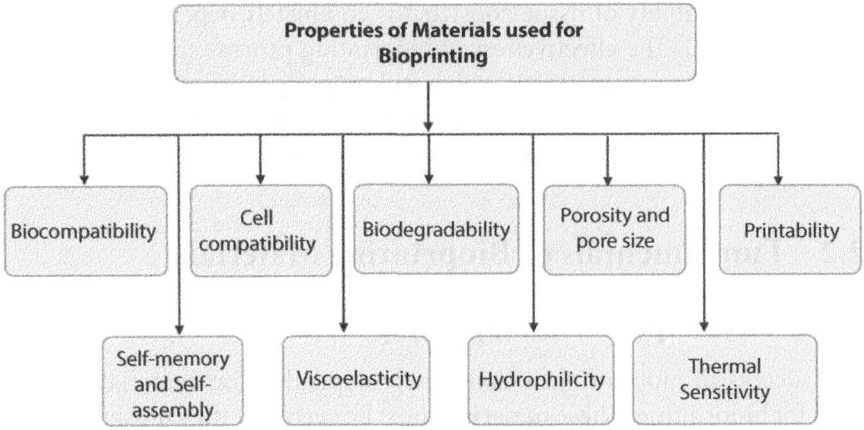

Figure 2.1 Properties of materials used for bioprinting.

mechanical property of the hydrogels. This property is useful in variety of applications [19]. Cross-linking is essential in transforming bioinks from sol to gel state. This helps the tissue construct to maintain its shape after printing. Crosslinking is done to achieve stable network of molecular bonds. This crosslinking can be induced by light, heat, enzymes, and ions [17]. While going for crosslinking, care must be taken that there is balance between the stiffness and self-healing property. The biomaterials used for extrusion based bioprinting must have shear-thinning ability. This ensures that cells are printed properly, and the shape of cell is also maintained after bioprinting [19]. They can mimic the natural cells and thereby support in cell function. This is achieved as they have ability to modify their mechanical property and rigidity [20]. The properties of bioprinting materials are illustrated in Figure 2.1. There is a rapid advancement in the field of bioprinting. But there are several challenges prevailing in this field. There is a need to optimize the formulation for different tissues, as it will help to ensure uniformity and consistency among batches. There is also need for developing proper regulatory guidelines [17].

2.2.2 Classification of Bioinks

A. Classification by Origin

Natural-Origin Bioinks: Collagen, gelatin, and alginate are the natural polymers used in composition of bioink. These natural polymers have good compatibility with body. They also mimic the native extracellular matrix. But they may face many challenges. The challenges are mostly associated with their mechanical property [17].

Synthetic-Origin Bioinks: Synthetic polymers offer great mechanical property and there is consistency in every batch. Synthetic polymers are used instead of animal-based natural polymers, as animal-based natural polymer causes adverse immune responses. Poly(α-amino acids) is example synthetic polymer used to prepare these bioinks [21].
Hybrid Bioinks: Both natural polymers and synthetic polymers are used to formulate hybrid bioink. They have advantages and best characteristics of both natural and synthetic polymers. These bioinks have good mechanical property. They also have good biological functionality [17].

B. Classification by Composition
Hydrogel-Based Bioinks: Bioinks can be formulated by using hydrogels. The constructs created by using hydrogel are strong as the hydrogels undergo cross-linking. Light, temperature, heat, and ion-tropic reactions can induce crosslinking in hydrogel [17].
Nanocomposite Bioinks: The biological and mechanical property of bioinks can be improved by using nano-composite bioinks. These bioinks are very sensitive to external stimuli like heat, ion, pH, and light. This enhances the functionality of the tissue constructs [22].

C. Classification by Functional Properties
Scaffold-Based Bioinks:
They are designed in such a way that they will act as a scaffold. They provide good mechanical strength and durability.

Scaffold-Free Bioinks:
They rely on assembling of the cells to form the tissue. So there is no need of supporting scaffold. They are used for creating tissue construct that can mimic the native tissue [23].

2.2.3 Challenges in Material Selection

Efficacy and functionality of bioprinted tissues is affected by various challenges that are faced during material selection. The challenges include material property, optimization process and regulatory issues. These challenges must be properly addressed to enhance the performance of bioink in tissue engineering can be enhanced by overcoming the challenges. The bioink should support cell viability and proliferation of the cells. The biomaterials derived from plants have good biocompatibility. However, they face several challenges related to mechanical integrity and printing resolution [24]. Printability is also affected by viscosity of bioink. The biofluids show non-Newtonian flowing behaviour so it becomes to predict the viscosity. Advanced machine learning algorithms can be used to predict the viscosity. This can bring consistency in printing result [25]. Crosslinking methods are adopted to improve the physical property of the material. The extent and timings of crosslinking should be monitored. If it is not monitored properly then it may affect the printing process [26]. The mechanical property of the tissue construct can be improved by mixing different types of biomaterial, but must be done cautiously. There is need to develop good quality control guidelines to maintain the batch-to-batch uniformity. Regulatory hurdles are the major issue as there are no standardized bioinks [17].

2.3 Materials for 3D Bioprinting

2.3.1 Natural Polymers

Natural polymers promote cell growth because they have good biocompatibility. Alginate and collagen are some of the commonly used natural polymers. These polymers form a hydrogel structure. The cells are encapsulated into these hydrated gel like structure. The liquid environment is similar to that of cell. This enhances cell viability, cell growth, and cell proliferation. Some bioinks are formulated by blending natural polymers with other material. These bioinks can place the cells properly in the matrix during 3D bioprinting.

This is method is one step and is very efficient than the traditional method which is multi-step and very tedious. Chitosan, alginate, gelatin, and collagen are some of the best examples of natural polymers. Each polymer has its own special characteristics, so they can be used for different application in tissue engineering. The performance and stability of bioinks formulated from natural polymers can be improved by cross-linking techniques. cross-linking can be done by using heat, enzyme, light, and ion. The crosslinking enables the bioink to transform from liquid to solid after printing [17].

2.3.2 Synthetic Polymers

Synthetic polymers have gained immense importance in the field of bioprinting. PCL, PEG, and PLA are some of the most commonly used synthetic polymers in bioprinting. They have many biomedical applications. These synthetic polymers are compatible with the body. They have good mechanical strength [28].

2.3.3 Composite and Hybrid Bioinks

The hybrid bioinks take the best properties of natural and synthetic polymers. These combination produces bioinks that offer several benefits. They have good mechanical strength. They have good compatibility with body. The degradation properties of these bioinks can also be tailored. These property helps in creating a scaffold that can mimic the native extracellular matrix and promote cell growth. Extrusion-based printing, laser-assisted printing, and inkjet-based printing use hybrid bioinks. They can precisely deposit the cell-encapsulated material, and hence, it helps in the creation of complex tissue structures. This is achieved due to rheological properties of the hybrid bioinks. Stabilization of hybrid material is achieved by cross-linking. Cross-linking can be done by means of heat, temperature, ion and enzyme. These help the bioink to transform from sol to gel, thus maintain the structural integrity of bioprinted tissue [17].

2.4 Materials for 4D Bioprinting

2.4.1 Stimuli-Responsive Biomaterials

These materials can change its shape and properties on application of stimuli. The external stimuli can be light, temperature, and pH. The tissue structures created using these materials can adapt to environmental change. These materials have wide applications in the field of regenerative medicines. These materials can be used to create novel drug delivery system. they can be also used to create small implants.

I. Thermoresponsive Hydrogels:
The changes in the temperature can bring change in the shape of these materials. These materials can be stimulated by changing the temperature, which will bring change in the morphology [32, 33]. PNVCL, PEG, PNIPAM, collagen, and gelatin are some of the examples of thermoresponsive polymers. They are most commonly used polymers, as they have good printing ability [31].

II. pH-Sensitive Polymers

These polymers have many functional groups present in them. The functional groups include pyridine, carboxyl, sulfonic phosphate, or acid. These material change in response to change in pH. The functional chemical groups present in material enable the material to respond when there is change in pH [34]. The chemical groups can undergo protonation and deprotonation, when the pH is changed [35]. This brings change in the structure. These materials are divided into two types, i.e., acidic polymers and basic polymers. The acidic polymers have $-SO_3H$ and $-COOH$ groups present in them. Basic polymers have $-NH_2$ group present in them. Collagen, hyaluronic acid, alginate, keratin, chitosan, PAA, PASP, and PHIS are some of the examples pH-sensitive polymers [31, 36].

III. Biological-Responsive Materials

Enzymes, glucose, and other biomolecules react with these polymers and produce a response. The peptide sequence in the polymer is cut by various enzymes, this can bring change in the structure of the polymer and can cause swelling of that polymer [31]. These materials have the ability to respond to biological stimuli. Biomolecules, bioenzymes and glucose act as the stimuli and can bring change in these biomaterials. These polymers are widely used in tissue engineering and regenerative medicines [31].

IV. Moisture-Responsive Materials

These materials have ability to adsorb or lose moisture in cell culture medium or humid environment. Swelling, twisting, folding or deformation are the morphological changes that can be achieved in case of these polymers. They function by swelling and deswelling mechanism. They have wide range of applications in soft robotics [31, 37].

V. Magneto-Responsive Materials

These materials have capability to respond to magnetic field. Magnetic field can trigger response in these biomaterials. They have many applications in biomedical field [38].

Paramagnetic and ferromagnetic nanoparticles can be incorporated into these magneto-responsive material, and they can be used to design novel drug delivery system. When such materials are used in drug delivery systems, they provide controlled release of drugs [31].

VI. Electro-responsive materials
These biomaterials can be conducted electrically. Bioelectric stimulation can control these polymer biomaterials. These materials can shrink, lengthen, or bend in different directions.

The electro-responsive property of these materials is enhanced by encapsulating them in nanomaterials like carbon nanotubes and graphene nanoparticles that are conductive in nature. Researchers have used 4D bioprinting to develop Origami robot that uses polypyrrole-based polymer. These polypyrrole-based polymers are electro-sensitive in nature. They have good biocompatibility and printing ability [31, 39].

2.4.2 Shape-Memory Polymers

When these materials are exposed to stimuli like light or heat, they have the ability to return to their original shape. They have many applications in 4D bioprinting. When there is change in moisture, temperature or light, these materials return back to their original form [27, 28].

2.4.3 Smart Hydrogels for Dynamic Structures

Smart hydrogels can create tissue structures that can respond to environmental stimuli. They have wide variety of applications in tissue engineering. These hydrogels are designed in such a way that it can product tissue structure that can mimic the native tissue. Hydrogels can be engineered such that they can change the shape in response to cellular force and external stimuli. The external stimuli include heat and light. This helps in developing complex tissue structures. The structures created using these smart hydrogels are precise. Smart hydrogels are designed in such a way that they can change their shape due to cell generated forces and external stimuli [7]. PEGDA and sodium alginate can be combined to produce the hydrogel of great mechanical strength. These materials have good biocompatibility. These properties are important to maintain cell viability during and after bioprinting [29]. Cross-linkers like poly(N-isopropylacrylamide) can be used in temperature-responsive hydrogels to obtain complex that can undergo change in shape [30]. Smart hydrogels can produce tissue constructs that can integrate easily with body systems [5].

2.5 Fabrication Techniques and Material Considerations

2.5.1 Extrusion-Based Bioprinting

Extrusion-based bioprinting is the most commonly used method for bioprinting. It has many advantages. It is very economical. It has very easy operating procedures.

Bioink is deposited through nozzle to create three-dimensional structures. This technique is widely used in tissue engineering and regenerative medicines. But these technique faces some challenges. One of the challenge is that these bioinks are complex. Another challenge is related to reproducibility of printed structures [40]. AI is used to overcome these challenges. Conventional Neural Network is used adjust the extrusion parameters and correct errors. This improves the accuracy of bioprinting [41]. The most critical aspect is the flow of non-Newtonian bioinks. The non-Newtonian bioinks show complex viscoelastic and thixotropic behaviour [42]. Issues, like tissue mismatch and complexity in operations, are solved by using robot assisted *in situ* bioprinting systems. Such systems have proper control over extrusion parameters [43]. Bioprinting process is influenced by extruder geometry and type of bioink. Innovative design tools are used to plan the process properly and have proper control on printing pressure and cell viability [44].

2.5.2 Inkjet-Based Bioprinting

Inject-based bioprinting is used to create complex 3D tissue structures. This technique has played a major role in the field of tissue engineering. In this technique, cells are deposited precisely. The bioink contains living cells. The bioink is deposited precisely by using thermal and piezoelectric printing modalities. This technique faces many challenges related to cell viability, bioink viscosity, and shear stress during droplet formation.

In this technique, biomaterials are created very precisely. This technique allows proper placement of cells and proteins. This technique has applications in drug delivery, bioelectronics and tissue engineering [49]. Very intricate tissue structures can be designed by using this technique [45]. Cell viability can be increased by using techniques like pneumatic conveying printing [46]. This technique can encapsulate many materials within hydrogel [47]. Bioinks need to have low viscosity, so this can bring limitations on the materials that are used.

The cells also get affected by the force exerted during droplet formation. So there is a need to optimize the printing parameters [48].

2.5.3 Stereolithography

Bioprinting can be done by stereolithography (SLA). It is very useful for 4D bioprinting. This technique uses photocrosslinkable polymers. The scaffolds created by using this technology promotes cell viability. This technique offers great resolution. This technique can be used to design hydrogel scaffolds. They have very fine filaments. The mechanical property and biocompatibility with cell can be improved by adding bioresins. This technique is used to produce scaffold that is required as per the need of patient. This will help in constructing tissue structures according to the requirement of patients. So this technique has great importance in regenerative medicines [50].

2.5.4 Laser-Assisted Bioprinting

Intricate tissue structures are created using laser. This technique has great importance in regenerative medicines. The laser created cavitation bubbles. The living cells are carefully placed inside the cavity. This technique creates tissue structure that has high resolution in cell pattern. This technique also ensures controlled deposition of biomaterials. This technique is used to design tissue structure that are very complex and fragile. Hence, it plays a great role in regenerative medicines. In this technique, laser energy is used to create cavitation bubbles. The cells are then transported to desired location. This technique offers great precision. It can create tissue structures that mimic the native tissues [51].

2.6 Applications of 3D and 4D Bioprinting Materials

2.6.1 Applications of 3D Bioprinting

Tissue Engineering: Complex tissues can be created by using this technology. The bioink is deposited in layers to form tissue constructs. Viable cells and other complementary materials are present in the bioink. They must promote cell proliferation and differentiation. 3D bioprinting helps in creating complex tissue constructs [52].

Regenerative Medicine:
3D bioprinting has important role in regenerative medicines. 3D bioprinting offers alternatives for organ transplant. This increases success of transplant and reduce risk of organ rejection. 3D bioprinting can produce structures that promote cell growth, cell colonization and cell regeneration [52].
Organ Transplantation: 3D bioprinting is widely used in tissue engineering to overcome the hurdles of organ transplantation. This technique created bioengineered organs that are similar to native organs. These bioengineered organs are composed of cells and biomaterials. They are prepared by depositing layers of bioink [52].

Wound Healing:
Bioengineered skin substitutes are developed using 3D bioprinting technology. This is bought great advancement in wound healing. Biomaterials are explored for their applications in tissue engineering. Nanocellulose based bioinks have promising applications in wound healing.

Advanced bioprinting techniques are used to create tissue structure that are used in wound healing [52]. Certain biomaterials have the ability to promote tissue healing and they can also restore the function of that tissue. The printing must be accurate and consistent so that the tissue structure generated functional and enhances wound healing [53].

2.6.2 Applications of 4D Bioprinting

Smart Implants: 4D bioprinting can create implants that can respond to body changes. These implants have brought a great revolution in fields of tissue engineering and regenerative medicines. The implants can respond to change in pH and temperature. The developed smart implants can mimic the native tissue [54].

Responsive Drug Delivery Systems:
Some biomaterials are designed in such a way that they will form a scaffold and deliver the drug to desired site, in response to stimuli. This stimuli can be triggered by thermal fluctuations or humidity. This has brought advancement in drug delivery system [51].
Organ Regeneration: 4D bioprinting can develop bioengineered organs. These bioengineered organs are similar to the native organs. They have same function as the native organs. 4D bioprinting uses biomaterial that responds to stimuli [54]. The applications of 3D bioprinting and 4D bioprinting are illustrated in Figure 2.2.

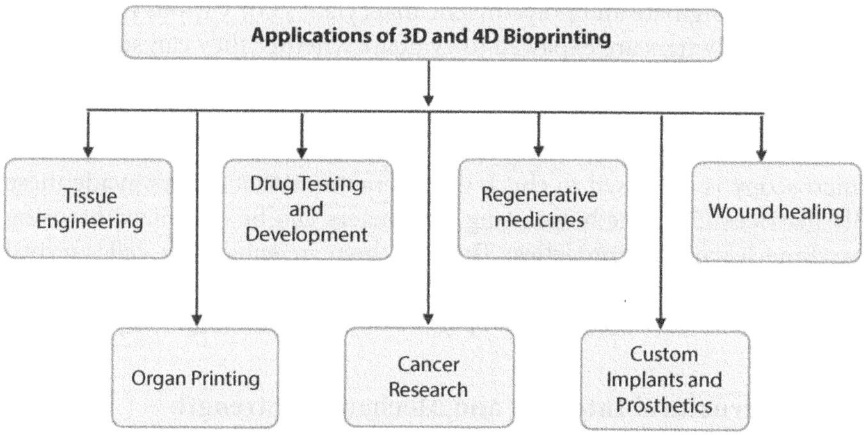

Figure 2.2 Applications of 3D and 4D bioprinting.

2.7 Challenges and Future Directions

2.7.1 Biocompatibility and Cell Viability Issues

Biocompatibility and cell survival are the major issues associated with bioprinting. The challenges associated with Bioprinting are illustrated in Figure 2.3. The bioinks must be selected carefully and the printing conditions must be optimized. This will promote the function and survival of the cells. This can help in overcoming the challenges associated with bioprinting. Cell viability can be promoted by using biocompatible materials

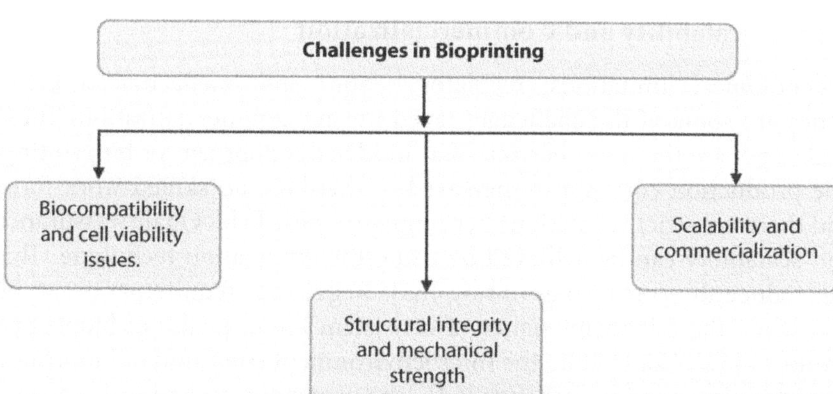

Figure 2.3 Challenges in bioprinting.

like sodium alginate and polyethylene diacrylate [55]. Various natural and synthetic polymers are explored to evaluate whether they can support cell viability [12]. The condition of cell inside the 3D complex can be evaluated using automated confocal microscopy. These advanced technology is far better than the conventional cell viability assays. Automated confocal microscopy is also used to check the interactions that occurs inside these 3D matrices [56]. The bioprinting parameters can be optimized by using machine learning approaches. This can help in enhancing cell survival during the bioprinting process [57]. The success of bioprinting is mainly dependent on the viscoelastic properties of the bioink [58].

2.7.2 Structural Integrity and Mechanical Strength

Structural integrity and mechanical strength is the one of the major challenge in bioprinting. Hydrogels have good biocompatibility, that is why they are most widely used in bioprinting process. But these hydrogels do not have good mechanical properties. The hydrogels aresoft, that is why they have poor mechanical strength. This challenge affects the printability. It also hinders the ability of these scaffold to function *in vivo* [59].

Norbornene-functionalized collagen is a photo-crosslinkable material. This material has good rheological properties and mechanical properties. The constructed tissues have greater stability and structural integrity [60]. The tensile strength of scaffold can be increased, when cells are infiltered and proliferated within the hydrogel matrix [61]. The mechanical properties of the tissue constructs improve, when cell begin to grow and proliferate within these tissue constructs [62].

2.7.3 Scalability and Commercialization

Technological limitations, regulatory issues and intellectual property issues are some of the challenges faced during commercialization. These challenges are the main barriers that hinder development of bioprinting. The production cost is also increased as there are no skilled work-force and there are strict regulations pertaining to these field. Quality assurance and scalability can be achieved by using the automation technique. This can reduce the cost of bioprinting [63]. Regulatory requirements can be met when there is maintenance of consistency and quality of bioprinted products [64]. Controlling the microenvironment condition of bioprinted tissues is also a major challenge [64]. The ownership and ethical management of bioprinted materials is properly addressed by current intellectual property laws [65].

2.7.4 Emerging Trends and Innovations

Complex tissue structures can be created by combining the techniques of biology and technology. There is rapid development in biomaterials and printing process. This has created many innovations in the field of regenerative medicines. Materials like alginate, collagen and hydroxyapatite can ensure cell viability [66]. Rapid advancement is made in the field of neuroscience. Novel bioinks are developed to achieve accuracy in bioprinting. These bioinks can develop tissue structures that have same biological property as native brain tissues [67].

Complex tissues and organs can be developed by integrating bioprinting, AI and advanced computational modeling [68].

2.8 Conclusion

There is rapid advancement in the field of regenerative medicines and tissue engineering due to emergence of novel materials in 3D and 4D bioprinting. Polymers, ceramics, and hydrogels can be used to design tissue structure that can change their characteristics on application of stimuli. Dynamic and responsive tissues can be created by using smart materials. Complex tissue structures are designed using biomaterials like polymer, ceramics, and hydrogels. The tissue structures created have customized properties. Dynamic and responsive tissue structures can be developed by using 4D bioprinting. 4D bioprinting uses smart materials to create such structures. There are many advancements made in this field. But there are still a few challenges prevailing in this field. The researchers should focus to develop materials that will create bioink, which will maintain cell viability and functionality for long duration. These materials should have structure and properties of the native structure. There is rapid advancement in the medical field due to this bioprinting technology. Patients are getting personalized treatment. Bioprinting has brought a rapid development and innovation in the field of drug discovery, tissue engineering and regenerative medicines. The development in science and technology will bring great advancement in the field of bioprinting. This will help in developing implants that will improve the health and quality of human life.

AI disclosure statement: The author acknowledges the use of AI-based tools for minor language editing or grammatical correction. However, all intellectual contributions, analyses, and conclusions presented in this chapter are entirely the author's own.

References

1. Gungor-Ozkerim, P.S., Inci, I., Zhang, Y.S., Khademhosseini, A., Dokmeci, M.R., Bioinks for 3D bioprinting: an overview. *Biomater. Sci.*, 6, 5, 915–46, 2018.
2. Groll, J., Boland, T., Blunk, T., Burdick, J.A., Cho, D.W., Dalton, P.D., Derby, B., Forgacs, G., Li, Q., Mironov, V.A., Moroni, L., Biofabrication: reappraising the definition of an evolving field. *Biofabrication.*, 8, 1, 013001, 2016 Jan 8.
3. Murphy, S.V. and Atala, A., 3D bioprinting of tissues and organs. *Nat. Biotechnol.*, 32, 8, 773–85, 2014 Aug.
4. Adhyapak, H.M., Gaonkar, S.L., Namratha, B., Overview of 3D Bioprinting Technology, in: *Compendium of 3D Bioprinting Technology*, pp. 27–46, CRC Press, Boca Raton, Florida, USA, 2025 Feb 12.
5. Imrie, P. and Jin, J., 4D Bioprinting: Keeping the Technology Alive. *Macromol. Mater. Eng.*, 2400386, 2025.
6. Karthi Kumar, R., Samyuktaa, V., Indhu, S., Harini, R., Advances in 4D Bio-Printing Technology for Enhanced Drug Delivery Systems. *J. Comput. Mech. Manage.*, 3, 4, 01–06, 2024, doi: 10.57159/jcmm.3.4.24142.
7. Pramanick, A., Hayes, T., Sergis, V., McEvoy, E., Pandit, A., Daly, A.C., 4D bioprinting Shape-Morphing tissues in granular support hydrogels: sculpting structure and guiding maturation. *Adv. Funct. Mater.*, 35, 5, 2414559, 2025 Jan.
8. Rana, A., Mittal, A., Vashist, C., Rajput, S., Sridhar, S.B., Malviya, R., Exploration of 4D Printing and its Applications in the Biomedical Sciences. *Curr. Pharm. Des.*, 2025.
9. Morouço, P., Lattanzi, W., Alves, N., Four-Dimensional Bioprinting As a New Era for Tissue Engineering and Regenerative Medicine. *Front. Bioeng. Biotechnol.*, 5, 61, 2017, https://doi.org/10.3389/fbioe.2017.00061.
10. Kim, H., Hu, G., Graham, A.J., Gu, G.X., Gartner, Z.J., Four-Dimensional Bioprinting: Harnessing Active Mechanics to Build with Living Inks. *Cold Spring Harbor Perspect. Biol.*, a041557, 2024 Nov 25.
11. Vo, P.H., Do, D.Q., Vu, B.T., Tang, T.N., Doan, H.N., Tam, P.T., Nguyen, T.H., Preparation and Characterization of the Hydrogel System N, O-Carboxymethyl Chitosan/Oxidized Xanthan Gum, in: *International Conference on the Development of Biomedical Engineering in Vietnam*, pp. 261–272, Springer Nature Switzerland, Cham, 2022 Dec 27.

12. Assad, H. and Kumar, A., Materials for 3D Bioprinting, in: *Compendium of 3D Bioprinting Technology*, pp. 177–190, CRC Press, Boca Raton, Florida, USA, 2025 Feb 12.
13. Schwab, A., Levato, R., D'Este, M., Piluso, S., Eglin, D., Malda, J., Printability and Shape Fidelity of Bioinks in 3D Bioprinting. *Chem Rev.*, 120, 19, 11028–11055, 2020 Oct 14. doi: 10.1021/acs.chemrev.0c00084. Epub 2020 Aug 28. PMID: 32856892; PMCID: PMC7564085.
14. Sonkar, A.K., Kundu, A., Sharma, D., Agarwal, V., Chanda, A., Polymeric Biomaterials for Bioprinting Applications, in: *Materials for Biomedical Simulation: Design, Development and Characterization*, pp. 171–191, Springer Nature Singapore, Singapore, 2023 Sep 27.
15. Yogeshwaran, S., Hosseinabadi, H.G., Gendy, D.E., Miri, A.K., Design considerations and biomaterials selection in embedded extrusion 3D bioprinting. *Biomater. Sci.*, 12, 18, 4506–18, 2024.
16. Fritschen, A., Acedo Mestre, M., Scholpp, S., Blaeser, A., Influence of the physico-chemical bioink composition on the printability and cell biological properties in 3D-bioprinting of a liver tumor cell line. *Front. Bioeng. Biotechnol.*, 11, 1093101, 2023 Feb 23.
17. Shukla, P., Ghosh, S., Ghosh, S., Saumya, S.K., Kiranmai, G., Bhavsar, A., Meenu, T.S., Pati, F., Bioinks for 3D Bioprinting, in: *Compendium of 3D Bioprinting Technology*, pp. 191–220, CRC Press, Boca Raton, Florida, USA, 2025 Feb 12.
18. Ballard, A., Patush, R., Perez, J., Juarez, C., Kirillova, A., Bioprinting: mechanical stabilization and reinforcement strategies in regenerative medicine. *Tissue Eng. Part A*, 30, 13-14, 387–408, 2024 Jul 1.
19. Karvinen, J. and Kellomäki, M., 3D-bioprinting of self-healing hydrogels. *Eur. Polym. J.*, 23, 112864, 2024 Feb.
20. Wierzbicka, A., Bartniak, M., Grabarczyk, J., Biernacka, N., Aftyka, M., Wójcik, T., Bociaga, D., Improvement of Mechanical Properties of 3D Bioprinted Structures through Cellular Overgrowth. *Appl. Sci.*, 14, 19, 8977, 2024 Oct 5.
21. Golunova, A., Dvořáková, J., Velychkivska, N., Strachota, B., Dydowiczová, A., Trousil, J., Proks, V., Fully synthetic, tunable poly (α-amino acids) as the base of bioinks curable by visible light. *Biomed. Mater.*, 19, 3, 035035, 2024 Apr 26.
22. Salahshour, P., Abdolmaleki, S., Monemizadeh, S., Gholizadeh, S., Khaksar, S., Nanobiomaterials/bioinks based scaffolds in 3D bioprinting for tissue engineering and artificial human organs. *Adv. Biol. Earth Sci.*, 9, 2024 Jan 2.
23. Benwood, C., Chrenek, J., Kirsch, R.L., Masri, N.Z., Richards, H., Teetzen, K., Willerth, S.M., Natural Biomaterials and Their Use as Bioinks for Printing Tissues. *Bioengineering, Basel, Switzerland),* 8, 2, 27, 2021, https://doi.org/10.3390/bioengineering8020027.

24. Hasan, M.M., Ahmad, A., Akter, M.Z., Choi, Y.J., Yi, H.G., Bioinks for bioprinting using plant-derived biomaterials. *Biofabrication.*, 16, 4, 042004, 2024 Aug 22.
25. Xu, Y., Sarah, R., Habib, A., Liu, Y., Khoda, B., Constraint based Bayesian optimization of bioink precursor: a machine learning framework. *Biofabrication.*, 16, 4, 045031, 2024 Aug 28.
26. Carpentier, N., Parmentier, L., Van der Meeren, L., Skirtach, A.G., Dubruel, P., Van Vlierberghe, S., Optimization of hybrid gelatin-polysaccharide bioinks exploiting thiol-norbornene chemistry using a reducing additive. *Biomed. Mater.*, 19, 2, 025025, 2024 Feb 5.
27. Zhu, Y., Guo, S., Ravichandran, D., Ramanathan, A., Sobczak, M.T., Sacco, A.F., Patil, D., Thummalapalli, S.V., Pulido, T.V., Lancaster, J.N., Yi, J., 3D-Printed Polymeric Biomaterials for Health Applications. *Adv. Healthcare Mater.*, 14, 1, 2402571, 2025 Jan.
28. Chauhan, A., Rajput, S., Malviya, R., Sridhar, S.B., Production and Multidisciplinary Applications of Bioplastic: A Biodegradable and Eco-Friendly Alternative to Plastic. *Curr. Green Chem.*, 2025.
29. Greco, I., Machrafi, H., Iorio, C.S., Double-Network Hydrogel 3D BioPrinting Biocompatible with Fibroblast Cells for Tissue Engineering Applications. *Gels.*, 10, 11, 684, 2024 Oct 23.
30. Nakamura, K., Di Caprio, N., Burdick, J.A., Engineered Shape-Morphing Transitions in Hydrogels Through Suspension Bath Printing of Temperature-Responsive Granular Hydrogel Inks. *Adv. Mater.*, 36, 47, 2410661, 2024 Nov.
31. Arif, Z.U., Khalid, M.Y., Zolfagharian, A., Bodaghi, M., 4D bioprinting of smart polymers for biomedical applications: Recent progress, challenges, and future perspectives. *React. Funct. Polym.*, 179, 105374, 2022 Oct 1.
32. Li, J., Wu, C., Chu, P.K., Gelinsky, M., 3D printing of hydrogels: Rational design strategies and emerging biomedical applications. *Mater. Sci. Eng.: R: Rep.*, 140, 100543, 2020 Apr 1.
33. Sharma, R., Malviya, R., Singh, S., Prajapati, B., A critical review on classified excipient sodium-alginate-based hydrogels: modification, characterization, and application in soft tissue engineering. *Gels.*, 9, 5, 430, 2023 May 22.
34. Ngan Giang, N., Kim, S.G., In, I., Park, S.Y., Real-time wireless monitoring of cell proliferation and detachment based on pH-responsive conductive polymer dots. *Anal. Chem.*, 93, 24, 8638–46, 2021 Jun 10.
35. Zhang, Z., Demir, K.G., Gu, G.X., Developments in 4D-printing: a review on current smart materials, technologies, and applications. *Int. J. Smart Nano Mater.*, 10, 3, 205–24, 2019 Jul 3.
36. Lai, J., Ye, X., Liu, J., Wang, C., Li, J., Wang, X., Ma, M., Wang, M., 4D printing of highly printable and shape morphing hydrogels composed of alginate and methylcellulose. *Mater. Des.*, 205, 109699, 2021 Jul 1.
37. Maraveas, C., Bayer, I.S., Bartzanas, T., 4D printing: Perspectives for the production of sustainable plastics for agriculture. *Biotechnol. Adv.*, 54, 107785, 2022 Jan 1.

38. Chu, H., Yang, W., Sun, L., Cai, S., Yang, R., Liang, W., Yu, H., Liu, L., 4D printing: a review on recent progresses. *Micromachines.*, 11, 9, 796, 2020 Aug 22.
39. Ding, H., Zhang, X., Liu, Y., Ramakrishna, S., Review of mechanisms and deformation behaviors in 4D printing. *Int. J. Adv. Manuf. Technol.*, 105, 4633–49, 2019 Dec.
40. Tseng, W.C., Liao, C.Y., Chassagne, L., Cagneau, B., An ink-insensitive deep learning model for improving the printing quality in extrusion-based bioprinting. *Int. J. Bioprint.*, 18, 6701, 2024 Dec.
41. Kelly, D., Sergis, V., Ventura i Blanco, L., Mason, K., Daly, A.C., Autonomous control of extrusion bioprinting using convolutional neural networks. *Adv. Funct. Mater.*, 12, 2424553, 2024 Dec.
42. Gharraei, R., Bergstrom, D.J., Chen, D., Extrusion bioprinting from a fluid mechanics perspective. *Int. J. Bioprint.*, 3973, 2024, https://doi.org/10.36922/ijb.3973.
43. Wang, Z., Wu, Y., Cui, Y., Wang, J., Developing a Robot-assisted in Situ Bioprinter Using Extrusion Technology, in: *2024 World Rehabilitation Robot Convention (WRRC)*, pp. 1–5, IEEE, 2024 Aug 23.
44. Chirianni, F., Vairo, G., Marino, M., Influence of extruder geometry and bio-ink type in extrusion-based bioprinting *via* an *in silico* design tool. *Meccanica*, 59, 8, 1285–99, 2024 Aug.
45. Zhu, H., Li, R., Li, S., Guo, K., Ji, C., Gao, F., Zheng, Y., Zhu, R., Wang, H., Zhang, L., Liu, L., Multi-physical field control piezoelectric inkjet bioprinting for 3D tissue-like structure manufacturing. *Int. J. Bioprint.*, 10, 3, 2120, 2024 Feb 29.
46. Bożek, J., Kurchakova, O., Michel, J., Groß, I., Gerhards, L., Zhang, Y., Brand, I., Bräuer, A.U., Pneumatic conveying inkjet bioprinting for the processing of living cells. *Biofabrication.*, 17, 2, 025003, 2025 Jan 24.
47. Suntornnond, R., Ng, W.L., Shkolnikov, V., Yeong, W.Y., A facile method to fabricate cell-laden hydrogel microparticles of tunable sizes using thermal inkjet bioprinting. *Droplet.*, 3, 4, e144, 2024 Oct.
48. Ng, W.L. and Shkolnikov, V., Optimizing cell deposition for inkjet-based bioprinting. *Int. J. Bioprint.*, 2024.
49. Cheng, C., Williamson, E.J., Chiu, G.T., Han, B., Engineering biomaterials by inkjet printing of hydrogels with functional particulates. *Med-x.*, 2, 1, 9, 2024 Jul 3.
50. Viray, C.M., van Magill, B., Zreiqat, H., Ramaswamy, Y., Stereolithographic visible-light printing of poly (l-glutamic acid) hydrogel scaffolds. *ACS Biomater. Sci. Eng.*, 8, 3, 1115–31, 2022 Feb 18.
51. Hakobyan, D., Kerouredan, O., Remy, M., Dusserre, N., Medina, C., Devillard, R., Fricain, J.C., Oliveira, H., Laser-assisted bioprinting for bone repair, in: *3D Bioprinting: Principles and Protocols*, pp. 135–44, Springer Nature, New York, NY, USA, 2020.

52. Sahoo, J., Prakashan, D., Gandhi, S., 3D Printing of Biomaterials, in: *Compendium of 3D Bioprinting Technology*, pp. 238–255, CRC Press, Boca Raton, Florida, USA, 2025 Feb 12.
53. Md Fadilah, N.I., Yamagishi, R., Zainuddin, A.N., Mustakim, N.N., Rosemanzailani, F.A., Fauzi, M.B., 3D Bioprinting of Nanocellulose Scaffolds for Tissue Engineering Applications.
54. Abolhassani, S., Fattahi, R., Safshekan, F., Saremi, J., Hasanzadeh, E., Advances in 4D Bioprinting: The Next Frontier in Regenerative Medicine and Tissue Engineering Applications. *Adv. Healthcare Mater.*, 2024, https://doi.org/10.1002/adhm.202403065.
55. Kim, J., Characterization of Biocompatibility of Functional Bioinks for 3D Bioprinting. *Bioengineering (Basel).* 10, 4, 457, 2023 Apr 9, doi: 10.3390/bioengineering10040457. PMID: 37106644; PMCID: PMC10135811.
56. Avnet, S., Di Pompo, G., Borciani, G., Fischetti, T., Graziani, G., Baldini, N., Advantages and limitations of using cell viability assays for 3D bioprinted constructs. *Biomed. Mater.*, 19, 2, 025033, 2024 Feb 13.
57. Mohammadrezaei, D., Podina, L., De Silva, J., Kohandel, M., Cell viability prediction and optimization in extrusion-based bioprinting *via* neural network-based Bayesian optimization models. *Biofabrication.*, 16, 2, 025016, 2024 Feb 15.
58. Jergitsch, M., Alluè-Mengual, Z., Perez, R.A., Mateos-Timoneda, M.A., A systematic approach to improve printability and cell viability of methylcellulose-based bioinks. *Int. J. Biol. Macromol.*, 253, 127461, 2023 Dec 31.
59. Chen, A., Deng, S., Lai, J., Li, J., Chen, W., Varma, S.N., Zhang, J., Lei, C., Liu, C., Huang, L., Hydrogels for Oral Tissue Engineering: Challenges and Opportunities. *Molecules.* 28, 9, 3946, 2023 May 7. doi: 10.3390/molecules28093946. PMID: 37175356; PMCID: PMC10179962.
60. Guo, K., Wang, H., Li, S., Chang, P., He, C., Sun, Q., Zheng, X., Bioprinting of light-crosslinkable neutral-dissolved collagen to build implantable connective tissue with programmable cellular orientation. *Biofabrication.*, 15, 3, 035007, 2023 Apr 12.
61. El-Sherbiny, I.M., Yacoub, M.H., Hydrogel scaffolds for tissue engineering: Progress and challenges. *Glob. Cardiol. Sci. Pract.*, 2013, 3, 316–42, 2013 Nov 1. doi: 10.5339/gcsp.2013.38. PMID: 24689032; PMCID: PMC3963751.
62. Rajput, S., Malviya, R., Uniyal, P., Advances in the treatment of kidney disorders using mesenchymal stem cells. *Curr. Pharm. Des.*, 30, 11, 825–40, 2024 Mar 1.
63. Lindner, N. and Blaeser, A., Scalable biofabrication: A perspective on the current state and future potentials of process automation in 3D-bioprinting applications. *Front. Bioeng. Biotechnol.*, 10, 855042, 2022 May 20.
64. Zhang, X., Zhao, G., Ma, T., Simmons, C.A., Santerre, P., A critical review on advances and challenges of bioprinted cardiac patches. *Acta Biomater.*, 2024 Oct 5.

65. Kantaros, A., Ganetsos, T., Petrescu, F.I.T., Alysandratou, E., Bioprinting and Intellectual Property: Challenges, Opportunities, and the Road Ahead. *Bioengineering (Basel)*, 12, 1, 76, 2025 Jan 15. doi: 10.3390/bioengineering12010076. PMID: 39851350; PMCID: PMC11761581."
66. Sousa, A.C., Mcdermott, G., Shields, F., Alvites, R., Lopes, B., Sousa, P., Moreira, A., Coelho, A., Santos, J.D., Atayde, L., Alves, N., Innovative Ink-Based 3D Hydrogel Bioprinted Formulations for Tissue Engineering Applications. *Gels.*, 10, 12, 831, 2024 Dec 17.
67. Bocheng, X. and França, R., Innovative 3D bioprinting approaches for advancing brain science and medicine: a literature review. *Biomed. Phys. Eng. Express*, 2024 Sep 11.
68. Yüksel, F., Advanced 3D Bioprinting Techniques for Regenerative Medicine: Exploring Applications in Tissue Engineering and Personalized Therapeutics. *Next Front. Life Sci. AI*, 8, 1, 2024.

3

3D Printing Biocomposites: Fabrication Methods and Process Optimization

N. Selvasudha[1], A.M. Renosa Sarun[1], K. Subalakshmi[1], V.G. Akilandeshwari[1], G. Geetha[1] and U.M. Dhanalekshmi[2*]

[1]*Department of Pharmaceutics, Kamarajar College of Pharmacy, Tamil Nadu, India*
[2]*Department of Pharmacology and Biological Sciences, National University of Science and Technology, College of Pharmacy, Muscat, Oman*

Abstract
Additive manufacturing (AM), commonly referred to as 3D printing, rapid prototyping, or solid freeform technology, enables the customized fabrication of objects with tailored mechanical and electrical properties. This approach offers advantages, such as moldability, geometric customization, simplicity, and reduced waste, compared to traditional subtractive manufacturing methods. 3D printing does not require any phase transition of device and equipment after creating the design in computer aided software. The materials used for printing are composite polymer, metals, and ceramic matrix composites. The materials used for 3D printing should have mechanical properties like flexibility or rigidity based on application, thermal stability, resistance to solvents, acids, and environmental degradation, compatibility with other materials for composite designs, safe for use in implants, prosthetics, and other biomedical devices, ease of processing, sustainability, compatibility with various 3D printing technologies, recyclable, and environmentally friendly options and the weight ratio are optimized based on the technology. Commonly, 3D printing fabrication methods are fused filament fabrication, fused deposition modeling, stereolithography, selective laser sintering, bioprinting, inkjet bioprinting, extrusion bioprinting, binder jetting (BJ), material extrusion (ME), material jetting (MJ), and many others are still under developments. Selecting fabrication techniques should be based on material expenses, resolution, complexity of geometry, it has its own leads and confrontation with fabrication of many composites. In each case, different AM processes, material and data acquisition techniques are

**Corresponding author*: dhanalekshmi@nu.edu.om

Rishabha Malviya and Mukesh Roy (eds.) Biocomposites in 3D and 4D Printing: Materials, Techniques, and Applications in Regenerative Medicine, (63–90) © 2026 Scrivener Publishing LLC

used before deriving the 3D printed. Process optimization includes parameter optimization, material selection, fiber orientation, and layer thickness. Conventional techniques of fabrication methods, like phase inversion, solvent casting, solute leaching, and electrospinning, often lack in their ability to incorporate fine internal architectural details and controlled porosity while ensuring reproducibility. 3D printing methods have started to address these issues by regulating precise pores sizes and their uniform distribution. It has to begin to revolutionize manufacturing, especially in biomedical engineering with increasing need for patient-specific medical devices. This chapter highlights the biocomposite and its properties, fabrication methods, and process optimization of 3D printing.

***Keywords*:** Sustainability, fused filament technology, inkjet, extrusion, electrospinning

3.1 Introduction

Three-dimensional (3D) anatomical models and customized physical implants are made using additive manufacturing (AM) techniques. The two most common ways to replicate biomodels from real anatomy are by using a computer simulation or by casting a chosen model into a physical form. Analysis of the processes and components using the virtual model is thought to be more difficult and less successful because research has revealed that recognition performance was noticeably better for real items than for a virtual image. Additionally, the fabrication of a casted polymer model is more time consuming and more expensive to customize. AM technology has made it possible to create complex, precise, and patient-specific models without the need for trained sculptors or artists. 3D printing uses sophisticated material expansion to produce tangible components from a geometrical interpretation. The procedure is rapid. These days, 3D printing is widely used globally. Large-scale customization and the creation of open-source blueprints in the fields of agriculture, medical, transportation, railroads, and construction are the main uses for 3D printing. According to the requirements, 3D printing, also known as AM, is a designing process used to deliver a 3D component layer by layer from a computerized 3D model. AM fabricating being faster, there is growing interest in fast prototyping and the creation of modifiable parts are 3D printing's advantageous features. Materials used by various kinds of 3D printers include metal, composites, ceramic, and polymers. Fused filament fabrication, a popular 3D printer, for instance, employs polymers and polymer-based composites along with several kinds of fillers, such as ceramic and metal nanoparticles.

With the exception of directed-energy deposition, a polymer is one of the most popular feedstocks for all forms of 3D printing techniques practically. The manufacturing process can be adapted to use various polymer types, such as thermoplastics and photopolymers, depending on the printing techniques and applications.

3.2 3D Printing Fabrication Methods

3.2.1 Extrusion-Based 3D Printing

This technology has started to gain popularity within the industry over the past few years through its ability to create high fidelity functioning products without moulds or inner tooling. This technology of additive manufacturing has discovered numerous implementations in various aspects of production, including aerospace, automotive, and biomedical industries, which make use of intricate parts such as engines, implants, and surgical instruments. Among these key advantages of extrusion-based additive manufacturing is those techniques are able to create geometrically complex and detailed features while achieving high strength and stiffness for many applications. Extrusion-based 3D is the most versatile method because it allows the incorporation of additional components that can be made of plastic, metals or even ceramics, each with their distinct properties.

3.2.2 Fused Deposition Modeling or Fused Filament Fabrication

Fused filament fabrication (FFF) also referred fused deposition modeling (FDM). Recently, FDM gained a lot of traction among industry users because it is effective in making complex geometrical shapes and highly functional products with great detail and accuracy [1]. FDM is done by first melting a thermoplastic filament, then the molten material is pumped through a heated nozzle which moves side to side to the desired position. This is known as 3D modeling and it is widely used because of its ease of application. It is a technique adopted and used across the globe [2]. Once the material has been applied to the upper side of a 3D model base, it adheres to the previous layer and thus attaches itself strongly to this layer that prevents it from breaking off.

One of the main advantages of using FDM technology is that parts are strong and stiff which makes this method beneficial in many fields such as medicine, the military, and cars [3]. Other methods of manufacturing, which include the use of FDM process include not only ABS and nylon but also PLA, broaden the usage of thermoplastics and provide advantages to FDM [4]. New materials and techniques like carbon fiber reinforced polymers (CFRP) and metal filled filament are furnishings of the new FDM technology. (CFRP) and metal filled filaments (MFs) were the new materials introduced that improved the fused deposition modeling (FDM) technology [5]. The acquisition of new materials allowed the FDM method to be able to parts of superior mechanical strength and thermal conductivity.

The development of FDM multi-material printing methods, as well as 4D printing, has brought new possibilities for product design that are both rich in detail and complex [6]. The geometric complexity and functional gradients of parts can be achieved through the multimaterial printing method which facilitates the simultaneous extrusion of different materials [7]. In contrast, 4D printing is the process of embedding materials and polymeric composites into structures that remodels according to changes in the environment. Such changes may include temperature and humidity [8]. From aerospace and automotive to biomedical and even consumer goods, it seems like the possible applications of FDM technology are vast and jaw droppingly impressive in scope [1]. Either way, FDM technology is useful when making lightweight components and satellite parts, as well as airplane interiors and other parts [3].

3.2.3 Direct Ink Writing

Direct ink writing (DIW) 3D printing possesses high design flexibility and functional gradient fabrication capability, it is useful for many. Within the sector of energy storage, high-efficiency batteries and supercapacitors have been made using DIW 3D printing technology [9]. It is well known that DIW 3D printing can create complex geometries with functional gradients, so it can be used in different fields [10]. Not only that, high numbers of inks and pastes of different viscosities and non-Newtonian [11] properties can be used with DIW 3D printing.

Numerous phenotypes, including biomedical engineering, energy storage, and soft robotics, are among the many uses for DIW 3D printing [11]. Custom-fit implants, prostheses, and surgical models have been produced in the biomedical field using 3D DIW printing [12].

3.3 Powder-Based 3D Printing

3.3.1 Selective Laser Sintering

Selective laser sintering (SLS) was used in numerous industries, including aerospace, automotive, and biomedical, have used it to manufacture complex parts like dashboard trim, engine components, and customized implants [13]. SLS also produces components with exceptional mechanical strength and stiffness, which is very beneficial for many applications [5]. Metals, ceramics, and polymers are just a few of the powdered materials that can be thinned and employed thanks to SLS; each of these materials has unique dynamics and features [9]. New powdered materials and methods, like the utilization of nanomaterials and hybrid powders, have been developed as a result of recent developments in SLS technology, increasing the technology's possibilities [14]. Applications for SLS are numerous and diverse, spanning industries like consumer products, automotive, medicinal, and aerospace [15]. SLS is used to make customized implants, prosthetics, and surgical models in the biomedical field [5]. However, a number of obstacles still exist, including as the post-processing required for SLS products and the resolution and accuracy limitations of SLS systems [15]. Further developments in SLS technology are expected to get over these challenges and expand its uses even more.

3.3.2 Electron Beam Melting

The other manufacturing process of 3D printing which will revolutionize the manufacturing industry gradually. This technology is melting metal powder by the energy of electron beam. A trailing electron beam is the main driver of the entire process inside a vacuum chamber. With the vacuum creating an oxygen-free environment and the low hydrogen risk, high purity is assured. To reduce the relapse tension and as a result create distortion and deformation, the high temperature of 700°C keep being inside the chamber when the part is created. First, EB pre-heats the powder bed with a high-feed rate, then liquefies based on the geometry defined *via* the CAD file. In electron beam melting (EBM) processing, every layer of the cut section in split into two steps. First it defines the outer shape of the part, a step called forming, and the powder in the mold is fused in this manner to create a layer. This process continues until the desired 3D object is completely finished [16].

EBM is widely used in the biomedical, for the production of complex parts implants [5]. The ability of EBM to produce components with remarkable mechanical strength and stiffness is a key benefit, making it appropriate for many benefits [15]. Furthermore, a broad range of powdered materials with unique qualities and attributes, including metals, ceramics, and composites, can be used using EBM [9]. Recent advancements in EBM technology have increased the potential of EBM by enabling the development of new powdered materials and techniques, such as the incorporation of nanomaterials and hybrid powders [14]. Applications of EBM are diverse and can be found in consumer products, medicinal, and aerospace industries, as reported by Kang *et al.* [13].

3.3.3 Binder Jetting

Binder jetting is a modern additive manufacturing method that builds complex three-dimensional objects by stacking powdered materials with a liquid binder [17]. Recently, this method has gained attention because it can make functional items with high accuracy and precision without using regular molds or tools [15]. The binder jetting method is commonly employed in different fields, including biomedical, for creating complex components like implants, and surgical tools [18]. In this advanced additive manufacturing process called "binder jetting," a liquid binder is used to firmly bond powdered materials together in a layered fashion to produce detailed three-dimensional objects [17]. This technology has recently attracted much interest due to its ability to create functional items with remarkable precision and without needing standard tooling or molds [6]. In the biomedical field, it is used to manufacture customized surgical tools, implants, and prosthetic devices [17].

3.4 Resin-Based 3D Printing

3.4.1 Stereolithography

Stereolithography (SLA), a resin-based 3D printing technology, has been successfully implemented in various fields particularly health care [19] stereolithography technique was familiarized in the last part of the 1980s. Although many other methods have been developed since, stereolithography is the most powerful and versatile technique. It offers the most notable creation correctness and there is an expanding number of materials that can be handled. Stereolithography is, like most advanced, flexible fabrication processes, an additive manufacturing process, which means that it allows for the construction of parts from a computer-aided design (CAD) file.

Stereolithography is based on the photopolymerization of a fluid tar, which is spatially controlled so as to generate 3D substances. Using a PC controlled laser bar or rational light projector with a PC-driven structure stage, a picture is illuminated on the surface of a sap. So, in the case above, the pitched is anchored to a defined depth, causing it to adhere to a substrate. After proven polymerization of the principal layer has occurred, the stage containing the puller and paddles is pulled away from the surface, and the assembled, polymerized layer is recoated with fluid tar. They are recycled to create vivid, three-dimensional writing. As a consequence of an abundant pitch, involving the washing system, a structure with depleting and manufactured structure is got.

SLA ideology in healthcare through preparing custom prosthetics for patients as dental implants and hip replacement. The advantages of SLA [20] are high resolution, high precision, and high speed (fast print speed). Nonetheless, SLA has some drawbacks such as small build largeness (space), expensive apparatus, and a few materials [21]. SLA, or stereolithography, is part of the oldest resin additive manufacturing and is used in numerous industries. With high resolution, accuracy, and print speeds, dirt-cheap KODAK system is an attractive proposition for creating complex geometries and custom components. Two of its biggest drawbacks are its small build size and high cost (and sometimes limited materials). These constraints need to be tackled, and additional investigation is necessary to reach out SLA's optimum performance.

3.4.2 Digital Light Processing

Digital light processing (DLP) 3D printing technology is the resin based printing technology and its most commonly used in the industrial applications due to its low price, high speed, and high resolution [22]. This approach, known as digital light processing (DLP), utilizes a projector to rapidly project a light on a liquid resin to cure it layer by layer, so that complex geometries and products can be manufactured [23]. Digital light processing (DLP) was introduced in the 1980s by the American technology firm Texas Instruments (TI, 1987). From the beginning, DLP was focused on display applications such as projectors and televisions. DLP has recently been adapted to 3D printing (e.g., resin 3D printing) [22]. Another widely employed principle in this domain is DLP, which is founded on the use of a projector for the curing of the liquid resin [23].

3.4.3 Continuous Light Interface Production

Continuous liquid interface production (CLIP) technology is a resin-based 3D printing technology, which makes the technology enter a new era of additive manufacturing. They recount the CLIP that the field uses and, its pros and cons. CLIP an abbreviation for continuous liquid interface production was also developed, and was originally introduced to the world, in 2015, by Carbon3D a firm founded by a group of scientists under Joseph DeSimone, a professor. This is done by (I) dispensing a layer of ℓ in resin on the build platform, (II) polymerizing ℓ into the correct shape with focused light, and (III) using oxygen to inhibit curing so that complex geometries can be made.

Factors influencing quality of the CLIP prints many process parameters needs to be considered for having a good quality of CLIP prints, such as light intensity, oxygen flow rate, and layer thickness. In medical care, CLIP offers personalized implants, whether they are dental implants or a hip replacement. In addition to speed and resolution, it can also produce complex geometries. In medical care, CLIP offers personalized implants, whether they are dental implants or a hip replacement. In addition to speed and resolution, it can also produce complex geometries [24]. For this reason, DeSimone and company devised CLIP in order to eliminate the slow printing times and limited build volumes of conventional 3D printing technologies. CLIP utilizes photopolymerization, where light and oxygen solidifies a liquid resin. A photo initiator created from light and oxygen leads to hardening of the liquid resin, enabling extremely complex geometries and custom, patient-specific devices to be produced at speeds never seen before.

Tip: The curing time is a function of the illuminating power, means the light intensity increases, the curing time decreases. The higher the oxygen flow rate, the more complex the geometries are obtained. The thicker the layers, the slower the build process, this is how quickly the build rises up. One of its most significant limitations is that CLIP has a small build size, it is expensive to produce, and has limited available options for materials. We all remember CLIP, the breakthrough freeform resin-based 3D printing process that revolutionized the field of additive manufacturing. It also generates a potential high-speed, high-resolution, high-geometric-complexity printing technique for unique product(s). All of that good stuff in terms of a relatively small build size, high cost, and low availability of materials for a lot are monumental limitations. And still some boundaries should be solved to improve CLIP efficiency and there is more work to be done for that [25].

3.5 Process Optimization

3.5.1 Parameters Optimization

Nevertheless, on the other side, optimization of 3D printing parameters are required for the significant improvement of mechanical properties, surface quality and performance of prepared biocomposites. These parameters are nozzle temperature, speed of the printing, infill density, thickness of layer and bed temperature. These parameters can be fine-tuned to achieve good results.

For instance, Kovan *et al.* [26] evaluated residual tensile strength of polyamide-66 Subjected to permanent deformation through a Taper Point Method. The impact of the printing limits also examined. The potential of biocomposites in the additive manufacturing process was like best mechanical and surface-quality at a nozzle temperature, printing speed, and infill density of 220°C, 40 mm/s, and 100%, respectively, were reported. Their findings also demonstrated how misconfigured parameters, like an overly high temperature for the nozzles, could destroy the material, and highlighted how low infill densities negatively impacted structure integrity. Similarly, Chen *et al.* [27] studied the printing speed and extrusion temperatures as one of the main parameters to reduce cases of warping and delamination during the 3D printing of natural fiber-reinforced biocomposites. They suggested systematic investigation of the interaction effects of parameters: process optimization methods like Design of Experiment (DoE).

3.5.2 Material Optimization

Optimisation of materials means the right combination of matrix materials, the right reinforcement and additives to withstand environmental effects of biocomposites [28]. investigated the polymer matrix compatibility of polylactic acid (PLA) in the composite with the natural fibers in 2024. Polylactic acid (PLA), a biodegradable polymer with good printability, is one of the principal candidates for sustainable 3D-printed biocomposites. Their mechanical performance was significantly shown for increases of cellulose loading with good interfacial bonding with cellulose fibers which reached impressive tensile strengths of 37.2 MPa and tensile modulus of 2.8 GPa, which means better 35% increase of tensile strength and stiffness of PLA with the use of cellulose fibers (Jour. Of Reinforced Plastics & Comp.).

A significant contribution is the work of who reported that the increase in the graphene oxide content leading to formation of stable PLA bionic Graphite nanoparticles, and also mechanical properties of the bio composite improved significantly [23].

Moreover, even hybrid enforcements have evidenced their effectiveness. In particular, they incorporated both natural and synthetic fibers into the PLA, in order to increase strength, stiffness, and biodegradability properties.

3.5.3 Fiber Orientation

Fiber orientation takes part in a major role in the anisotropic mechanical qualities of the additive manufacturing biocomposite. The fibers having the largest sturdiness and firmness in both direction along with the direction of the applied force makes them exhibit isotropic orthotropic for all other directions. It showed multi axis 3D printing for 283 continuous filament reinforced composite with precise fiber alignment control. They found that aligned fibers improved tensile strength by up to 50% relative to the randomly oriented fibers. According to the researchers, advanced printing processes that allow control over the placement or arrangement of fibers are needed to develop these biocomposites mechanically [29]. Another study by Liu *et al.* [30] reported PLA and flax fibers based biocomposite fiber with effective fiber orientation scalars and quentals forces.

3.5.4 Post-Processing and Finishing

Relevance of post-processing processes is to augment the machine like qualities, surface state of being 3D biocomposites. Post-processing consists of some practices similarly to sanding, heat therapy, chemical smoothing, coating, etc. Based on research conducted in a thesis from Malmö University by on post processing systems for sanding had the most dramatic effect on reducing surface roughness, and chemical treatment (acetone vapor smoothing) both improved surface finish and inter-layer adhesion. Post-processing is especially important when an application calls for aesthetics or tight tolerances, noted [31].

The studied effects of annealing on biocomposite mechanical properties. A comparison study of PLA with hempfiber, polymers and composites intensive investigation has shown that a two-hour annealing of the PLA at 80°C can eliminate all internal tensions and increase crystallinity, resolving the problem mechanical and thermal stability [32].

3.5.5 Layer Thickness

Once more, in the world of 3D printing, layer thickness is a critical variable that influences the resolution, surface finish and mechanical properties of components. Thin layers offer greater detail but take more time to print, while speedy prints feature thicker layers that sacrifice quality for time.

The effect of thickness and material combination on the mechanical properties of polylactide biocomposites structural development-based on layered construction. As reported by the authors, decreasing the layer thickness from 0.3 mm to 0.1 mm renders a smoother surface and higher interlayer adhesion of PLA biocomposites. Thinner layers also reduced printing time by as much as 60%, suggesting a trade off between quality and speed [26].

The influence of layers thickness on the mechanical behavior of wood-filled PLA biocomposites, For this trade off, we found that the optimal layer thickness was 0.15 mm in terms of tensile strength vs dimension accuracy [26].

Optimizing all the settings necessary for proper 3D printing of biocomposites is a combinatorial challenge that touches on bottom-up contributions associated with material type, fiber orientation, and post-processing and top-down contributions related to layer thickness and other factors. All of these are important factors in appropriately achieving mechanical properties, surface properties and sustainability [33].

3.6 Recent Advancement of 3D Printing

3.6.1 Tissue Engineering and Regenerative Medicine

An instrument for recapitulating the vascular physiology is offered by 3D printing. Using 3D printing, discrete units of print materials can be piled into 3D structures according to a predetermined path. As 3D printing technology advances quickly, complicated vascular scaffolds can be created by identifying print materials, enhancing printing equipment, and streamlining printing procedures. Print materials decompose, are biocompatible, and are beneficial to cells. Hydrogel materials, including methylated gelatin (GelMA), alginate hydrogel, and gelatin, are frequently utilized in 3D printing because of their biocompatible qualities and strong crosslinking. There are comparable 3D printing procedures for various print materials [34].

3D printing techniques make it simple to use computer-assisted TE scaffold design techniques. A wide range of 3D printing methods are currently being used for TE applications. These facilitate the creation of TE scaffolds from various materials, such as hydrogels, metals, polyesters, and ceramics. The usage of 3D printers in tissue engineering has also increased due to their improved ultra-high resolution. These days, 3D printing techniques allow for the creation of TE structures that are employed to regenerate various tissues, including skin, cartilage, vascular networks, and whole organs [35].

The process of designing human tissue structures *via* 3D bioprinting is more intricate than the development of simple cell layers and natural extracellular matrix. In contrast to 2D structures, some studies have demonstrated a clear influence on cell characteristics as migration, differentiation, and proliferation. It was proposed that in order to activate native extracellular matrix, 3D bioprinted structures were necessary. The study indicates that the two biocompatible kinds that are most thoroughly look over for 3D printed structures are hydrocolloid and permeable structures. Furthermore, it has previously been demonstrated that 3D of physiologically adequate tissue reconstructions is a crucial instrument for prosthesis and surgery. With the use of 3D printing technologies, complex assorted materials implants with patient-precise geometries for tissue engineering medical prostheses can be made to order. Material creation is one of the main obstacles on the topic. Rheological properties are necessary for printable materials to enable extrusion and solidification when they are expanded into mechanically robust 3D printed structures [36].

Tissue engineering, which seeks to produce functioning tissues for use in drug testing and advancing medicine, depends heavily on three-dimensional bioprinting. The possibility of replacing or mending damaged tissues and organs may become possible with tissue regeneration and reconstruction. Because synthetic tissue engineering products do not require scaffolding or biomaterials that are absent from native tissues, their adoption can result in shorter organ donor waiting periods and fewer animal testing, which is its indisputable social benefit. During the first decade of the twenty-first century, 3D bioprinters were successfully used to create the first organs that were purely lab-grown. In this area, organs for numerous body sections, tissue-engineering structures, and cell therapies have been created [37].

In addition to tissue/organ transplantation, one of the main problems facing biomedical research groups is comprehending the cellular and molecular causes of human disease in order to create novel treatments, preventative measures, or diagnostic tools. Two needs have fueled recent developments in tissue engineering and regenerative medicine (TERM): creating bioartificial tissue or organs and *in vitro* tissue/organ equivalents to maximize bioavailability while reducing immunological response to save humans. The unit's manual method of performing traditional tissue engineering was technically complex. Additionally, a significant obstacle to conventional TERM approaches is the replication of organ complexity and revascularization of implanted tissue/organ at the human scale [38].

3.6.2 Implants and Prosthetics

3D printing offers a lot of potential for producing useful and reasonably priced goods. Prosthetists still design and fabricate a prosthesis socket for a particular patient using their expertise, experience, and craftsmanship. Another issue is the subjective sensation of the test prosthesis fit (comfort, stability, etc.), which is difficult to express. It has been demonstrated that this workshop's trial-and-error method, which involves regular prosthesis modifications and pricey materials, is time-consuming, material-wasting, and ultimately costly. In the first five years following amputation, the cost of constructing patient-specific prosthesis sockets ranges from approximately USD 6000 to nearly USD 20,000, according to prior studies. 3D-printed components can be utilized as final goods in addition to prototypes. Low-cost 3D-printed transtibial prostheses manufactured from PLA filament using the FDM method were examined in a recent observational cohort research. Results from an evaluation of the prostheses in a rural Sierra Leonean community demonstrated their functionality and safety. PLA is the most popular plastic filament for FDM 3D printing because it is affordable, biodegradable, and exhibits good printability from renewable resources. Functional items like 3D-printed PLA sockets have shown a lot of promise, but they still require experimental and numerical testing to demonstrate their suitability during the prosthesis's lifetime [39].

The surgical correction of bone abnormalities in different anatomical sites or in situations where traditional implants are not suitable can be accomplished with 3D printed prostheses. By applying silver ions to the surface of the intended implants, infection rates can be decreased, depending on the patient's condition. A surface coating of hydroxyapatite applied to the bone–implant contact zones can also speed up the osseointegration process. Implants intended for joint replacement can use an interface between metal and polyethylene. In addition to having the structural geometry to meet each patient's unique surgical needs, 3D-printed personalized implants enable biomechanical evaluation under patient-specific loading conditions for design modification before the actual implant fabrication and application to minimize pain, speed up recovery, and improve osteointegration if needed. This workflow for 3D printing necessitates the use of several distinct engineering software packages to master several steps of the working process under the surgeon's supervision. Since each of the many materials—including metallic alloys, polyetheretherketone, polycaprolactone, bioceramics, and ceramic scaffolds—has unique biomechanical and biocompatibility characteristics, it is possible to customize implant designs to each patient's requirements The purpose of this study is to describe the phases of design, modeling, and manufacture for four distinct customized 3D printed prostheses and their use in clinical settings for patients who received treatment in diverse anatomical regions with a well-defined indication for implantation [40].

3D printing technology was used to create a breast mold. demonstrates the idea behind the construction of a drug-loaded implanted prosthesis made using 3D printing to prevent breast cancer from returning after conserving surgery. The printed breast mold was used to create a polydimethysiloxane (PDMS) prosthesis that was loaded with PTX and DOX microspheres. The produced prosthesis's physiochemical characteristics, such as its shape, drug loading and encapsulation efficiency, and drug release *in vitro*, were assessed. The prosthesis's systemic toxicity and therapeutic efficacy were then assessed in a mouse model of breast cancer metastasis and local recurrence. The outcomes showed that the 3D-printed prosthesis containing PTX and DOX microspheres could deliver the medications continuously for over three weeks and prevented cancer from recurring [41].

3.6.3 Drug Delivery and Release

3D printing has drawn interest from a variety of industries, providing new manufacturing paradigms for producing new or existing product designs. As a revolutionary technology, it has the ability to digitalize pharmaceutical production by dispense individual units of customized pharmaceuticals in a flexible method, shifting therapies away from one-size-fits-all dosage. It is possible to use 3D printing technologies to construct 3D items that are often difficult to make using traditional manufacturing methods because of their special qualities, such as their additive nature and ability to be combined with 3D imaging. They are therefore especially promising for moving the pharmaceutical industry into digital healthcare through the invention and evaluation of novel drug-eluting dosage forms. A 3D printing technique that uses a power source to selectively combine particles into 3D objects. Selective laser sintering/melting, direct metal laser sintering,

EBM, and multi-jet fusion are some of the derivatives of PBF. These technologies are ideal for creating custom drug-loaded formulations, devices, and implants because they offer a number of advantages. They can precisely build complex goods because of their improved printing speed and resolution [42].

3D printing has been investigated for their potential to create pharmaceutical formulations: nozzle-based deposition systems, printing-based inkjet systems, and laser-based writing systems. Among, three systems the laser-based writing system works by using a laser to solidify a layer of photosensitive liquid resin. This process is then repeated to harden more layers, creating a three-dimensional structure. 3D printing and offers a current assessment of its role in medication delivery. The first step will be to identify and explain the basic ideas behind the many forms of 3D printing that have been investigated for the creation of dosage forms. 3D printing technology creates 3D items from computer blueprints layer by layer. Inkjet, extrusion, or laser-based systems are the most common types of 3D printers, depending on how the materials are layered [43].

3D printing technology has been increasing, as has interest in 3D printed pharmaceutical products. The most efficient method for developing pharmaceutical products with a patient-focused approach is thought to be 3D printing. Because 3D printing may be used to adapt therapy to match patient needs, its application in the pharmaceutical industry has been a growing area of research. 3D printing technology is growing from its early stages of development to the manufacturing of finished items with streamlined design and production cycles because of its versatility and flexibility. In on-demand production, 3D printing is becoming more and more popular than traditional manufacturing methods due to its automated process and low operating costs. Extrusion-based systems, liquid solidification, and powder solidification are the three primary mechanisms used in 3D printing processes. Personalization, increased product complexity, and on-demand manufacturing are some advantages of 3D printing over traditional tablet manufacturing techniques. As a result of this technique, drug formulations with customized release profiles and designs can be produced, and dosages can be customized based on the patient's body mass index, metabolism, genetic variants, and other comorbid disorders. It makes it possible to combine the patient's daily prescriptions into a single polypill or multi-drug dose. Particularly for young and elderly patients, this tailoring may increase treatment adherence [44].

(3DP) is anticipated to be among the pharmacy's most innovative developments. Due to the potential benefits of 3D printing over tailoring medications in individually tailored dosages, its usage in drug delivery systems has grown rapidly in recent years. Drug configuration, manufacture, and use may alter as a result of 3DP's ability to precisely deposit medications and excipients. It can include frontline medical care, preclinical events, clinical preliminary findings, and drug improvement measures. Even though 3DP technology has financial and clinical benefits, its use with pharmaceutical products is restricted by some administrative and specialist issues. Therefore, it is necessary to continuously develop and improve 3DP techniques in order to surpass existing limitations and integrate patients' specific medical services with the use of tailored medications [45].

Hydrogel-based scaffolds that can deliver drugs on demand have drawn more attention for use in tissue engineering and localized cancer treatment. However, due to the uncontrolled diffusion of medicines from the swelling hydrogels, the majority of drug-loaded hydrogels are typically not appropriate for long-term drug administration. Thus, a core/shell fiber scaffold was created in this study by covering the 3D printed alginate-gelatin hydrogel scaffolds with a uniform layer of polycaprolactone (PCL). Drugs from the core gels may diffuse less freely thanks to the PCL coatings. The Gel/PCL core/shell scaffolds were then covered with polydopamine (PDA), which gave them strong photothermal properties. Thus, the thermally induced sol-gel transition of the core gels made it possible to achieve near-infrared (NIR) laser triggered on-demand drug release in this system. "Doxorubicin and photothermal therapy, two released medications, have the potential to successfully prevent or eradicate tumors both *in vitro* and *in vivo*. Three-dimensional porous scaffolds that may release drugs under regulated conditions have been utilized extensively in tissue engineering and have recently attracted more attention in the realm of cancer treatment. In particular, scaffolds that can administer drugs on-demand in response to external stimuli have been crucial for topical cancer chemotherapy and tissue regeneration. Drug-loaded scaffolds for localized chemotherapy might significantly lessen the adverse effects of chemotherapeutic medications on healthy tissues and organs since the chemicals were delivered precisely at the tumor site in a manner that was controlled by time and dose [46].

3.6.4 Case Studies and Industrial

3D printing technologies now on the market, 3D printing earth materials like cob still poses a number of difficulties. This essay outlines the steps used to create a cob 3D printing system that satisfies the modern demands of digital building. The study began by outlining the process for creating an updated cob recipe that could be printed using 3D technology. Following extensive research into the characteristics of three different extrusion system types using both atmospherically and electromagnetic techniques, the study ultimately produced a unique custom dual-ram extruder. The design and construction industries now have a plethora of potential uses thanks to 3D printing (3DP) technology for huge sheets. Concurrent studies emphasize the benefits of 3D printing in construction to get a greater level of process optimization (e.g., cost, construction time, and personnel resource), as well as the rise of new digital procedures related to building information. The life cycle of 3D printed items and construction components can help the environment, and recent modeling offers the possibility of mass customization. This suggests that 3D printing of cob is a promising road toward more environmentally friendly 3DP practices and that it has a lower environmental impact than concrete [47].

Industry 4.0 has seen a huge increase in interest in 3D printing. With the growing number of users requesting 3D printing services, it is becoming increasingly difficult to manage the supply and demand of 3D printing resources in an earthly dispersed setting. As a recently developed service-directed manufacturing paradigm, cloud manufacturing offers a promising solution to the problem by incorporating 3D printing into a cloud-based environment. This could foster the growth of intelligent virtual 3D printing cloud networks and encourage the creation of a new 3D printing service business model to accomplish mass customization. Additive manufacturing (AM), another name for 3D printing, is seen as the beginning of "a third industrial revolution. These days, additive manufacturing is used in many different industrial applications, such as robots, medical gadgets, building, airplanes, and weaponry. In addition to offering manufacturing solutions at the industry level, 3D printers have been filling the market with consumer-level solutions during the past five years. This trend has made 3D printers, particularly the small desktop models, more widely available to the general public, which enables individuals to easily produce customized 3D printed goods the accomplishment of widespread personalization [48].

3D printing in the food business by comparing its present potential to meet the SDGs. With a focus on customized nutrition, food waste management, and the creation of intricate, individualized edible patterns, the study examines current advancements in additive manufacturing in a range of applications. This demonstrates how the technology has a lot of advantages over traditional methods of producing food. Limitations on the technology's performance are also examined in order to close the gap between market demand and 3D printing. With its patented technique, 3D food printing (3DFP) offers a practical substitute for current food production

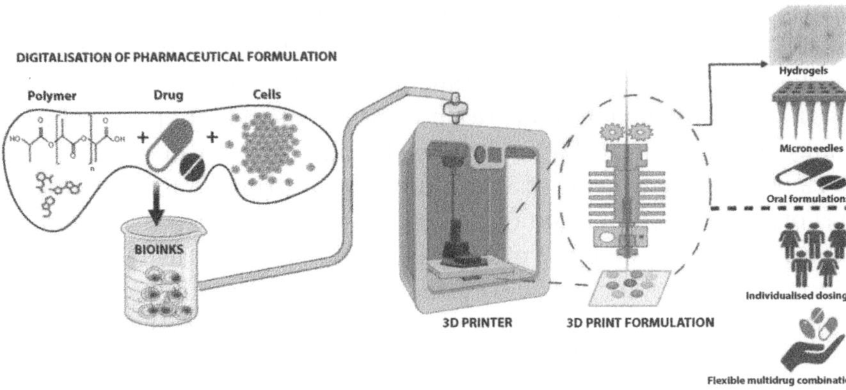

Figure 3.1 Role of 3D printing in the drug delivery system.

methods, ensuring a sustainable and effective food supply in the future while also delivering a novel and fascinating culinary experience. By simulating a layer-by-layer deposition process using a computer-aided design (CAD) model-based printer, additive manufacturing (AM) enables the production of 3D printed food goods. Even though the equipment is still in its early stages of commercialization, it has been effectively used in a variety of market environments, including the medical area. As illustrated in Figure 3.1, the evidence of numerous and current studies published as primary and secondary sources supports 3DFP. It is evident that throughout 2018–2022, the visualization map gave the field of 3D printing a high level of characterisation, with strong associations to topics related to food, rheological qualities, sustainability, individualized nutrition, and customized food production [49].

3.7 Application of 3D Printing Biocomposites Fabrication Methods and Process Optimization

With the potential applications for 3D printing increasing, engineering businesses are beginning to explore ways to build new business models and opportunities with the methods. 3D printing technology plays a key part in the big three engineering industries, aerospace, automotive, and biomedical. The applications of 3D printing technology in these three fields are increasing every day and have several advantages. For functional goods with desired mechanical qualities, such as ultra-lightweight

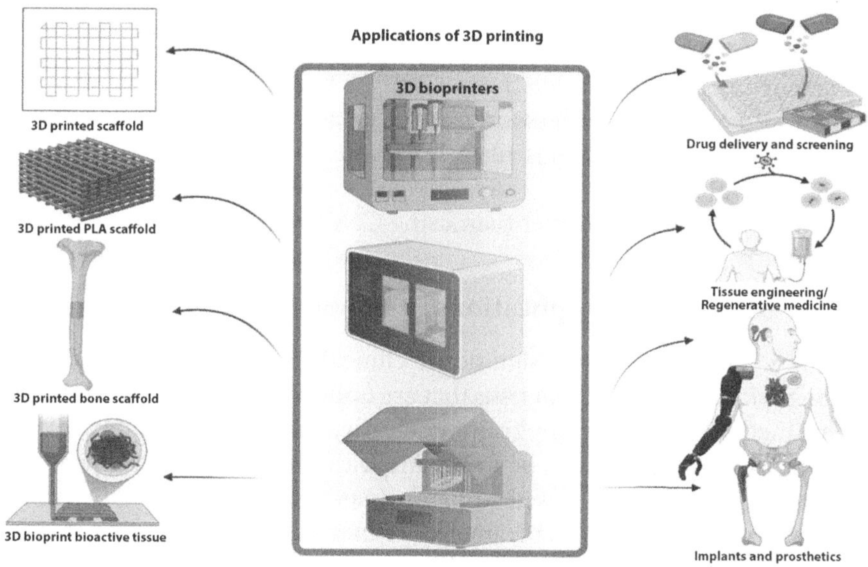

Figure 3.2 Application of 3D printing.

or damage-tolerant structural materials for storage, micromechanical sensors, and biomedical devices, there is a growing need for novel materials (Figure 3.2). Research in the past has concentrated on creating and altering materials through chemical composition manipulation, but microstructure design of materials at the micro and/or nanoscale is now becoming commonplace. One of the possibilities is architected materials, which are multiphase materials with a regulated and tailored microstructure and topology for physical qualities that are not possible with just constituent materials. Although these advantages make the designed materials promising, their multi-scale structure and intricate design make it challenging to create them using traditional production methods. One manufacturing technique that can be used to conquer challenges and leads to fabrication of novel methods.

3.7.1 3D Printing Applications in an Industrial Fixtures

The advantages of cheap cost, quick response, and flexible manufacturing have made fixtures and tooling a significant application area for 3D printing in the industrial sector. The majority of short fiber reinforced composites and 3D printed thermoplastic polymers, however, have comparatively poor mechanical qualities, which limit their use to the fabrication of specific prototype parts. Continuous Fiber Reinforced Polymer Composites that are 3D printed are just as strong as machined aluminum and can be used as finished components right away. CFRPCs that are 3D printed are also lighter and won't harm the workpiece's surface. When the Dixon Valve Company used a continuous fiber 3D printer from Markforged to construct a robot end effector fixture, the cost was 96.8% lower and the production time was 87% shorter than using CNC machining [50].

3.7.2 3D Printing Applications in Biomedical Field

Donor inadequacies are a significant clinical diagnostic worldwide for organ replacement. Potential risks that are undoubtedly present with traditional methods include complications, elective injuries, and limited source donors. These obstacles can be overcome with the help of the 3D printing technique, which can be used to print tissue and organs directly, create personalized tissue building frameworks in a short amount of time, and 6x tissue surrenders *in situ* with cells. Biomedical applications, including tissues, structures, and obsession devices, have specific production components that should be considered. In biological applications, it is strongly advised to employ these 3D printing procedures without rigid support structures. Current advancements in tissue design and recovery, remedial delivery, clinical device production, employable administration, and planning ensure that 3D printing will continue to assume an unstoppable job in healthcare. Medical 3D printing was once a hopeful but unfeasible dream. The effort and time make it modest. Today, the 3D printing procedure is a leading way to help pharmaceutical and clinical organizations make more visible pharmaceuticals, allowing for the rapid creation of clinical embeds, and changing the way experts and specialists organize their methodology. Patient-specific 3D-printed anatomical models are becoming increasingly useful tools in the current paradigm of precision medicine and customized medications. 3D-printed implanted organs will eventually be available, reducing the amount of lives saved and reducing the holding up archives. Evolution is still the main source of further substance production for human services. Nevertheless, it is being used in the clinical setting from a variety of perspectives that were previously unexplored under tremendous tension concerning the ideal execution and decreased costs will raise extraordinary benefits from this great as-gold novelty [16].

3.7.3 Applications in Organ Transplantations

Current methods of tissue or organ transplantation in clinical medicine, such as auto-transplantation, xenotransplantation, and artificial organ implantation, continue to present challenges because of their high cost, scarcity of donors, and immunological rejection. The ability to rapidly produce customized tissues and organs using biocompatible polymers, such as synthetic polymers (poly(ethylene glycol) diacrylate, PVA, poly(D,L-lactic-co-glycolic acid), PCL, and PLA) and natural polymers (gelatin methacrylate [GelMA], alginate, and collagen), is one of the many ways to manufacture artificial organs.100 Polymers for 3D bioprinting should have high mechanical and structural qualities and be biocompatible, in contrast to those used for nonbiomedical printing. For bone tissue engineering scaffolds to regenerate, they must have the right mechanical qualities, biocompatibility and degradability of the material, the right structure (shape, pore size, and porosity), and the right cellular connections [51].

3.7.4 Applications of Parts and Equipment for Medicine

Cellular viability and biocompatibility were the main topics of recent AM research. Using AM, porous scaffolds can be created and manufactured. They can then be used as bone grafts thanks to 3D interconnectivity, which can support a more abundant supply of blood, oxygen, and nutrients for the healthy ingrowth of bone cells. The physical, mechanical, and biological characteristics of the materials are essential for producing high-quality medical parts. Additive manufacturing can create porous scaffolds with a shape memory effect that can be utilized as self-fitting implants to replace lost bone. Other potential scientific and medical uses include the production of medical instruments, orthopedic devices, and prostheses. In terms of AM's commercial success, dental applications have been growing [52].

3.8 Global Market for 3D Printed Biocomposite Products

3D-printed biocomposites have revolutionized the global market recently [53]. Due to their environmental impact and innovative approaches, these biocomposites are directly related to the Global Sustainable Development Goals (SDGs) or Global Goals. In recent years, two keywords that have become increasingly popular are sustainability and biodegradability (the requirement for biodegradable composites) [54]. Hence, it is unsurprising that they are ruling the international market with burgeoning market supply and demand. The worth of 3D printed composites in the global market was USD 324.79 million in 2024, increased to USD 448.60 million in 2025, and is expected to reach around USD 8,206.65 million by 2034, with a growth rate of 38.12% from 2024 to 2034 [55]. Federal regulations, green environment sustainability awareness, and innovations in 3D printing technologies are major contributing factors to biocomposite products' compound annual growth rate (CAGR).

In the healthcare sector, biocomposites have transformed the manufacturing of surgical instruments, hard and soft tissue applications, implants, and prostheses. They also play a significant role in patient-specific applications because of their biodegradability, biocompatibility, and flexibility for customization, which are required for the patient's needs. Recent biomimetics (prosthetic limbs) developments have sparked an essential breakthrough in the global biocomposites market [56]. International market demand always depends on the end-user requirements. 3D-printed biocomposites are playing a significant role in healthcare, automotive, aerospace and defence, and others, and the CAGR of the health sector is projected to rise from 2024 to 2034. In the health sector, 3D-printed composites create lightweight, convenient, and durable prosthetic limbs. They are more convenient for the manufacturers because of the flexibility to develop unique limbs depending on the individual patient's requirements, which improves the market demand. They are also used to create customized surgical equipment and fixtures. They are customized in such a way as to fit the anatomy of a patient, which improves the accuracy, comfort, mobility, and efficiency of the surgeries. In 2023, North America held the largest share (38.1%) of the global market of 3D-printed composites. The driving force for this increase in market share is the adoption of advanced technology across various industries, including health care. The healthcare market for 3D printing is driven by the lightweight, flexible, and other characteristics of 3D biocomposites, especially in producing customized implants and prosthetics. In recent years, the healthcare industry has been transformed by using 3D-printed biocomposites to create unique, precise, patient-specific parts with intricate geometrics and great precision, improving patient outcomes and lowering costs [55]. The Asia-Pacific region is expected to experience rapid market growth of 3D print biocomposites in the upcoming years, driven by government policies that promote advanced technologies and a growing demand for personalized products by the public. There is rising government support in China, India, and South Korea to boost the growth of 3D printing, and opportunities are flourishing for 3D printing market players. As per the market size analysis report 2030, the estimated market size of 3D printing (application, technology, and processes) USD will be 37.4 billion by 2029 with a CAGR of 16.4% [57]. The key players in the global market for 3D printed biocomposite products include reputed firms like Nature Works LLC and BASF SE (leading suppliers of biopolymers used in biocomposites) and Stratasys, Ultimaker, and Formlabs are integrating biocomposites into their material portfolios and product manufacturing companies like Stryker. As mentioned before, the demand for sustainability holds untapped potential for biocomposites. The customization and flexibility characteristics of biocomposites provide a unique edge over traditional manufacturing methods. Despite their advantages, biocomposites often face challenges, especially in the cost of production, material properties, availability, and regulatory hurdles (lack of standard regulations) that may impede the market growth of 3D printed biocomposites [58, 59]. 3D printed biocomposites global market is the convergence of sustainability and technological innovation. The use of these products in healthcare is expected to increase as consumers and entrepreneurs are moving toward Global Sustainable Development Goals (SDGs) or Global Goals. There are challenges, but technological advancements will overcome these challenges to make a substantial contribution of 3D printed biocomposites to a sustainable future with sustainability in market growth.

3.9 Conclusion of Future Prospects of 3D Printing Biocomposites

The future of 3D printing biocomposites is promising, with continuous advancements in materials, technology, and applications. As the demand for sustainable and high-performance materials grows, biocomposites offer an eco-friendly alternative to traditional synthetic composites. Innovations in material science, including the development of novel bio-based polymers and reinforcements, will further enhance the mechanical, thermal, and biodegradable properties of these composites. With ongoing research and development, 3D printing biocomposites are expected to find applications across various industries, including healthcare, aerospace, automotive, and construction. In healthcare, biocomposites can be used for customized prosthetics, implants, and scaffolds for tissue engineering. Challenges such as material consistency, scalability, and cost-effectiveness must be addressed to fully integrate biocomposites into mainstream manufacturing. However, advancements in additive manufacturing processes, such as multi-material printing and nanotechnology integration, will help overcome these barriers. Hence, the combination of sustainability, versatility, and technological innovation ensures that 3D printing biocomposites will play a significant role in the future of manufacturing. As industries shift toward greener solutions, the potential for biocomposites in additive manufacturing will continue to expand, fostering a more sustainable and efficient production landscape.

AI disclosure statement: The author acknowledges the use of AI-based tools for minor language editing or grammatical correction. However, all intellectual contributions, analyses, and conclusions presented in this chapter are entirely the author's own.

References

1. Gao, H. and Duan, H., 2D and 3D graphene materials: Preparation and bioelectrochemical applications. *Biosens. Bioelectron.*, 65, 404–419, 2015, Elsevier.
2. Chua, C.K. and Yeong, W.Y., *Bioprinting: principles and applications*, vol. 1, World Scientific Publishing Co Inc, Singapore, 2014.
3. Ahn, Lee, E.-H., *et al.*, Effect of post-curing time on the color stability and related properties of a tooth-colored 3D-printed resin material. *J. Mech. Behav. Biomed. Mater.*, 126, 104993, 2017, Elsevier.
4. Wang, Q., *et al.*, 3D printing with cellulose materials. *Cellulose*, 25, 4275–4301, 2017, Springer.
5. Li, M., *et al.*, Continental-scale mapping and analysis of 3D building structure. *Remote Sens. Environ.*, 245, 111859, 2020, Elsevier.
6. Ge, G., *et al.*, 3D printing of hydrogels for stretchable ionotronic devices. *Adv. Funct. Mater.*, 31, 52, 2107437, 2019, Springer.
7. Kim, H. and Han, S., Interactive 3D building modeling method using panoramic image sequences and digital map. *Multimedia Tools Appl.*, 77, 20, 27387–27404, 2018, Springer.
8. Momeni, A., *et al.*, 3D pore network modeling and simulation for dynamic displacement of gas and condensate in wellbore region. *Int. J. Multiphase Flow*, 97, 147–156, 2017, Elsevier.
9. Wang, B., *et al.*, Current progress on the 3D printing of thermosets. *Adv. Compos. Hybrid Mater.*, 3, 462–472, 2020, Springer.
10. Sun, F., *et al.*, Application of 3D-printed, PLGA-based scaffolds in bone tissue engineering. *Int. J. Mol. Sci.*, 23, 10, 5831, 2019, Elsevier.
11. Rana, A., Mittal, A., Vashist, C., Rajput, S., Sridhar, S.B., Malviya, R., Exploration of 4D Printing and its Applications in the Biomedical Sciences. *Curr. Pharm. Des.*, 2025.

12. Tang, T., Zhao, S., Joralmon, D., Poit, Z., Ahire, B., Li, X., Recent advancements and applications in 3D printing of functional optics. *Addit. Manuf.*, 52, 102682, 2019, Elsevier.
13. Kang, M.-J., *et al.*, Effects of post-curing light intensity on the mechanical properties and three-dimensional printing accuracy of interim dental material. *Materials*, 15, 19, 6889, 2019, MDPI.
14. Yan, *et al.*, A critical review of fused deposition modeling 3D printing technology in manufacturing polylactic acid parts. *Int. J. Adv. Manuf. Technol.*, 102, 2877–2889, 2019, Springer.
15. Gu, J., *et al.*, Characterization of particulate and gaseous pollutants emitted during operation of a desktop 3D printer. *Environ. Int.*, 123, 476–485, 2019, Elsevier.
16. Srinivasan, D., *et al.*, 3D printing manufacturing techniques, materials, and applications: an overview. *Adv. Mater. Sci. Eng.*, 2021, 1, 5756563, 2021, Elsevier.
17. Cima, J., Shakor, P., *et al.*, Review of binder jetting 3D printing in the construction industry. *Prog. Addit. Manuf.*, 7, 4, 643–669, 2016, Elsevier.
18. Sachs, E.M., *et al.*, Three-dimensional printing techniques, U.S. Patent No. 5,204,055, 2018, MDPI.
19. Bourell, D.L., Perspectives on additive manufacturing. *Annu. Rev. Mater. Res.*, 46, 1, 1–18, 2016.
20. Gibson, I., *et al.*, Photopolymerization processes, in: *Additive manufacturing technologies: Rapid prototyping to direct digital manufacturing*, pp. 78–119, New York, Springer Nature, 2010.
21. Bourell, D.L., Rosen, D.W., Leu, M.C., The roadmap for additive manufacturing and its impact. *3D Print. Addit. Manuf.*, 1, 1, 6–9, 2014.
22. Lee, K.-Y., *et al.*, Accuracy of three-dimensional printing for manufacturing replica teeth. *Korean J. Orthod.*, 45, 5, 217–225, 2015.
23. Wu, P., Wang, J., Wang, X., A critical review of the use of 3-D printing in the construction industry. *Autom. Constr.*, 68, 21–31, 2016.
24. Tumbleston, J.R., *et al.*, Continuous liquid interface production of 3D objects. *Science*, 347, 6228, 1349–1352, 2015.
25. Kumar, S., Malviya, R., Sundram, S., Management of peripheral nerve injuries using natural based biomaterials and their derivatives: Advances and prospective. *MedComm–Biomater. Appl.*, 3, 1, e72, 2024 Mar.
26. Kovan, V., *et al.*, Effect of printing parameters on mechanical properties of 3D printed PLA/carbon fibre composites. *Mater. Sci. Non-Equilib. Phase Transform.*, 4.4, 126–128, 2018.
27. Chen, T., *et al.*, Optimizing photocuring properties of ceramic slurry for 3D printing of ceramic membranes. *Ceram. Int.*, 50, 1, 1732–1741, 2024.
28. Zhang, Y., *et al.*, 3D-printed bioceramic scaffolds with antibacterial and osteogenic activity. *Biofabrication*, 9, 2, 025037, 2017.
29. Fang, Z.-C., *et al.*, Review on residual stress in selective laser melting additive manufacturing of alloy parts. *Opt. Laser Technol.*, 129, 106283, 2020.

30. Liu, B., *et al.*, Spherical path planning for multi axis support free additive manufacturing of truss structures. *J. Manuf. Processes*, 109, 198–212, 2024.
31. Zettersten, J., The processing of a 3d-printed biocomposite: A material driven study conducted in collaboration with Stora Enso, 2023.
32. Goh, G.D., *et al.*, Large-format additive manufacturing of polymers: a review of fabrication processes, materials, and design. *Virtual Phys. Prototyping*, 19, 1, e2336160, 2024.
33. Sharma, R., *et al.*, Bioceramic filled PLA based nanocomposites for biomedical application: An experimental investigation. *Ceram. Int.*, 50, 16, 28662–28681, 2024.
34. Wang, P., *et al.*, 3D printing of tissue engineering scaffolds: a focus on vascular regeneration. *Bio-Des. Manuf.*, 4, 2, 344–378, 2021.
35. Zaszczyńska, A., *et al.*, Advances in 3D printing for tissue engineering. *Materials*, 14, 12, 3149, 2021.
36. Mani, M.P., *et al.*, A review on 3D printing in tissue engineering applications. *J. Polym. Eng.*, 42, 3, 2022243-265.
37. Kantaros, A., Soulis, E., Alysandratou, E., Digitization of ancient artefacts and fabrication of sustainable 3D-printed replicas for intended use by visitors with disabilities: the case of Piraeus archaeological museum. *Sustainability*, 15, 17, 12689, 2023.
38. Jain, P., Kathuria, H., Dubey, N., Advances in 3D bioprinting of tissues/organs for regenerative medicine and *in-vitro* models. *Biomaterials*, 287, 121639, 2022.
39. Plesec, V., *et al.*, Numerical analysis of a transtibial prosthesis socket using 3D-printed bio-based PLA. *Materials*, 16, 5, 1985, 2023.
40. Safali, S., *et al.*, The possibilities of personalized 3d printed implants—A case series study. *Medicina*, 59, 2, 249, 2023.
41. Rajput, S., Sharma, P.K., Malviya, R., Biomarkers and treatment strategies for breast cancer recurrence. *Curr. Drug Targets*, 24, 15, 1209–20, 2023 Nov 1.
42. Awad, A., *et al.*, Advances in powder bed fusion 3D printing in drug delivery and healthcare. *Adv. Drug Delivery Rev.*, 174, 406–4245, 2021.
43. Mohammed, A.A., *et al.*, 3D Printing in medicine: Technology overview and drug delivery applications. *Ann. 3D Print. Med.*, 4, 1000375, 2021.
44. Dumpa, N., *et al.*, 3D printing in personalized drug delivery: An overview of hot-melt extrusion-based fused deposition modeling. *Int. J. Pharm.*, 600, 1205015, 2021.
45. Mohapatra, S., *et al.*, Approaches of 3D printing in current drug delivery. *Sens. Int.*, 3, 1001465, 2022.
46. Liu, C. and Shen, Y., Application of 3D U-Net-based neutral hydrogen foreground reduction method. *J. Alloys Compd.*, 968, 1720675, 2023.
47. Gomaa, M., *et al.*, 3D printing system for earth-based construction: Case study of cob. *Autom. Constr.*, 124, 1035775, 2021.
48. Kumar, S., Malviya, R., Sridhar, S.B., Bioresorbable Polymeric Scaffold: Advancing Minimally Invasive Surgical Procedure for Cardiovascular and

Its Applications, in: *Sustainable Nanocomposites with Green Biomaterials*, pp. 61–82, Springer Nature Switzerland, Cham, 2025 Mar 25.
49. Alami, A.H., *et al.*, 3D printing in the food industry: Recent progress and role in achieving sustainable development goals. *Ain Shams Eng. J.*, 15, 2, 102386, 2024.
50. Tian, X., *et al.*, 3D printing of continuous fiber reinforced polymer composites: development, application, and prospective. *Chin. J. Mech. Eng.: Addit. Manuf. Front.*, 1, 1, 100016, 2022.
51. Park, S., *et al.*, 3D printing of polymer composites: Materials, processes, and applications. *Matter*, 5, 1, 43–76, 2022.
52. Valino, A.D., *et al.*, Advances in 3D printing of thermoplastic polymer composites and nanocomposites. *Prog. Polym. Sci.*, 98, 1011625, 2019.
53. Banjanin, B., Vladić, G., Adamović, S., Bošnjaković, G., Global market structure, in: *Polymers for 3D Printing Methods, Properties, and Characteristics*, pp. 353–367, Plastics Design Library. William Andrew Applied Science publishers, Kidlington, Oxford, United Kingdom, 2022.
54. Santhosh, N., Praveena, B., Gowda, A., Duhduh, A., Rajhi, A., Alamri, S., Berwal, P., Khan, M., Wodajo, A., Innovative eco-friendly bio-composites: A comprehensive review of the fabrication, characterization, and applications. *Rev. Adv. Mater. Sci.*, 63, 1, 20240057, 2024.
55. Bidwai, S., 3D Printed Composites Market Size, Share, and Trends 2024 to 2034, Report Code : 2843, Precedence Research Category of Chemical and Material, 2024.
56. Varaganti, P. and Seo, S., Recent Advances in Biomimetics for the Development of Bio-Inspired Prosthetic Limbs. *Biomimetics (Basel)*, 9, 5, 273, 2024 Apr 30, doi: 10.3390/biomimetics9050273. PMID: 38786483; PMCID: PMC11118077.
57. Marketsandmarkets, 3D Printing Market Size, Share, Industry Analysis Report 2030, SE2936.3D Printing Market, 2024.
58. Shahrubudin, N., Koshy, P., Alipal, J., Kadir, M.H.A., Lee, T.C., Challenges of 3D printing technology for manufacturing biomedical products: A case study of Malaysian manufacturing firms. *Heliyon*, 6, 4, e03734, 2020 Apr 12, doi: 10.1016/j.heliyon.2020.e03734. PMID: 32322726; PMCID: PMC7160453.
59. Chen, L., Zhang, Y., Chen, Z., *et al.*, Biomaterials technology and policies in the building sector: a review. *Environ. Chem. Lett.*, 22, 715–750, 2024.

Part 2
SPECIALISED APPLICATIONS

4

Applications of Biocomposites in Vascular and Cardiac Tissue Engineering

S.K. Abdul Rahaman[1]*, V. V. Siva Krishna Pushadapu[2], P. Srinivasa Babu[2] and T. E. G. K. Murthy[3]

[1]Department of Pharmaceutical Sciences, Vignan Foundation for Science, Technology and Research, Vadlamudi, Guntur, Andhra Pradesh, India
[2]Department of Pharmaceutics, Vignan Pharmacy College, Guntur, Andhra Pradesh, India
[3]Department of Pharmaceutics, Bapatla College of Pharmacy, Guntur, Andhra Pradesh, India

Abstract

Biocomposites (comprising natural polymers, e.g., collagen, chitosan) and reinforcing agents (e.g., cellulose, bioactive fibers) are characterized by biocompatibility, mechanical strength, and versatility. They have applications in tissue engineering, regenerative medicine, implantable devices, and drug delivery systems, including implantable devices to conduct cardiovascular operations. Although challenges like regulation and the need to have standardized testing exist, the research at hand tries to enhance their qualities and expand their applications, and this shows that they are transformational in the medical field. The polymers, metal oxides, graphene derivatives, and organometallic composites are currently used in tissue engineering due to their distinctive properties, but the lack of biocompatibility and cell compatibility, as well as the presence of teratogenicity and carcinogenicity problems, restrict their clinical usage. Bacterial cellulose (BC), natural polysaccharides, and proteins have been known to be biocompatible, sustainable, biodegradable, and have potential applications in such fields as the delivery of specific drug types straight to the target, vascular grafting, bone implants, and personalized heart muscle. By delivering drugs or other therapeutic agents directly to a specific tissue, e.g., the bones or tissues in the mouth, using a bioconjugate delivery mechanism, it provides more specific treatments, which are also effective. It is possible to tune biopolymers to meet certain functional requirements of the

**Corresponding author*: rahamandr123@gmail.com

Rishabha Malviya and Mukesh Roy (eds.) Biocomposites in 3D and 4D Printing: Materials, Techniques, and Applications in Regenerative Medicine, (93–122) © 2026 Scrivener Publishing LLC

target tissue. The forms of biopolymers and composites and their interaction with particular cells and tissues are discussed in the chapter. It also researches cellular processes in tissue regeneration and highlights the role played by the biopolymers, focusing on the effects of these polymers on these processes and their advantages and disadvantages.

Keywords: Biocomposites, biopolymers, cardiac regeneration, regenerative medicine, tissue engineering, vascular grafts

4.1 Introduction

4.1.1 Overview of Biocomposites

A widely known material made up of one or more distinct materials or polymers with significantly varied chemical or physical properties is called a composite. By improving or adding characteristics that certain homogeneous materials cannot exhibit, composites are used to create specific attributes. Biopolymers are increasingly being used in place of petroleum-based polymers in the field of composite materials [1]. Biocomposites are composite materials made using natural fibers or bio-based matrix materials, with synthetic or bio-based reinforcement. They are materials designed to offer enhanced mechanical, thermal, and environmental properties, while also being environmentally friendly. Lightweight, biodegradable, and sustainable, biocomposites are usually used in the packaging, building, automotive, and medical industries. The constituent materials are the matrix and the reinforcement materials. The major load-bearing elements are reinforcing elements, and they may be fibers, whiskers, particles, or flakes. The matrix ensures that the reinforcing elements are supported and attached. Cellulose, hemicellulose, and lignin are the biggest biofiber (lignocellulose) constituents [2]. This material is gaining lots of interest nowadays because of the potential of its multifunctional and high-performance biocomposites with low environmental impact and special qualities, including lightweight, renewable, abundant, and eco-friendly. In biochemistry, biomedical engineering, and optics, biopolymeric composites should replace synthetic materials with a wide range of uses, and research and investment in these materials should rise dramatically [3].

4.1.2 Importance in Vascular and Cardiac Tissue Engineering

There is an increased appreciation of biocomposites in the context of possessing an extraordinary capability to be capable of copying the mechanical and biological characteristics of actual tissues, more so in cardiac and vascular tissue engineering. Tissue engineering is an attempt to prepare tissues and organs that imitate their functions in a way that could be used to repair or replace damaged biological structures. Biocomposites are very promising in this aspect [4]. The biocomposites are important in vascular and cardiac engineering because they combine the tissue-based material with biocompatibility, biodegradability, and multifunctionality. This substance could be in a position to develop working tissue to replace or mend the damaged tissues and blood vessels. Such materials could generate useful tissues that can be employed to regenerate or replace the damaged blood vessels and heart tissues [5].

4.1.3 Mimicking the Natural Environment

There is a possibility to create close to the extracellular matrix (ECM) of heart and vascular tissue biocomposites. A biochemical guidance and ECM direct the cell activity. The researchers are able to make the scaffolds with the required mechanical strength and the bioactivity of the tissue regeneration by using natural fibers and biopolymers (collagen, gelatin, or polylactic acid) [6].

4.1.3.1 Vascular Tissue Engineering

Blood vessels possess unique mechanical characteristics, which are elastic and tensile strength, that are needed to make them functional. When biocomposites are designed to become flexible and strong as vascular tissue, it is possible to consider them as an imitation of a synthetic blood vessel or stents that can fit in the body [7].

4.1.3.2 Cardiac Tissue Engineering

Cardiac tissue can be described as having structure and mechanical characteristics that enable contraction and expansion in response to electrical stimuli. It is possible to produce biocomposites using natural fibers and biopolymers to provide a system on which cardiac cells (like cardiomyocytes) can proliferate as a scaffold on which tissue regeneration can occur [8].

4.1.4 Biocompatibility and Biodegradability

Naturally, biocomposites are biocompatible and biodegradable because most of them are created using natural or renewable sources, which is important in the field of tissue engineering [9].

4.1.5 Controlled Release of Growth Factors

Growth factors are extremely important in tissue engineering. These factors facilitate the growth of cells, their differentiation, and eventually tissue formation. In order to promote the regeneration of blood vessels or heart tissue, biocomposite scaffolds can be designed to administer growth factors, cytokines, or other bioactive compounds in a regulated manner [10].

4.1.6 Customization for Specific Mechanical Properties

Development of materials with the required mechanical properties is one of the major problems of tissue engineering. The cardiac and vascular tissues need a fine ratio of stiffness, flexibility, and elasticity to be able to perform their work properly under the physiological conditions [11].

4.1.7 Enhanced Cell Attachment and Proliferation

The regeneration of tissues requires the capability of functionalizing the natural fibers and biopolymers employed in biocomposites to increase cell adhesion and proliferation. The bacteria can be integrated into the host tissue.

4.1.8 Integration with Host Tissues

Biocomposites facilitate better integration by providing a scaffold that mimics the native ECM, which can promote vascularization (the growth of blood vessels) and tissue remodeling. This is more so when it comes to vascular and cardiac tissue engineering, where the new tissue must be able to become fully functional and be able to blend in with the already existing biological systems of the patient [12].

4.2 Classification of Biocomposites in Vascular and Cardiac Engineering

4.2.1 Natural Biopolymers: Collagen, Chitosan, Cellulose

Natural biopolymers are macromolecules that come from microbes, plants, or animals. To increase their range of applications, these polymers can be manufactured through particular chemical alterations or employed directly. Natural biopolymers can be produced from substrates that are high in lipids, proteins, or carbohydrates. The types and uses of naturally occurring biopolymers can be used to categorize them [13]. Figure 4.1 illustrates the process of tissue regeneration using biopolymers, including cell isolation, expansion, and implantation.

4.2.1.1 Collagen

Collagen is the most abundant and important natural biopolymer occurring in the human body and many other organisms. Three polypeptide chains or alpha chains wound together to form a triple-helix structure make up the fibrous protein known as collagen [14]. It is this very structure that confers strength and stability to collagen. It contains mainly the amino acids glycine, proline, and hydroxyproline. The extracellular protein

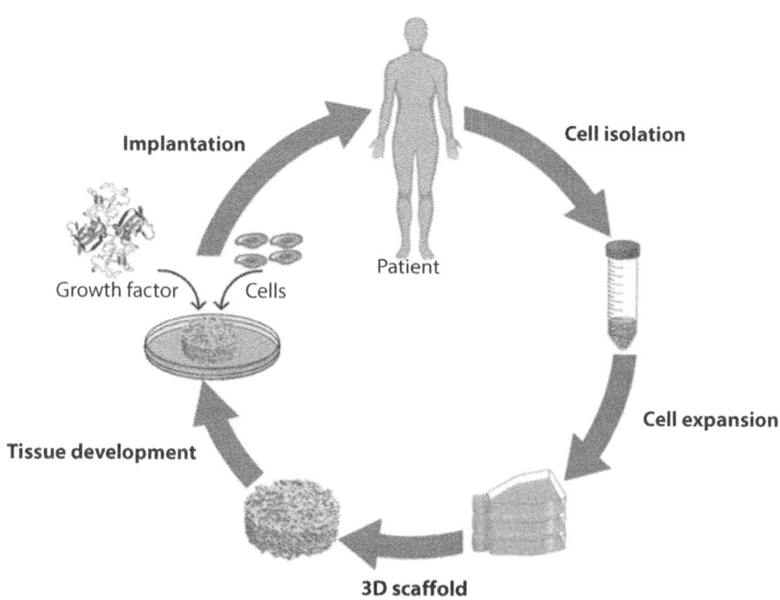

Figure 4.1 Various tissue regeneration from biopolymers.

that most mammals contain is called collagen. In particular, the biomass of fish and by-catch animals such starfish, undersized fish, sharks, sponges, and jellyfish has higher collagen levels. Since it promotes tissue growth, it is mostly employed for wound healing applications. It is extensively utilized in many biological applications, including as the treatment of burns and wounds, bone and cartilage repair, hernia, and dentistry applications. Collagen has a big drawback: it might trigger allergic responses. Furthermore, the primary source of collagen is bovine or porcine, which increases the risk of disease transmission. The introduction of recombinant collagen does not ease the problem either, because it lacks posttranslational changes and hence is not physiologically active [15]. Collagen is a remarkable natural biopolymer with exceptional potential for use in tissue regeneration. Its biocompatibility, enhancing cell growth, and ability to support the healing of tissues make it an ideal material for applications across many medical and regenerative programs. Whether used alone or in combination with other biomaterials, such scaffolds play a vital role in developing efficient therapies targeting wound healing, bone and cartilage regeneration, skin repair, among many other more applications [16].

4.2.1.2 Cellulose

It is a polysaccharide with a highly crystalline structure made up of lengthy chains of glucose units connected by β-1,4-glycosidic linkages. The textile, paper, and pharmaceutical Sectors are only a few of the many industrial uses for cellulose, which gives plant cells structural support and stiffness. Due to its biodegradability, renewability, and mechanical properties, cellulose is also gaining significant attention in biomaterial and tissue engineering applications. The primary sources of cellulose are plants, such as wood and cotton. Furthermore, it can also be acquired from some animal sources. Plant cellulose is still a significant part of plant cell walls, even though bacterial cellulose (BC), which is produced by bacteria, is chemically identical to plant cellulose. Advances in material modification and fabrication techniques continue to unleash the promise of cellulose in the production of sustainable and effective biomaterials for medical use, despite obstacles like improving cell adhesion and processing [17].

4.2.1.3 Chitosan

Chitosan (CS) is a polysaccharide with numerous benefits, including biocompatibility, biodegradability, low immunogenicity, and excellent antibacterial activities. It finds utility in bone tissue engineering because it increases osteoblast development and mineral-rich matrix deposition. Chitosan, composed of 2-amino-2-deoxy- d-glucopyranose units, constitutes the second biggest copolymer after cellulose. It belongs to a different class of biopolymers that are exclusively derived from the chitin molecules. It forms through the partial deacetylation of the chitin material in an alkaline environment. Chitosan is soluble in mildly acidic solutions and is modifiable. When chitosan is broken down by human enzymes, it promotes hemostasis and accelerates the process of tissue regeneration, which aids in wound healing. Chitosan can be used to create a range of unique bioscaffold materials by lyophilizing and altering its surface. They are also a viable option for application in tissue engineering, wound healing or dressings, and ophthalmology because of their biocompatibility and biodegradability qualities [18]. Figure 4.2 shows the various biomedical applications of chitosan in tissue engineering, including bone, skin, and vascular applications.

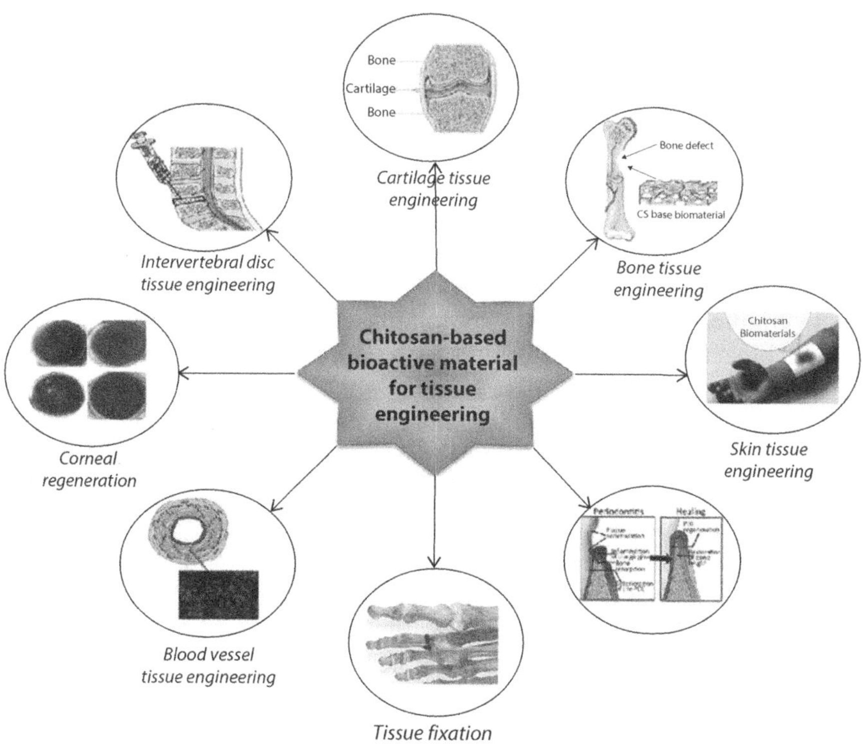

Figure 4.2 Applications of biopolymer chitosan.

4.2.2 Synthetic and Hybrid Biocomposites

However, poor mechanical properties in natural polymers limit their applicability, therefore they are being blended with others and synthetic polymers to elevate their mechanical property. Hybrid-based wound dressings are made by combining natural and synthetic polymers. Numerous hybrid-based wound dressings, including foams, hydrogels, hydrocolloids, nanofibers, films, and membranes, have been developed. The most commonly used synthetic biopolymers include polylactide (PLA), poly (glycolic acid) (PGA), poly (lactide-co-glycolide) (PLGA), Polycaprolactone (PCL), and poly (para-dioxanone) (PPDO) [19].

PLA, a traditional bioplastic made from biomass, can be produced in three steps: fermentation, separation, and polymerization. Using monomers made from renewable sugarcane, cornstarch, and other plentiful natural resources, PLA is a biodegradable polyester polymer. The reason why PLA is the perfect polymer for biomedical applications like clips, sutures, and drug administration systems is that it does not poison nor is carcinogenic to humans. The breakdown products are mainly H_2O and CO_2. Due to superior mechanical qualities and biocompatibility, PLA is more suitable for load-bearing applications, especially scaffolds that connect the tendon, ligament, and tendon/ligament-bone contact. Nevertheless, this biopolymer has several disadvantages for this use, such as a slower rate of breakdown, weak ductility, and hydrophobicity. Ongoing research focuses on addressing the limitations of PLA, such as improving its hydrophilicity, accelerating its degradation rate, and enhancing its mechanical properties. Researchers hope to further enhance PLA-based scaffolds for a variety of tissue engineering applications by mixing PLA with other biomaterials, adding bioactive compounds, and using sophisticated production processes [20].

In comparison to PLA, PGA is more hydrophilic and stronger, and its self-reinforced form is much stiffer than any other therapeutically useful biopolymer that may be used on the bone-side bone-unit scaffolds. PGA degrades fast because it breaks down hydrolytically in aqueous environments. It is useful for temporary scaffolds, in which rapid disintegration is required to keep up with the growth speed of the new tissue. Glycolic acid is one of the non-toxic breakdown products of PGA that can be easily metabolized by the body, resulting in being safer for therapeutic use. To improve its biological properties, PGA is frequently used in combination with a variety of other materials, collagen, for example. An example of these scaffolds is a collagen-PGA composite that will aid the supplementing of cell adhesion, proliferation, and osteogenic differentiation capacity because of the dual functionality of the biological functionality of collagen and the strength of PGA. The reason why PGA is commonly used in 3D printed scaffolds is because of its high processability. This provides exact regimentation of scaffold, pore size, and porosity. This is critical since it allows for vascularization and the assurance that the cells can enter and populate the scaffold so that better bone regeneration and tissue integration can occur [21].

PLGA is a widely used synthetic polymer in bone tissue engineering, and the popularity of this polymer is explained by its flexibility, biocompatibility, and biodegradability. To create PLGA, a copolymer, lactic and glycolic acids are combined, and the proportion of usage of these two monomers can change the breakdown rate. PLGA can be used in bone regeneration in the form of scaffolds, drug delivery systems, and also microspheres. The PLGA copolymers are unlike the other constituent homopolymers, PLA and PGA, in that they offer hindrance in the rate of degradation by monomer ratio control, which in turn regulates the hydrophilicity and crystallinity. The adaptability of the degradation profile of PLGA is one of the key elements that make it suitable for bone regeneration because it could be used in the short run as well as the long run. This proportion of lactic acid to glycolic acid determines the speed at which the degradation process will take place without hydrolysis. The ratio of 50:50, i.e., is worse than the other compositions. This allows the development of scaffolds to match the needs of the bone problem. Polyglycolic acid is one of the most useful biomaterials that can be used in tissue engineering. All these mechanical, biocompatible properties and biodegradable properties make them an excellent material to be used as bioprostheses in the repair of tissues. Nonetheless, the rate at which the material degrades and an inflammatory reaction is likely to occur should be taken into account. Ongoing research and development aim at breaking these limitations and finding additional uses of PGA in the regenerative medicine field [22].

With its mechanical dexterity, biocompatibility, and biodegradability, the FDA-approved semi-crystalline polyester PCL is a dependable scaffold substance to be used in any long-term scenario in load-bearing bone regeneration. The slow rate of deterioration is considered to be one of the main advantages of PCL; it ensures the growth of the bone and enables the preservation of the scaffold structure over a long time. The semi-crystalline aliphatic polymer (PCL) is non-toxic to cells or living organisms and biocompatible as well as bioresorbable, and has been widely utilized in tissue engineering, drug delivery, and administration. Nevertheless, PCL may render cell growth and adhesion inactive due to its hydrophobicity. To beat this, bioactive materials, such as hydroxyapatite (HA) or ceramic micro powder, are usually blended with PCL. PCL is a flexible, lower Young's modulus, tensile strength, and higher tensile strength at breakage material compared to PGA and PLA. To fabricate the scaffolds, which are just appropriate in treating the bone defects, frequently, PCL is frequently employed in 3D printing. Precise control of the scaffold architecture and pore size of 3D-printed PCL scaffolds enhances their mechanical properties and ensures complete vascularization and cellular penetration, which are essential to bone regeneration [23].

PPDO is a synthetic, biodegradable polymer, and over the last several years, it has attracted much attention in the area of tissue engineering. It is a biocompatible, histoconductive, and resorbable semi-crystalline polymer of thermoplastic type, which is based on polyether-esters and possesses good biocompatibility. The degree of crystallinity of PPDO is considerably higher than compared of other biodegradable thermoplastic polymers. This may modify the mechanical properties of the material by changing the quality of crystalline structure with respect to the non-isothermal cooling profile. Due to its special properties, it is a convenient source of scaffolds and implants that facilitate the process of tissue growth. Another biodegradable polyester, which has outstanding mechanical properties, high biodegradability, bio absorbability, and biocompatibility, PPDO is used both in soft and hard tissue engineering. PPDO finds a wide range of applications in the industrial production of surgical sutures, based on its incredible bio absorbability and mechanical flexibility [24].

4.2.3 Biopolymer-Based Composites for Tissue Engineering

A new person is added to the US national transplant waiting list around every 10 minutes due to the prevalence of organ or tissue loss from illness or trauma. Because organ donation appears to be rare, some 20 Americans die every day due to a lack of donor organs. Tissue engineering and regenerative medicine enable biomedical engineers and doctors to efficiently replace damaged tissues or organs using biomaterials. Polymers' special qualities that can provide the required performance are a major factor in tissue engineering. This material can be formed and designed into a range of microstructures to do that. In tissue engineering, they are typically employed as a supportive substrate that transports the cell and therapeutic material to the intended area. Biopolymers show excellent tissue biocompatibility and can be degraded within the body with minimum toxicity [25]. The degree of biological interaction serves as the foundation for the ideas that support the significance of mechanical and physicochemical properties while creating biomaterials, especially biopolymers.

The biological interactions of biopolymers must be understood for desired tissue regeneration. Cell adhesion, as a part of the interactions of cells with substrates and microenvironments, is one of the important factors for tissue regeneration. The attachment of cells to the scaffolds is significantly influenced by the way proteins bind to external surfaces. Protein-biopolymer interactions are mediated by ionic/electrostatic interactions, surface energy, hydrophobicity, and intermolecular forces. By showing their modes of action, biomaterials can be controlled through machining, alloying, and other methods to reach the desired performance for biomedical engineering [26].

4.2.3.1 Metal Oxide-Based Composites

Metal oxide composites are also one of the materials that have emerged as a promising line of materials in the field of tissue engineering. The materials that contain the individual properties of the metal oxides, along with the adaptability of the other biocompatible systems, are very promising in producing scaffolds with improved capacity to recap the natural ECM and consequently shape regeneration of tissues. Due to their mechanical strength, anti-bacterial properties, biocompatibility with living tissue, and ability to promote cell adhesion and proliferation, metal oxide-based compounds have produced interest in tissue engineering. The key categories of polymers, ceramics, or other bioactive agents which are generally used with metals are: Metal oxide-based composites are a promising category of materials used in the development of tissue engineering, which is a class of compounds that combines metallic features to meet a wide variety of clinical applications [26, 27].

4.2.3.2 Graphene Derivatives-Based Composites

Graphene composites of very high mechanical strength, electrical conductivity, biocompatible, and cell proliferation, as well as differentiation, are products of the ever-growing niche of graphene derivatives in the area of tissue engineering. These are polymers, ceramics, or other bioactive molecules combined with graphene or graphene derivatives to make multifunctional scaffolds in a very diverse selection of applications in tissue engineering. Graphene derivatives, including graphene oxide (GO) and reduced graphene oxide (rGO), have much better mechanical and electrical properties, and since the last several years, they have attracted particular attention in tissue engineering. These materials can be used together with other biocompatible materials to generate composites capable of being used to rejuvenate the natural ECM and stimulate tissue regeneration. The composites of graphene derivatives have the potential to emerge as multifunctional materials in tissue engineering, and future research and developments of the material have the potential to transform regenerative medicine [27].

4.2.3.3 Organometallic-Based Composites

Organometallic type of compounds that include a metallic-carbon bond have been identified to have a great potential in the future of tissue engineering. The organometallic compounds have interesting property combinations, such as tunable physical and chemical properties, which may be tailored to produce responsive and bioactive scaffolds. A new round of materials in tissue engineering is the organometallic-based composites, which have peculiar strengths as a result of their hybridization, lying between organic and metallic ones. Composites that are oriented on organometallic elements combine the biocompatibility and responsiveness of organic molecules as well as the real-life attributes of metals, such as catalytic efficiency, electrical conductivity, and mechanical properties. Their tunable chemical composition and functional responsiveness render them an ideal group of materials for an extensive variety of uses in tissue engineering and regenerative medicine [27].

4.2.3.4 Nanocomposite-Based Hydrogels

The latest forms of tissue engineering biomaterials are hydrogel nanocomposites. When the combined properties of nanomaterials and those of hydrogel materials are used, the properties of the latter are normally enhanced in mechanical, biological, and functional characteristics. Hydro gels of nanocomposites show potential and enthusiasm in the field of tissue engineering due to their increased mechanical strength, bioactivity, and versatility in functions. The introduction of nanoparticles or other nanostructured materials allows the formation of these composites that mimic the extracellular matrix and induce cell adhesion, proliferation, and tissue reconstruction. Hydrogels based on nanocomposites create a novel category of transformative material in tissue engineering, which is capable of better recreating and repairing natural tissue environments with improved regenerative properties [28].

4.3 Interactions between Biopolymers and Cells

An array of biomolecules on the surface of cells allows them to constantly interact with their immediate environment, which includes nutrients, signaling chemicals, extracellular matrix, and other cells. Cells are constantly gathering many kinds of information from the local environment and responding accordingly. Indeed, any alteration of the microenvironment experienced by a cell can disrupt the subtle equilibrium between a cell and its surrounding environment with biological consequences such as protein expression [29].

4.3.1 Cellular Mechanisms in Tissue Regeneration

The biological process of tissue regeneration is incredibly complex and depends upon the cooperation of many types of cells and signaling networks to occur. The extracellular matrix (ECM) is simulated by natural or synthesized materials known as biopolymers, and they are indispensable for stimulating and directing several cellular processes. Some of the important cellular mechanisms involved in biopolymer-mediated tissue regeneration include. Figure 4.3 represents the cellular proliferation triangle involving cells, scaffolds, and growth factors.

4.3.1.1 Cell Adhesion and Migration

- **Biopolymer-Cell Interactions:** The surface chemistry and topography of biopolymers affect cell adhesion. Biopolymers containing specific functional groups (e.g., RGD peptides) can interact with cell surface receptors (e.g., integrins) and stimulate cell attachment and spreading.
- **Cell Migration:** Biopolymers can help in the migration of cells. They support the process of tissue repair by guiding the cells to the area of damage. The biopolymer's porosity and rate of breakdown can have an effect on how cells migrate.

4.3.1.2 Cell Proliferation and Differentiation

- **Growth Factor Delivery:** Growth factors, or medications, can be incorporated into biopolymers to functionalize them and stimulate cell division and proliferation. The controlled release of these molecules from the biopolymer will also be close to the natural healing process.

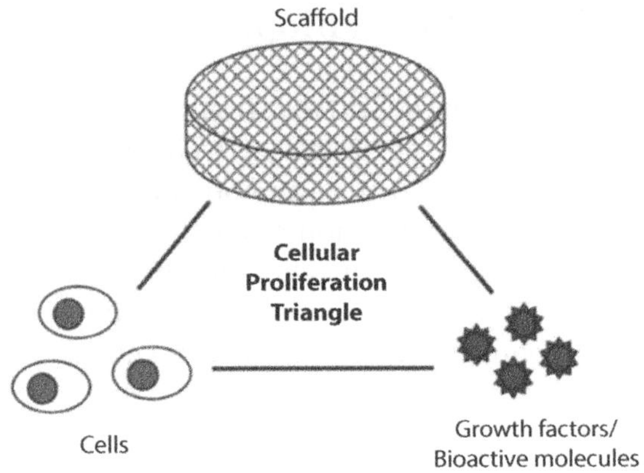

Figure 4.3 Cellular proliferation triangle.

- **Mechanical Cues:** The mechanical properties of biopolymers, their stiffness and elasticity them might be a factor in cell behavior. For example, the stiffness difference makes it possible to differentiate the cells according to their environment.

4.3.1.3 Angiogenesis

- **Vascularization:** New blood vessel formation, or angiogenesis, is a critical component of tissue repair and regeneration. By offering a scaffold for endothelial cells to migrate and multiply, biopolymers can promote angiogenesis.
- **Growth Factor Release:** Biopolymers can be engineered to release pro-angiogenic factors, such as vascular endothelial growth factor (VEGF), to stimulate blood vessel formation.

4.3.1.4 Immunomodulation

- **Inflammation:** Inflammatory response is an essential phase of the repair of tissues. Biopolymers can regulate the inflammatory response by interacting with the immune cells and releasing anti-inflammatory substances.
- **Immune Cell Recruitment:** There are some immune cells, e.g., macrophages, that facilitate the removal of debris and stimulate tissue regeneration, and these can be attracted to the location of the wound by biopolymers.

4.3.1.5 ECM Remodeling

- **Degradation:** The novelly created extracellular matrix (ECM) must be biodegradable by having its rate of breakdown such that tissue regeneration is made possible, so that biopolymers are slowly replaced.
- **ECM Deposition:** Cells that are found inside the biopolymer scaffold are capable of depositing their ECM proteins, e.g., collagen and elastin, which further stimulate the regeneration of tissues [30].

4.3.2 Cytocompatibility and Biocompatibility

Both cytocompatibility and biocompatibility play a vital role in assessing the polymers in the development of tissue regeneration, as they determine how the material interacts with the cells and the tissue itself. Cytocompatibility means that a substance has the capacity to promote growth, proliferation, and the functioning of cells that come into direct contact with the substance without any inherent toxic and harmful effects. To simplify the word, it explains the extent to which cells can live and work at and around the polymer surface. In case of tissue regeneration with the help of polymers, cytocompatibility is the most important factor since the material, in this case, must be able to permit adhesion, growth, and even proliferation of cells, both the stem cells and the differentiated cells. Hydrophobicity, surface charge, roughness, and porosity are important characteristics of the polymer surface that are relevant to the interaction of the material with cells. One method of cytocompatibility is *in vitro* cell culture in which the cells are cultured on the material, and different features of the cells are evaluated, including cell viability, cell morphology, and rates of cell proliferation [31].

Biocompatibility is a more general term describing the ability of a material of a living being to perform its functioning without causing an undesirable immunological response, toxicity, or chronic problems. It entails the interaction of the material with the different biological systems, including cells, tissues, and fluids, and whether the material generates any detrimental systemic inflammatory or immune reactions. In the case of polymers to be used in tissue engineering, biocompatibility should ensure that it does not cause rejection and other unpleasant responses in the body. It is essential to use temporary or long-term implants or scaffolds that facilitate the regeneration of the tissue. Biocompatibility is usually measured by using *in vivo* testing (animal models), where the material is inserted into a living body. The researchers can see how the body has reacted to the material in the long run and whether it has an inflammatory reaction, an immune reaction, or toxicity manifestations. A histological examination of tissue around the implant is usually done to check for any undesired effects [31].

4.4 Applications of Biopolymeric Composite Materials

4.4.1 Vascular Engineering: Grafts and Stents

The creation of new blood vessels is crucial in tissue engineering and many vascular diseases. It was discovered that blood vessels derived from BC were biocompatible and stable vascular conduits. The area where BC-based blood arteries and the surrounding media come into touch underwent structural changes, according to morphological review. The presence of fibroblasts in this area indicated that integration without degradation had occurred. When these devices were implanted *in vivo*, the fractional shortening and ejection fraction were much higher than in partially untreated hearts.

Stents are implantable medical devices used to restore the shape of damaged or constricted arteries. They are delivered inside the coronary artery with the use of an inflatable balloon, attached to a catheter. Aside from diminishing systemic toxicity and hastening the repair of the damaged veins upon implantation of stents, this biodegradable polymeric drug-eluting stent further resists restenosis. Through the employment of poly-lactic acid, the first ever biodegradable polymeric drug-eluting stent was derived [32].

4.4.2 Cardiac Engineering: Heart Muscle Regeneration, Patches, and Valves

Biocomposites are receiving much attention in the area of cardiac engineering due to their ability to replicate the structural, mechanical, and functional properties of natural cardiac tissues. These materials, which are mixtures of biopolymers and additives with compatibility in biomedical applications, are promising solutions for the regeneration of the heart muscle, cardiac patches, and valve replacement or repair. Biocomposites are becoming interesting materials in the context of cardiac engineering due to their unique combination of properties able to mimic the complex environment of the heart. These materials consist of a natural or synthetic

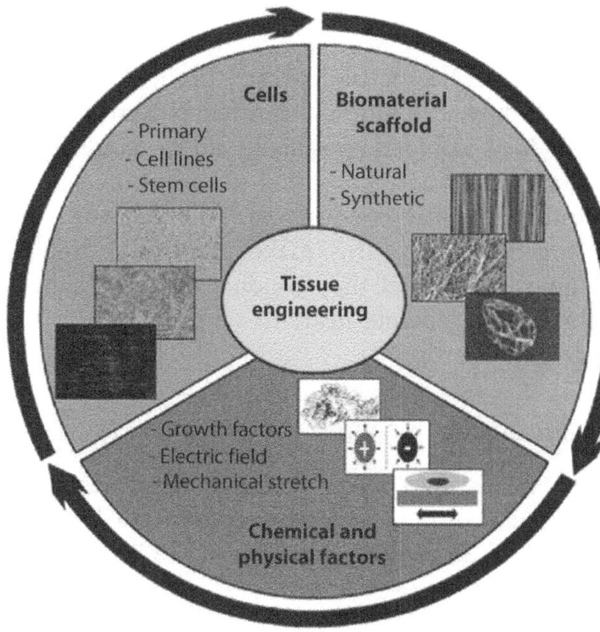

Figure 4.4 Cardiac tissue engineering.

polymer matrix reinforced with biological or synthetic materials such as fibers, particles, or even cells. This could be used for the development of scaffolds, which can have tailored mechanical, biological, and electrical properties that are suitable for different applications in the cardiac area [32]. Figure 4.4 depicts the components and factors involved in cardiac tissue engineering.

4.4.2.1 Heart Muscle Regeneration

A myocardial infarction or heart attack is linked with the death of some cell of heart. It leads to the formation of scars as well as poor heart performance. Biocomposites can help the damaged heart tissue to regenerate in a favourable environment by:

- Mimicking the extracellular matrix (ECM): ECM provides cells the metabolic and structural signals that they need to develop and differentiate. In order to address the need of better cell adhesion, proliferation, and differentiation, biocomposites may be designed to mimic the composition,

topography, and mechanical properties of the extracellular matrix.
- Bioactive molecules can be delivered through Biocomposites. Growth factors, cytokines, and other bioactive substances can be added to biocomposites to aid in tissue repair and regeneration.
- Providing electrical conductivity: The heart is an electrically active organ. This makes it possible to engineer Biocomposites so that they conduct electrical signals and thereby help regenerate heart tissue in sync with the surrounding myocardium.

4.4.2.2 Cardiac Patches

Cardiac patches are used to repair damaged heart tissue, such as that resulting from MI or trauma. Compared to conventional synthetic patches, biocomposite patches can provide a number of benefits, such as:

- **Improved biocompatibility:** Natural-based Biocomposites can elicit a less inflammatory response and promote better integration with the host tissue.
- **Enhanced mechanical properties:** The incorporation of reinforcing fibers or particles can improve the strength and durability of the patch, allowing it to withstand the mechanical stresses of the heart.
- **Tailored functionality:** Biocomposites can release therapeutic drugs, such growth factors or anti-inflammatory drugs, into damaged tissue.

4.4.2.3 Cardiac Valves

The heart's cardiac valves control blood flow. When valves become diseased or damaged, they may need to be replaced. The possibility for tissue regeneration, particularly that of more complicated tissues like heart valves, has been made possible by tissue-engineered scaffolds. A few days after various human cells were seeded onto the biohybrid scaffolds, a notable amount of extracellular matrix deposition was seen, suggesting that each cell had grown and proliferated. Biocomposite valves offer several potential advantages over traditional mechanical and biological valves:

- **Reduced risk of thrombosis:** The use of biocompatible materials can minimize the risk of blood clot formation, a major complication of mechanical valves.
- **Improved durability:** The incorporation of reinforcing fibers or particles can enhance the durability of the valve, reducing the need for future replacements.

4.4.3 Targeted Drug Delivery Systems

One of the most important applications of biomaterials is drug delivery systems. These materials can be utilized in the phasing of pharmacological materials in a controlled and accurate release. There are several benefits of this way, including better drug absorption, reduced side effects, and slow absorption of long-term diseases.

Sustained Release: Therapeutic agents that are released over time can be engineered in Biodegradable Biocomposites. This kind of profile would decrease the rate of administration, maximize the drug effect, and decrease the adverse effects of the drug.
Ligand-Conjugated Biocomposites: Biocomposites can be functionalized with ligands, i.e., peptides or antibodies that have recognizing power to attach to receptors on target cells or tissues. Such a technique of individual delivery may enhance treatment efficacy and enhance the presence of medication in the desired site.
Scaffolding for Tissue Growth: It is also possible to develop biocomposites that can be used in the form of a scaffold in tissue regeneration to provide a three-dimensional environment that promotes cell proliferation and differentiation. The biocomposites can then accelerate repair and regeneration of tissues by the controlled release of growth factors and other bioactive substances by the biocomposites.
Improved Biocompatibility: Natural-based biocomposites often display excellent biocompatibility, which minimizes the chances of an undesirable immune response and complicates tissue integration [33].

4.4.4 Bone and Dental Tissue Engineering

In Bone Tissue engineering, an alternate method for making bone uses biomaterials alone or in association with cells. Bone tissue engineering has focused its attention on the creation of 3D scaffolds with the desirable and sufficient pore size to offer support and mechanical reinforcement, as well as order tissue regeneration and replacement in physiological conditions.

Biodegradable spongy structures with selected properties like osteogenicity and osteoconductive, good biocompatibility, and eventually biodegradation are often picked for BTE.

Mimicking Natural Bone: Biocomposites can closely resemble the mechanical characteristics and biochemical makeup of natural bone because they frequently combine polymers (like collagen and chitosan) with ceramics (like hydroxyapatite). This promotes better integration and bone regeneration.
Controlled Drug Delivery: In order to promote bone cell growth and differentiation, biocomposites can be designed to release growth factors and other bioactive compounds gradually. This enhances healing and reduces the need for multiple surgeries.
Improved Biocompatibility: Natural-based biocomposites frequently have high levels of biocompatibility, which reduces the possibility of negative immune responses and speeds up the healing process.

4.4.5 Dental Tissue Engineering

Bone Grafting: Biocomposite scaffolds can be used to fill out a bone defect in the jaw, aiding in the placement of a dental implant, or to support complicated reconstructions of a jaw. Figure 4.5 shows the classification of scaffolds used in bone tissue engineering based on composition and geometry.

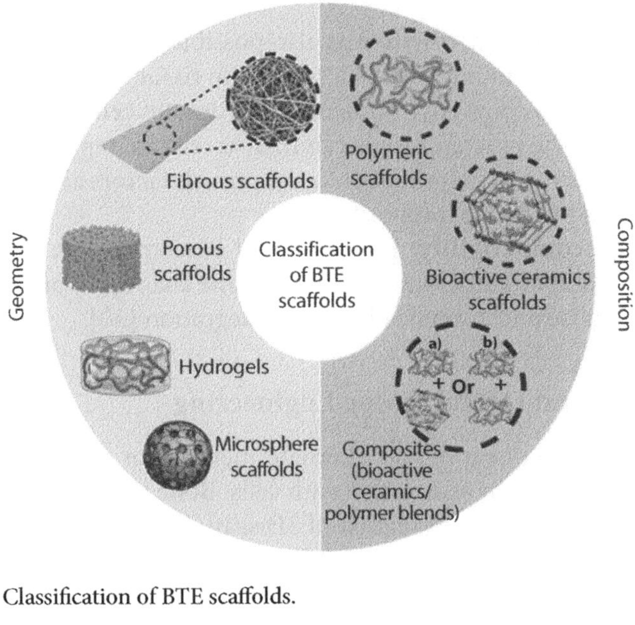

Figure 4.5 Classification of BTE scaffolds.

Periodontal Regeneration: The biocomposites can be used for the regrowth of the lost periodontal tissues, such as bone, ligament, and cementum, which aids and enhances the stability and function of the teeth.
Dental Implants: Coatings of biocomposites on dental implants can be used to improve osseointegration (contact between bone and implant), which will improve the stability and longevity of the implant [33].

4.4.6 Pros of Biopolymers in Tissue Engineering

1. Biocompatibility
   Collagen, fibrin, and hyaluronic acid are examples of natural materials from which many biopolymers are made. This natural source often means that the material will have superior biocompatibility, which reduces the risk of a dangerous immunological response or internal toxicity. Our bodies are already used to these materials, making the rate of rejection less likely.
2. Biodegradability
   These substances can spontaneously decompose in the body with the course of time. As they degrade, they form space for the development of new tissue, meaning that they do not need to be surgically removed after they have served their purpose. The fact that many biopolymers are manufactured so that they degrade slowly within the body is a huge plus. This obviates the necessity for secondary surgeries to remove implants once the tissue has regenerated. The degradation products are often not harmful and can easily be eliminated by the body.
3. Tunable Properties
   Biopolymers are extremely versatile. Researchers can tune their attributes to suit some of the demands of tissue engineering by altering chemical composition, structure, and processing methods.
4. Mimicry of Natural Tissues
   Collagen and alginate are two examples of biopolymers resembling the extracellular matrix (ECM) present in tissues in nature. This will induce tissue regeneration and increase cell proliferation under a more natural environment.
5. Encouraging Cellular Growth
   Biopolymers are used to deliver important components, such as cell attachment sites or growth factors, that can enhance

cell growth and differentiation and are required to achieve successful tissue engineering.

6. Renewability and Sustainability
   Biopolymers are usually derived from renewable, natural resources; thus, making them environmentally friendly than synthetic materials.
7. Scaffolding Potential: Biopolymers are very good at creating three-dimensional scaffolds that mimic the extracellular matrix (ECM), which is the natural habitat of cells. These scaffolds offer:
   - Structural support: A structure on which cells can get attached and grow.
     Guidance for cell migration: On the way to where they are needed: Directing cells to the right places.
   - Control over tissue architecture: Controlling the formation of the regenerated tissue into the required shape.
   - Drug Delivery: It is possible to produce biopolymers that act as growth factors, medications, or other therapeutic materials with controlled release. It would, therefore, maintain the delivery of such substances for maximum expression of their therapeutic effect and with a minimum of side effects [34].

4.4.7 Cons of Biopolymers in Tissue Engineering

1. Variability
   Natural biopolymers, such as those from animal sources, can be prone to variation in their properties from batch to batch. This inconsistency can impact their performance and reproducibility, which is difficult to achieve in tissue engineering applications, where it is important to have consistent results.
2. Mechanical Weakness
   Many biopolymers do not have the strength and stiffness required for some types of body tissues. Some of the limitations of natural biopolymers are that the material can be tunable and may not be favorable in terms of mechanical strength. This may be a concern for applications where there is a great deal of load-bearing ability, such as bone or cartilage regeneration.

3. Immunogenicity
   Although mostly biocompatible, there is always a slight risk of an immunological response to certain biopolymers, even those of natural origin. This risk is considered to be generally low, but requires careful consideration.
4. Inconsistent Properties
   Biopolymers may vary in their quality depending on where they are sourced from which may make their properties inconsistent. This can mean that it is more difficult to accurately control material properties such as strength or rates of degradation.
5. Slow or Uneven Degradation
   While biodegradability is a plus, some biopolymers can degrade too slowly, which can hinder the process of tissue regeneration. Alternatively, if the scaffold is unevenly degraded, it could impact the scaffold's function.
6. Potential Immune Response
   Despite being natural, certain biopolymers (like those derived from animals) may still trigger immune responses, especially if not properly purified or if sourced from non-human animals.
7. Manufacturing Challenges
   Processing biopolymers into the desired scaffold forms can be more complex than working with synthetic materials. Ensuring consistency, uniformity, and desired properties in the final product is often more difficult.
8. Limited Suitability for Some Tissues
   Biopolymers may not be able to deliver the mechanical properties required for all kinds of tissues. For example, bone tissue needs a good and stiff scaffold, which cannot be provided by many natural biopolymers alone.
9. Cost
   Certain types of biopolymers (especially those isolated from natural sources) may be quite expensive to produce and purify. This may add to the cost of tissue engineering therapies as a whole [35].

4.5 Future Directions

Over time, the method for creating scaffolds for tissue engineering has been progressively improved. Scaffolds have been effectively produced utilizing natural precursors in numerous recent attempts to prevent any potential toxicity, either short-term or long-term. To advance toward safe clinical trials and authorized goods devoid of adverse effects based on these superior biomaterials, more research must be done. Certain biopolymers, such as silk fibroin, have been shown to pose a risk to scar tissue when used as tissue scaffolds for wound and skin regeneration. Future tissue engineering technology is expected to use 4D printing of biopolymer-based scaffolds in addition to standard 3D printing over time to overcome some of the drawbacks of 3D printing, such as the intricate dynamics of native tissue fabrication and the optimization of cell–construct interaction functional responses. Because of their great potential for use in future 4D bioprinting of tissue engineering and regenerative medicine, biopolymers may therefore be crucial biomaterials for bioink formulations [36].

Recent decades have witnessed a high increase in research in this field, as seen by the upsurge of newly published papers. Packaging, automobiles, construction, electronics, and—above all—medicine are just a few of the industries that use biopolymer composites on a regular basis. For the synthesis of biopolymer composites, injection, extrusion, and *in situ* synthesis are the most used biochemical recipes. The addition of graphene and its derivatives as fillers enhances the mechanical properties of biopolymers, including tensile, effect, flexural, and other structural properties. Future directions are toward developing new biomaterials that overcome the above-mentioned issues, are eco-friendly, highly recyclable, and economically viable [37].

4.6 Challenges and Future Developments

Tissue engineering challenges can be divided into two categories: universal bioink for all tissues and the regulatory category, or innovative bioink research and development for particular tissues. A universal bioink should ideally be a combination of biomaterials that support neuronal intercalation and angiogenesis of natural tissues, chemical markers, and growth hormones. New technologies that enable the creation of intricate textiles, including additive manufacturing, can help overcome these obstacles. Vascularization impedes the development of sustainable angiogenesis treatments that include bone marrow clots, platelet additions, angiogenic growth agents, and bioreactors.

Graphene and its composites, along with metal nanoparticles, have become crucial fillers into biopolymers to develop the necessary biomaterials. These are used in medical applications to reinforce tensile, effect, bending, and other structural properties of biopolymers. The significant problems while using these biopolymers to prepare Biocomposites are mechanical behaviors and insufficient dispersion. The formation of agglomerates with a biopolymer matrix leads to poor mechanical characteristics from the fillers and a weak interfacial connection with defective structural harmonics. This leads to many other unique characteristics, including low impact strength, shelf life, and vulnerability to high temperatures and humidity. Future regulations, therefore, result in new biomaterials that are recyclable, eco-friendly, and commercially successful [38].

In the future, biocomposites will be essential to the creation of multipurpose scaffolds for tissue engineering and regenerative medicine. As biopolymers are applicable to tissue engineering, their biological interactions, characteristics, advantages, and disadvantages support this contribution toward the development of Biocomposites in the future. In addition, particular attention should be paid to the contribution of machine learning algorithms in the design and fabrication of intelligent Biocomposites [39].

4.7 Conclusions and Prospects

Tissue engineering is a promising discipline that regenerates injured tissues using bioactive chemicals, scaffolds, and cells. Scaffolds should have morphology, biochemistry, and mechanics similar to ECM in order to support cell adhesion, proliferation, and differentiation. Numerous natural and synthetic polymers are used to make scaffolds. Given the benefits and drawbacks of each natural and synthetic polymer, combining them is a good option. Smart hydrogels, which can be employed as injectable *in situ* forming hydrogels with a less invasive procedure, are among the crucial structures. Consequently, the right biomaterials and scaffold-building technique can be developed based on the target tissue. The best alternative to synthetic petroleum polymers is biopolymers, which have significant renewable, biodegradable, and ecologically friendly properties. Biopolymers lack mechanical characteristics such as strong tensile strength, impact strength, bending force, and thermal stability. They may, however, use their ceramic composites' mechanical strength to carry out load-bearing tasks. The usual mixing processes and the adaptation of

Biocomposites microstructural characteristics through the use of reinforced materials require more attention and innovation. The development of tissue engineering scaffolds utilizing nanoscale fabrication techniques should focus on biopolymer scaffolds, which are the biomaterial of the future. The main problems with inorganic scaffolds and synthetic polymers, such as cytotoxicity and biocompatibility, can be resolved by these novel scaffolds. For both present and future tissue engineering scaffolds, biopolymer scaffolds show great promise as the primary precursor [40].

AI disclosure statement: The author acknowledges the use of AI-based tools for minor language editing or grammatical correction. However, all intellectual contributions, analyses, and conclusions presented in this chapter are entirely the author's own.

References

1. Hsissou, R., Seghiri, R., Benzekri, Z., Hilali, M., Rafik, M., Elharfi, A., Polymer composite materials: A comprehensive review. *Compos. Struct.*, 262, 113640, 2021 Apr 15.
2. Vinod, A., Sanjay, M.R., Suchart, S., Jyotishkumar, P., Renewable and sustainable biobased materials: An assessment on biofibers, biofilms, biopolymers and biocomposites. *J. Cleaner Prod.*, 258, 120978, 2020 Jun 10.
3. Xiong, R., Grant, A.M., Ma, R., Zhang, S., Tsukruk, V.V., Naturally-derived biopolymer nanocomposites: Interfacial design, properties and emerging applications. *Mater. Sci. Eng.: R: Rep.*, 125, 1–41, 2018 Mar 1.
4. Rajput, S., Malviya, R., Sridhar, S.B., Osteoporosis Risk Assessment and Individualized Feature Analysis Using Interpretable XAI and RAI Techniques, in: *Explainable and Responsible Artificial Intelligence in Healthcare*, pp. 89–113, Elsevier, Amsterdam, Netherlands, 2025 Mar 13.
5. Al Mahmud, M.Z., Exploring the versatile applications of biocomposites in the medical field. *Bioprinting.*, 36, e00319, 2023 Dec 1.
6. Shapira, A., Feiner, R., Dvir, T., Composite biomaterial scaffolds for cardiac tissue engineering. *Int. Mater. Rev.*, 61, 1, 1–9, 2016 Jan 2.
7. Goonoo, N., Bhaw-Luximon, A., Bowlin, G.L., Jhurry, D., An assessment of biopolymer-and synthetic polymer-based scaffolds for bone and vascular tissue engineering. *Polym. Int.*, 62, 4, 523–33, 2013 Apr.
8. Rajput, S., Malviya, R., Srivastava, S., Ahmad, I., Rab, S.O., Uniyal, P., Cardiovascular disease and thrombosis: Intersections with the immune system, inflammation, and the coagulation system. *Ann. Pharm. Fr.*, 83, 2, 228–250, 2025 Mar 1. Elsevier Masson.
9. Luzi, F., Puglia, D., Torre, L., Natural fiber biodegradable composites and nanocomposites: a biomedical application, in: *Biomass, Biopolymer-Based*

Materials, and Bioenergy, pp. 179–201, Woodhead Publishing, Cambridge, UK, 2019 Jan 1.
10. Lee, K., Silva, E.A., Mooney, D.J., Growth factor delivery-based tissue engineering: general approaches and a review of recent developments. *J. R. Soc. Interface*, 8, 55, 153–70, 2011 Feb 6.
11. Chapekar, M.S., Tissue engineering: challenges and opportunities. *J. Biomed. Mater. Res.*, An Official Journal of The Society for Biomaterials, The Japanese Society for Biomaterials, and The Australian Society for Biomaterials and the Korean Society for Biomaterials 53, 6, 617–20, 2000.
12. Nambiar, J., Jana, S., Nandi, S.K., Strategies for Enhancing Vascularization of Biomaterial-Based Scaffold in Bone Regeneration. *Chem. Rec.*, 22, 6, e202200008, 2022 Jun.
13. Kaplan, D.L., Introduction to biopolymers from renewable resources, in: *Biopolymers from renewable resources*, pp. 1–29, Springer Berlin Heidelberg, Berlin, Heidelberg, 1998.
14. Nimni, M.E. and Harkness, R.D., Molecular structure and functions of collagen, in: *Collagen*, pp. 1–78, CRC Press, Boca Raton, FL, USA, 2018 Jan 18.
15. Kumar, S., Malviya, R., Sridhar, S.B., Biomaterials for Tissue Engineered Cartilage Product for Orthopedic (Osteoarthritis) Reconstruction, in: *Sustainable Nanocomposites with Green Biomaterials*, vol. 25, pp. 45–60, Springer, Singapore, 2025 Mar.
16. Li, Y., Liu, Y., Li, R., Bai, H., Zhu, Z., Zhu, L., Zhu, C., Che, Z., Liu, H., Wang, J., Huang, L., Collagen-based biomaterials for bone tissue engineering. *Mater. Des.*, 210, 110049, 2021 Nov 15.
17. Chaudhary, S., Jain, V.P., Jaiswar, G., The composition of polysaccharides: monosaccharides and binding, group decorating, polysaccharides chains, in: *Innovation in Nano-Polysaccharides for Eco-Sustainability*, pp. 83–118, Elsevier, Amsterdam, Netherlands, 2022 Jan 1.
18. Khan, A. and Alamry, K.A., Recent advances of emerging green chitosan-based biomaterials with potential biomedical applications: A review. *Carbohydr. Res.*, 506, 108368, 2021 Aug 1.
19. Aderibigbe, B.A., Hybrid-based wound dressings: combination of synthetic and biopolymers. *Polymers.*, 14, 18, 3806, 2022 Sep 12.
20. Taib, N.A., Rahman, M.R., Huda, D., Kuok, K.K., Hamdan, S., Bakri, M.K., Julaihi, M.R., Khan, A., A review on poly lactic acid (PLA) as a biodegradable polymer. *Polym. Bull.*, 80, 2, 1179–213, 2023 Feb.
21. Samantaray, P.K., Little, A., Haddleton, D.M., McNally, T., Tan, B., Sun, Z., Huang, W., Ji, Y., Wan, C., Poly (glycolic acid)(PGA): A versatile building block expanding high performance and sustainable bioplastic applications. *Green Chem.*, 22, 13, 4055–81, 2020.
22. Archer, E., Torretti, M., Madbouly, S., Biodegradable polycaprolactone (PCL) based polymer and composites. *Phys. Sci. Rev.*, 8, 11, 4391–414, 2023 Nov 20.

23. Rana, A., Mittal, A., Vashist, C., Rajput, S., Sridhar, S.B., Malviya, R., Exploration of 4D Printing and its Applications in the Biomedical Sciences. *Curr. Pharm. Des.*, 2025.
24. Yuan, L., Xiao, L., Zhang, J., Xiao, Y., Yu, L., Yang, K.K., Wang, Y.Z., Engineering Biodegradable Polyurethanes with Precisely Controlled Hierarchical Structures to Access Shape Memory Effect and Enhanced Bioactivities. *Biomacromolecules*, 2024 May 23.
25. Biswal, T., Biopolymers for tissue engineering applications: A review. *Mater. Today Proc.*, 41, 397–402, 2021 Jan 1.
26. Huang, H., Feng, W., Chen, Y., Two-dimensional biomaterials: material science, biological effect and biomedical engineering applications. *Chem. Soc. Rev.*, 50, 20, 11381–485, 2021.
27. Muthukrishnan, L., Sagadevan, S., Hossain, M.M., Metal oxide-based fiber technology in the pharmaceutical and medical chemistry, in: *Metal Oxides for Optoelectronics and Optics-Based Medical Applications*, pp. 259–300, Elsevier, Amsterdam, Netherlands, 2022 Jan 1.
28. Anwer, A.H., Ahtesham, A., Shoeb, M., Mashkoor, F., Ansari, M.Z., Zhu, S., Jeong, C., State-of-the-art advances in nanocomposite and bio-nanocomposite polymeric materials: A comprehensive review. *Adv. Colloid Interface Sci.*, 102955, 2023 Jun 25.
29. Cassim, S. and Pouyssegur, J., Tumor microenvironment: a metabolic player that shapes the immune response. *Int. J. Mol. Sci.*, 21, 1, 157, 2019 Dec 25.
30. Forbes, S.J. and Rosenthal, N., Preparing the ground for tissue regeneration: from mechanism to therapy. *Nat. Med.*, 20, 8, 857–69, 2014 Aug.
31. Komeri, R., Kasoju, N., Kumar, P.A., *In vitro* cytotoxicity and cytocompatibility assays for biomaterial testing under regulatory platform, in: *Biomedical Product and Materials Evaluation*, pp. 329–353, Woodhead Publishing, Cambridge, UK, 2022 Jan 1.
32. Zia, A.W., Liu, R., Wu, X., Structural design and mechanical performance of composite vascular grafts. *Bio-Des. Manuf.*, 5, 4, 757–85, 2022 Oct.
33. Ciolacu, D.E., Nicu, R., Ciolacu, F., Natural polymers in heart valve tissue engineering: strategies, advances and challenges. *Biomedicines.*, 10, 5, 1095, 2022 May 8.
34. Veeman, D., Sai, M.S., Sureshkumar, P., Jagadeesha, T., Natrayan, L., Ravichandran, M., Mammo, W.D., Additive manufacturing of biopolymers for tissue engineering and regenerative medicine: an overview, potential applications, advancements, and trends. *Int. J. Polym. Sci.*, 2021, 1, 4907027, 2021.
35. Ghosh, S., Chaudhuri, S., Roy, P., Lahiri, D., 4D Printing in biomedical engineering: A state-of-the-art review of technologies, biomaterials, and application. *Regener. Eng. Transl. Med.*, 9, 3, 339–65, 2023 Sep.
36. Abolhassani, S., Fattahi, R., Safshekan, F., Saremi, J., Hasanzadeh, E., Advances in 4D Bioprinting: The Next Frontier in Regenerative Medicine and Tissue Engineering Applications. *Adv. Healthcare Mater.*, 2403065, 2024.

37. Andrew, J.J. and Dhakal, H.N., Sustainable biobased composites for advanced applications: recent trends and future opportunities–A critical review. *Composites, Part C: Open Access*, 7, 100220, 2022 Mar 1.
38. Kim, S., Oh, T., Lee, H., Nam, J.M., Trends and perspectives in bio-and eco-friendly sustainable nanomaterial delivery systems through biological barriers. *Mater. Chem. Front.*, 6, 16, 2152–74, 2022.
39. Verma, D., Dong, Y., Sharma, M., Chaudhary, A.K., Advanced processing of 3D printed biocomposite materials using artificial intelligence. *Mater. Manuf. Processes*, 37, 5, 518–38, 2022 Apr 4.
40. Mabrouk, M., Beherei, H.H., Das, D.B., Recent progress in the fabrication techniques of 3D scaffolds for tissue engineering. *Mater. Sci. Eng. C.*, 110, 110716, 2020 May 1.

5

Cartilage and Soft Tissue Engineering Using Biocomposite

Prerna Mehta

Department of Biotechnology, GD Rungta College of Science & Technology, Bhilai, Chhattisgarh, India

Abstract

It was a significant advance in the tissue engineering of the cartilage and soft tissues using biocomposites which are synthetic and natural materials that are integrated to enhance strength and the biological tolerance. The subject of the modern studies is the manufacturing of the scaffolds with the help of bioceramics with hydrogel and polymer mixtures mainly to correspond to the key features of the extracellular conditions of the natural tissues. The works on biocomposite materials that have recently been published are viewed through the prism of the design strategy and manufacturing practices and their impact on the human body biologically. The text presents the 3D bioprinting and electrospinning since they are able to create complex structures that trigger cell binding and cell multiplication and regulate cell differentiation. The paper will examine the role of both bioactive molecule, such as growth factors and peptides in biocomposite activities, which mediate tissue healing activities. The chapter addresses the issues both in the context of the life expectancy and tissue–host integration at the border and the evidence of future application of individualized medicine. The use of biocomposite-based techniques has transformed cartilage and also the soft tissue engineering since it generates regenerative therapies that could be critical in the therapy of musculoskeletal diseases.

***Keywords*:** Biocomposites, cartilage engineering, regenerative medicine, soft tissue repair, tissue scaffolds

Email: prernamehta326@gmail.com

Rishabha Malviya and Mukesh Roy (eds.) Biocomposites in 3D and 4D Printing: Materials, Techniques, and Applications in Regenerative Medicine, (123–156) © 2026 Scrivener Publishing LLC

5.1 Introduction

The presence of soft tissues and cartilage is crucial to the human body since it forms the structure of the body and allows it to move [1]. The primary composition of cartilage includes an extracellular matrix (ECM) that contains collagen fibers together with proteoglycans and lacks direct blood circulation. The cartilage structure succeeded in bearing compressive forces because it serves vital joint functions and creates a smooth articulating surface. The regenerative therapies for cartilage become necessary after the occurrence of injuries and degenerative health issues, including osteoarthritis, which causes tissue damage. The biomechanical functions and physiological homeostasis heavily depend on soft tissues which include tendons along with ligaments and adipose tissues. Injuries within these tissues and their deficiencies weaken patient mobility and substantially decrease their life quality [2].

The field of tissue engineering presents itself as an effective method to address the treatment and improvement of damaged cartilage together with damaged soft tissues (Figure 5.1). Through the combination of three disciplines, which include biology and materials science and engineering, this research field develops structures that assist cell growth while

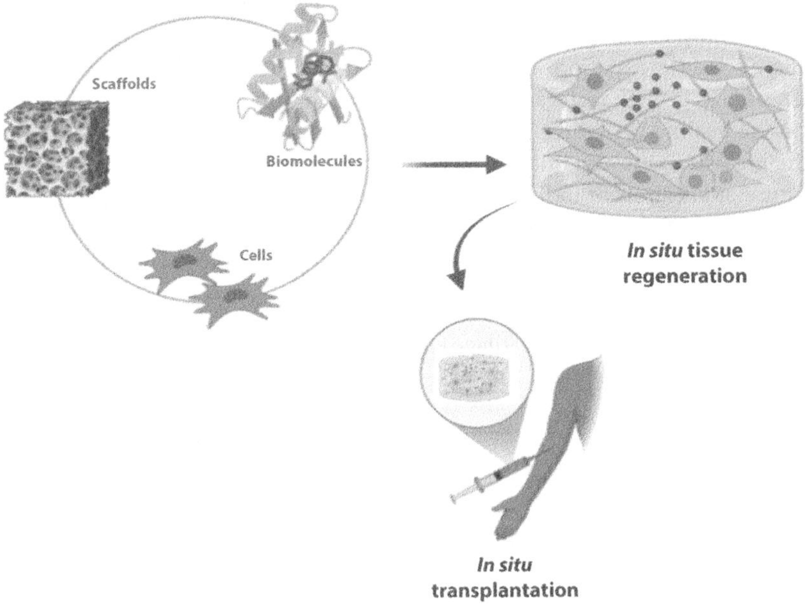

Figure 5.1 Schematic diagram of tissue engineering construct.

driving tissue replenishment. Recent research demonstrates that engineered scaffolds require structures which duplicate natural tissue architectures together with biochemical milieux to facilitate cellular attachment along with increase cell growth while enabling proper differentiation. Biocomposites represent a fundamental breakthrough because they merge natural compounds with synthetic elements into scaffolds which combine enhanced mechanical strength and biological acceptability [3].

5.1.1 Overview of Cartilage and Soft Tissue Engineering

Soft tissue engineering alongside cartilage creation deals with building functional tissue replacements for damaged and diseased organs. The use of autografts and allografts as traditional methods faces challenges due to their negative effects on donor sites and encounters problems with tissue rejection from the immune system. The engineered tissues are individual tissues solutions that are capable of addressing the needs of a specific patient case. The introduction of scaffolds becomes the key element of cartilage regeneration since such structures contribute to upholding the structure and supporting biological functions. Advancement in material science led to the development of biocomposite scaffold with a blend of bioceramics with biodegradable polymers and hydrogels that can be used to replicate the characteristics of the native cartilage in terms of structure and physiology [4].

5.1.2 Importance of Biocomposites in Tissue Engineering

The use of biocomposites has better advantages than traditional materials that are applied in cartilage and soft tissue engineering. The composite materials provide synergistic advantages on the basis of a high level of cooperation of all their natural and synthetic components. The mixture of natural polymers like chitosan or gelatin introduces biocompatible features with cellular interaction facilitation and synthetic polymers introduces mechanical strength and bio-degradability features to the system [5]. Hydroxyapatite is a bioceramic material that mimics the inorganic parts of the bone and cartilage and improves the osteoconductive qualities of the material. It is this property which has been supplemented with factoring in growth factors and peptides into biocomposites which have increased the functionality of biocomposites in supporting both cell differentiation processes and tissue regeneration processes. The biocomposites have a high level of performance which makes them significant building blocks of the future therapy founded on the healing of damaged soft tissues and cartilage [6].

5.1.3 Objectives of the Chapter

The chapter discusses in detail how the use of biocomposite materials has propelled cartilage and soft tissue engineering in the last years. This will accomplish four major things first, outline the basic principles of cartilage and soft tissue engineering and how biocomposites work in them; second, the many ways biocomposites can be designed and produced; third, how these living tissues actually react to biocomposite scaffolds, and what it entails to healing; and the fourth, what we currently do and what we can do next and how we can use these materials to make every patient unique. Combining what we already know and where we are headed, this chapter is meant to illustrate how biocomposites may help reverse the tide in helping the body to heal and make cure more successful on individuals struggling with muscle and bone problems.

5.2 Fundamentals of Cartilage and Soft Tissue

Engineers must understand much about how these tissues operate, and what they comprise in order to develop new cartilage and soft tissues. This involves knowing their composition, the various types that exist and the work they perform within the body. The role of the surrounding of cells which can be referred to as the extracellular matrix (ECM) in keeping tissues in a healthy and good condition should also be understood. This knowledge is important in the development of new methods of assisting tissues to regenerate.

5.2.1 Structure and Function of Cartilage

The special type of cartilage is known as connective tissue that has a firm type combined with a flexible one. Chondrocytes characterize the cartilage and are mostly located in a gelatinous framework consisting of water and collagen fibres and other macromolecules and proteoglycans. There are three types of cartilages, such as hyaline elastic and fibrocartilages that serve a certain purpose to the human body [7].

Table 5.1 Overview of soft tissue types and their functions.

Tissue type	Subcategories	Description	Function	References
Soft Tissues		Form a combination of several tissues that attach, support, or enclose other structures in the body.	Necessary to ensure body functions and structural support.	[9]
Muscle Tissue	- Skeletal	Voluntary muscle that causes movement and posture; striped in appearance.	Allows voluntary movements and helps with posture.	[10]
	- Cardiac	Involuntary, striated muscle adapted for heart contractions.	Enables rhythmic heart contractions for blood circulation.	[11]
	- Smooth	Involuntary muscle responsible for regulating several visceral functions; non-striated.	Responsible for controlling involuntary movements like digestion and constriction of blood vessels.	[12]
Connective Tissue	- Tendons	Attach muscle to bone; made of dense collagen fibers with very high tensile strength.	Enables movement and stability by attaching muscles to bones during physical actions.	[13]
	- Ligaments	Attach bone to bone at joints; contain a looser collagen structure to allow flexibility.	Supports joints and gives stability while permitting a level of movement.	[14]

(*Continued*)

Table 5.1 Overview of soft tissue types and their functions. (*Continued*)

Tissue type	Subcategories	Description	Function	References
	- Adipose tissue	Stores energy in the form of fat; insulates and cushions the organs.	Functions in energy storage, insulation, and protective cushioning; also plays a role in hormone secretion.	[15]
Other roles of adipose tissue		Has adipocytes that secrete hormones and signaling molecules.	Affects metabolism, immune function, and overall endocrine control.	[16]

Hyaline cartilage is the most common kind of cartilage which serves a surface of joint movement over the vast articular bone locations and trachea forms and rib cage parts. The thick type II collagen and proteoglycans create an interlacing which provides compressive strength as well as sustaining the framework. Tissue studies have validated the fundamental role of the aggrecan compound—a primary proteoglycan—as the main element determining hyaline cartilage compressive capabilities [8].

The elastic cartilage cells containing more elastic fibers for bending support are located in the ear and epiglottis. Fibrocartilage contains numerous dense type I collagen fibers that can be observed at intervertebral discs and pubic symphysis. The arrangement of fibers in this material enables it to resist stretch forces when located at points of mechanical strain.

The exclusive cell type of cartilage tissue called chondrocytes maintains all of the cartilage matrix structures. The cells produce and release the ECM components and control its remodeling process by enzymatic activity of MMPs and aggrecanases. Osteoarthritis occurs as a degenerative disorder upon the synthesis and degradation existing in an unbalanced situation.

5.2.2 Types of Soft Tissues

The various body structures are supported as required since the soft tissues can keep in touch with these parts and cover them. There are three different types of soft tissue components in the body each having its own functions to support the wellness of the body and the performance of the body. Table 5.1 shows an overview of soft tissue types and their functions.

5.3 Role of Extracellular Matrix in Tissue Health

The complex extracellular matrix of proteins and carbohydrates provides structural support and biochemical support to cells surrounding it. The extracellular matrix facilitates both integrity and normal functioning of any type of soft tissue cell such as cartilage. The ECM aids cell connection and their processes of movement and differentiation which reshape tissue structure and stabilization through its properties.

The mechanical properties of cartilage tissue are by default the outcome of ECM composition. By collagen fiber-proteoglycan binding cartilage made it possible to resist compressive stress without diminishing its elasticity. The hyaluronan that is present in the form of glycosaminoglycan in the cartilage ECM also retains high levels of water and also gives cartilage the elastic-plastic response required to absorb the shock [17].

EMC remodeling is an activity that requires the release of MMPs although ADAMTS (a disintegrin and metalloproteinase with thrombospondin motifs) also acts as a central enzyme. These enzymes degrade ECM structures that enables the tissue to repair itself and undertake its regular turnover functions. This imbalance between these enzymes also creates undesirable outcomes that result in the damage of the cartilage in case of inflammatory or injury conditions hence aggravation of osteoarthritic disorders [18].

ECM health status is a crucial condition on which tissue regeneration should take place. The progress of tissue engineering led to the introduction of biomimetic scaffolds that recapitulate the tissue elements and mechanics of natural ECM. This procedure causes the engineered tissue to be incorporated into the host tissue, hence, resulting in better host tissue regeneration processes. Scientific study is currently underway to add growth factors and bioactive molecules to scaffolds as these signals can help cells regenerate ECM, i.e., to build tissues [19].

The development of tissue engineering requires the fundamental knowledge about cartilage and soft tissue behavior and full knowledge about the role of the significant extracellular matrix. Research on tissue structure informs the development and improvement of restoration plans to functional tissue integrity of damaged tissues.

5.3.1 Biocomposites: Definition and Relevance

The industry of tissue engineering depends a lot on biocomposites as they serve as important components for repairing cartilage with soft tissues. Integration of biological components with synthetic elements aims to increase the biological characteristics of mechanical properties and regenerative drug scaffolds. Tissue engineering applications require better knowledge about how biocomposites have been defined with their structural structure and physical characteristics.

5.3.2 Definition of Biocomposites

The definition of biocomozites involves a mixture of natural (biological) components with synthetic elements for biological interactions. The engineer material is designed to create a particularly suitable cellular environment that enables cells to stick to the surfaces and multiply by differentiating the cells required for tissue regeneration. A co-interactive relationship that emerges from natural and synthetic components provides better biological abilities compared to single-domestic materials, as well as single-creating materials. Figure 5.2 shows the diverse potential applications of biopolymers combined with various fillers. (Source: Author's own work).

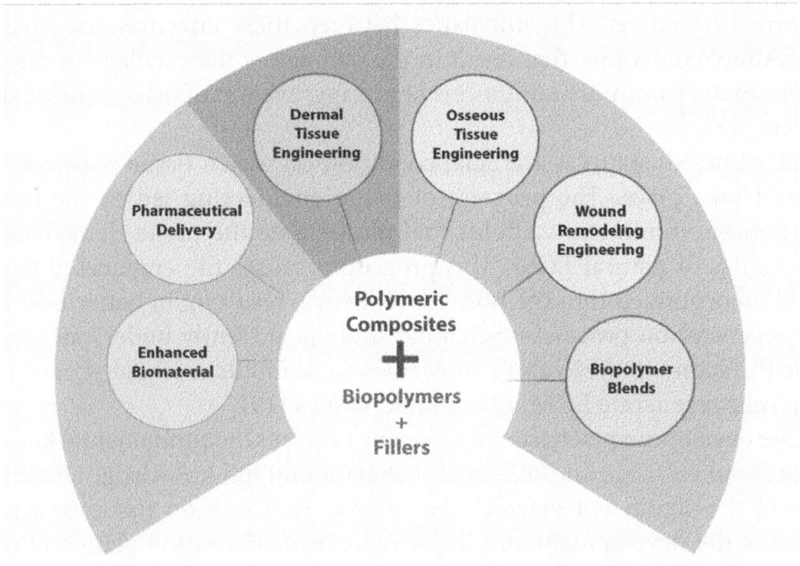

Figure 5.2 Diverse potential applications of biopolymers combined with various fillers.

Biocomogen development arises from a method that the benefits of natural ingredients are accompanied by increased durability and adaptability of synthetic materials with cellular support capabilities to merge the benefits of biocompatibility. Custom works for scaffolding designs that meet different requirements between different soft tissues and cartilage, which forms the necessary biocomposites in tissue engineering [20].

5.3.3 Biocomposites Composition: Materials-Natural vs. Synthetic

Separate format for biocomposites is present in association with the required specific properties. Natural components used in biocomogen formulations include proteins that include collagen and fibrin and gelatin, which contains polysaccharides that contain alginal and chitosan and hyaluronic acid and bioactive substances obtained from biological sources. These biometricals have both cell attachment abilities and biological reaction stimulation symptoms which are beneficial for medical recovery applications.

Engineer synthetic materials, including polycaprolactone (PCL) with polylalctic acid (PLA) and poly (lactic–coalic acid) (PLGA) (PCL), are available for their corresponding mechanical characteristics and declining behavior and biodegradability properties. Biocomposite can be adapted to the mechanical properties and structural quality of the scaffolding through synthetic components. Scientists have recently developed multi-coal polymer compounds and mixed through progression in biological tissue-compatible material design for better cellular infiltration capabilities [21].

Incorporating nanoparticles, like hydroxyapatite or bioactive glass into biocomposite formulations, have been shown to improve their bioactivity and mechanical performance. Such nanoparticles can encourage cellular response and induce osteointegration, which is considered critical for the regeneration process of bone and cartilage. Table 5.2 presents the ideal properties of biocomposites for cartilage and soft tissue engineering.

5.3.4 Ideal Properties for Tissue Engineering

Biocomposites are emerging as a very important component in the field of tissue engineering, especially in the upliftment of cartilage and soft tissue. The ideal biocomposite must have many properties, so that they can enable successful clinical results, repair, and integrate with damaged tissues.

Table 5.2 Ideal properties of biocomposites for cartilage and soft tissue engineering.

Property	Description	Importance for tissue engineering	Reference
Biocompatibility	Biocompatibility allows the biocomposite not to induce adverse immune responses upon implantation.	Most important for scaffolds' success by facilitating cellular functions and supporting long-term integration of tissues.	[22]
Mechanical properties	Mechanical properties similar to natural tissues, such as tensile and compressive strength, elasticity, and capability to withstand physiological loads.	Some tissue tolerate the mechanical needs of some tissues and meet the internal environment of cartilage and soft tissues.	[23]
Porosity and interconnectivity	The scaffold must be porous with connected pores to increase cellular infiltration, transport of nutrients, and waste elimination.	Enhances vascularization and integration into host tissues through facilitation of fibroblast and chondrocyte migration essential for regeneration.	[24]

(*Continued*)

Table 5.2 Ideal properties of biocomposites for cartilage and soft tissue engineering. (*Continued*)

Property	Description	Importance for tissue engineering	Reference
Degradability	Biocomposites must degrade at the same rate as new tissue development to offer structural support throughout healing.	Ensures that scaffolds dissolve without forming toxic residues, reducing inflammatory responses throughout healing.	[22]
Antimicrobial properties	Incorporation of antimicrobial agents (e.g., silver nanoparticles or essential oils) into the biocomposite matrix.	Prevents infection at critical phases of scaffold integration, minimizing postoperative complications and enhancing outcomes.	[25]

5.4 Design Approaches in Biocomposite Development

The exact method development for biocomposites for the purposes of cartilage and soft tissue engineering depends on the selection of appropriate materials, careful planning of scaffold architecture, and establishing means moving towards growth of biological compatibility. All these aspects represent important components that ensure the desired clinical results.

5.4.1 Selection of Materials

The choice of the material in biocomozit manufacturing is itself a fundamental need to attain the effectiveness and success in the application in the tissue engineering application environment. Resources for material selection depends on four important factors that have both biocompatible properties and mechanical properties combined with the decline profile and cell adhesion and proliferation capabilities.

Research has shown that natural polymers, including collagen, gelatin and chitosan, are mainly chosen because they are both biocompatibles by nature and are bioactive in their behavior. The most abundant protein substance in human tissue acts to support the body structure and in addition to the cellular reactions required to restore tissue. The development of bioactive Cross-linked collagen matrix represents the recent focus of research that aims to improve matrix structural power.

Research on chitosan has been intense due to its appropriate compatibility with its appropriate compatibility with biological systems. The beneficial charge of chitosan AIDS in its ability to bind with negative cell membrane surfaces is therefore running a cellular connection. Chitosan tissue shows suitability for engineering progress as its biodegradability allows medical professionals to accurately manage the process of decline.

Biocomposites regularly include synthetic polymer poly (lactic acid) (PLA) and polycaprolactone (PCL) because of their adjustable mechanical properties. The PLA stands out because it reflects rapid decomposition

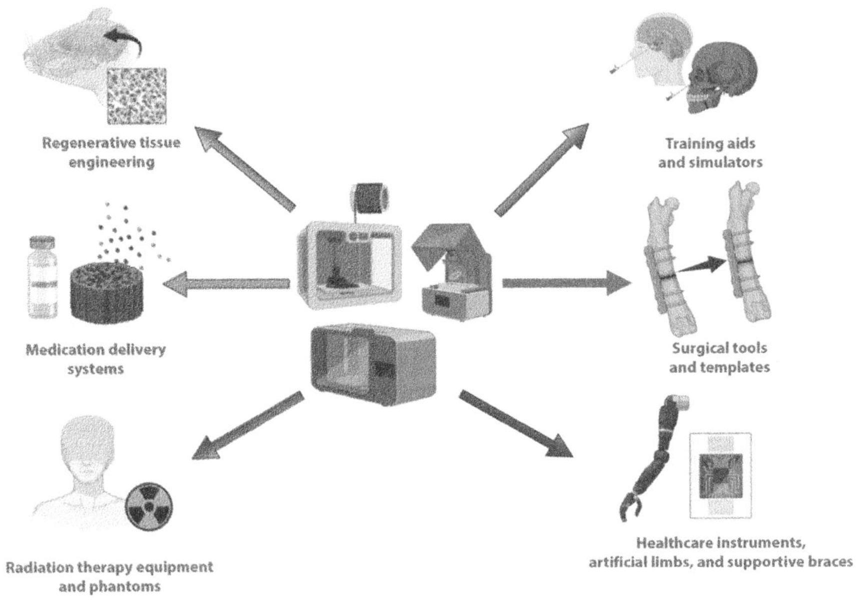

Figure 5.3 Biomedical and clinical applications of 3D-printed PLA.

and strong mechanical properties that make it suitable for weight-bearing applications. Figure 5.3 shows biomedical and clinical applications of 3D-printed PLA. By combining synthetic polymers with natural materials, researchers identify new compounds, which provide the benefits of both substances producing better performance results [26].

The 3D printing process has progressed to a stage where it enables scaffolding of overall bioinks and filaments that engulfs numerous material. The accurate management of materials during printing enables users to include bioactive compounds in addition to growth factors and enzymes to improve the biological performance of scaffolds manufactured.

5.4.2 Design Considerations for Scaffold Architecture

Scaffold architectural design elements determine how well tissue engineering scaffolds work and perform. The main factors to consider in scaffold design include openness between pores along with their dimensions and both geometrical volumes and surface shapes.

The direct relationship between scaffold porosity determines how nutrients spread throughout the structure and how waste will leave along with cell migration capability. An appropriate pore structure enables tissue growth through enough internal space while preventing weakness in the structure. The ideal porosity range for cartilage regeneration exists between 70% and 90% since it maintains appropriate osseous strength alongside biological tissue function according to research findings.

Porli, the dimensions of the cell, affects how the cells behave. Research confirms that tissue angiogenesis works best when the perforation diameter falls within 100 to 500 micrometer range as it allows adequate blood vessel growth and nutrient distribution in developed tissues. Very large hole dimensions compromise adequate mechanical stability in tissue.

The characteristics of the surface of a tissue between topographical size and hydrophilic property determine the way cells react to their environment. Creating surface features at a subtle or nanoscopic level increases cell adhesion functions by promoting their proliferation state with directional development. There are two fundamental approaches to develop electrospinning and foam templating scaffolds that use their related procedures to mimic the natural external matrix structure.

Research benefits from computational modeling can create simulation for scaffold design optimization before implementing scientific physical construction protocol. One of the available finite element analysis methodologies helps researchers in making foretell of mechanical results in the biological surroundings to inform the best design specifications of special works.

5.4.3 Strategies for Enhancing Biological Compatibility

Biological compatibility enhancement is an important consideration to make in cases where biocomposites are to be introduced into host tissue. There are methods of various kinds that can contribute to the achievement of successful cellular reactions with integration of tissues.

Coating or chemical remedies that facilitate protein absorption by surface modifying methods and then cell adhesion is possible to generate a superior biodiversity.

Cultural cells demand certain behavior of the treatment tissues in treatment, and thus the scientific surface modifications are applied, which entail the additional of bioactive molecules such as peptides and growth factors, to motivate the behavior.

The successful approach to improve osteochondral activity uses the application of bioactive semeics, such as hydroxyapatite to cover scaffolds due to its effects of assisting scaffold integration with nearby bone tissue. The bioconsciousness of scaffold material improves with hydroxyapatite coating that works to attract osteoblast cells required for simultaneous bone development.

The biological compatibility of bioocomposites reaches a widespread growth when growth factor or enzymes are involved in their composition. The cellular activities of proliferation and migration and differentiation process heavily depend on growth factors that include transforming growth factor-beta (TGF-β) and vascular endothelial growth factor (VEGF). The controlled release of biologically active molecules through biodegradable microspheres or hydrogels directs these materials to the target site over time which helps tissue regeneration [27].

Researchers have started exploring the therapeutic applications of extracellular vesicles derived from mesenchymal stem cells since these vesicles contain proteins and RNA and lipids that facilitate tissue repair along with tissue regeneration. Researchers enable regenerative properties in these vesicles when they integrate them into biocompogen structures.

Different methods used in biocomposite development require intensive evaluation of materials with scaffold design and biological growth measures. New development in the field of research will improve biocomogen properties that promote cartilage and other soft tissue regeneration possibilities.

5.5 Fabrication Methods for Biocomposites

The process of biocomposite fabrication is fundamental for tissue engineering because it determines how the scaffolds perform and how well they do. This section of the article reviews traditional construction strategies and innovative third generation construction approaches such as 3D bioprinting as well as electrospinning with other emerging techniques with recent developments in tissue engineering applications.

5.5.1 Traditional Fabrication Techniques

Older techniques developed to create biocomospices usually include solvent castings with freeze-drying and molding processes.

The solvent casting approach creates a homogeneous solution using suitable solvents to replace polymers unless the solution allows the mold solvent evaporation. This method helps produce materials that make movies or make coatings when producing scaffolds that can be controlled for their thickness and porous structure. The selection of solvent type with solvent concentration control and evaporation speed allows researchers to develop specific microstic features in their scaffold products. The solvent provides a means to improve biological properties of scaffold structures, in addition to bioactive components for casting solutions during casting methods.

Lyophilization produces porous structures through multiple stages where the polymer solution causes ice pronunciation before reducing pressure. As a result of the manufacturing process, scaffolds are formed through porous channels that enable the solution transportation as well as cell penetration. Modern freeze-drying technology enables researchers to regulate cold rates after which they can develop scaffolding structures with better power and adjustable fora structures. The manufactured scaffolds are based on the natural tissue structure which supports a stronger cell structure [28].

Biocomposite structures that can be reused are made using moulding techniques like compression and injection moulding. Two main advantages of using moulds are the ability to create complex shapes and meet precise physical fit requirements for specific medical uses. The particulates of bioactive glass or hydroxyapatite demonstrate high potential when incorporated in the polymer matrices during the moulding processes as they improve the biological and mechanical properties.

5.5.2 Advanced Fabrication Techniques

The integration of biological complication with scaffolding designs has now become quite popular with the improvement in technologies. These are far better and enable better incorporation of biological signals.

5.5.2.1 3D Bioprinting

The generation of biocomogeneous have now become feasible to affect through 3D bio-printing since this versatile technique enables layer-by-layer deposition for building three-dimensional complex tissue structures. There, instead presence of cells is employed in the approach to form complex 3D shape, and biocompatible materials are dispersed into the ink filament on printed duct to create complex 3D shape.

The principle benefit of 3D bioprinting is single scaffolds that are fitted with anatomical features to suit the body of the patient. One of the extrusion-based and the inkjet-based bioprinting technologies is THE added new physical diversity as a larger to get researchers install their own admix polymers of hydrogel or bioactive molecules. Besides these bio-aches have growth factors and cultured derived extracellular matrix proteins that also increase the cell binding even more and they induce cell differentiation [29].

Coupling many material printing capabilities, specialists are able to stack many substances (e.g., with inverse properties) into a single structure. When rigid and soft materials are structured in a gradient pattern the result becomes a scaffold that duplicates different tissue types' mechanical properties to advance implant integration capabilities.

Bioprinting technologies receive close scrutiny to determine their capability for building vascular networks in printed tissues which is essential to maintain survival of larger tissue constructs. The brain architecture looks promising because researchers developed microvascular channels through sacrificial materials that teams could eliminate following the printing process.

5.5.2.2 Electrospinning

The fabrication process called electrospinning enables producing nanofiber scaffolds with structures identical to natural extracellular matrix through extensive use. Application of high-voltage electricity during solution processing produces extremely thin fibers which get collected onto a substrate.

Electrospinning technologies produce fabricated scaffolds that feature elevated surface area-to-volume ratios, adjustable fibers diameters and better cell binding and cell mobility due to its uncommon capabilities. Researchers control the fiber diameter range from nanometers to micrometers through processing parameter adjustments of voltage strength and fluid flow rate in addition to nozzle-to-collector distance changes [30].

Scientists have utilized co-electrospinning technology to embed polymers and additives together during electrospinning processes which produces complex structures with upgraded mechanical alongside biological qualities. Bioactive molecules and nanoparticles added directly to electrospun fibers boost cellular functions when promoting differentiation together with proliferation. Hybrid electrospinning techniques along with other fabrication methods allow researchers to create multiscale and multifunctional scaffolds through the integration of electrospun fibers.

5.5.2.3 Other Emerging Techniques

Other than 3D bioprinting and electrospinning several new techniques are currently increasing their popularity in biocomposite fabrication.

Specific biomaterial designs through biomimetic methods use natural structures and functions for creating materials that mimic them. The biomedical field has discovered two significant approaches which enable scaffold production from emulsion templates and sol-gel processes to replicate natural tissue structures.

Microfluidics has emerged as a technological method which enables small-scale control of fluids for the creation of micro- and nanostructured scaffolds. Through microfluidic devices one can create intricate biomaterial structures with gradients that improve cellular responses including movement and development.

Decellularization provides techniques to maintain the original structure of extracellular matrix through the removal of tissue cells from naturally derived materials. The cell seeding may take place on the integral matrix following a method in which is obtained a highly tissue-friendly structure capable of both integration and proper function.

Self-organization leads to the spontaneous formation of ordered structures when biomaterials perform molecular recognition in self-assembly. Interestingly, as the scaffolds are made from these biomaterials they could respond to *in vivo* cues and create active cultivation environments for tissue growth and healing.

Biocomposite generation represents fast progress, since classic techniques are used but 3D bioinkings set new standards during scaffold design for electrospinning. Structural architectural features of the customable material structure enabled to generate possible and impossible previously for successful regeneration solutions to repair defects in cartilage and soft tissue.

Further research in the region will continue to increase the properties and properties of bionomist for better results with tissue engineering applications [31], which includes biological signals in the construction process.

5.6 Biological Responses to Biocomposites

Biocomposite interaction with biological systems represents the chorocate of effectiveness in tissue engineering problem. To customize the design and functional performance of these materials, one has to understand the biological response which are currently. This section considers etiquette in which cells interact with biocomostases, evaluation of biocompatibility, and in increasing tissue regeneration, the role of bioactive molecules including growth factors and peptides.

5.6.1 Interactions with Cells

The way biocomposites interact with cells, they stand as a serious position for the proper installation and functioning of engineer tissues. Cells are associated with transplanted biocomposites through many mechanisms that increase practical characteristics with attachment strength and cellular proliferation and cell number such as mobility and tissue-specific development. The physical chemical makeup of biocomogen material determines how the cells will interact through a combination of surface properties and a combination of topographic characteristics with hydrophilic character.

The way surfaces interact chemically with their environment, it proves necessary for cell binding processes. Some functional surface groups enable the absorption of scaffolds proteins that serve as an initial status for cell connections. Surfaces that enable fibronectin and vitronectin extracellular matrix proteins to attach effectively promote the adhesion and proliferation of fibroblast and osteoblast cells. Research has focused on surface functionalization through RGD peptides (arginine-glycine-aspartic acid) application to enhance cell-integrin receptor binding and support strong cellular attachment [32].

The behaviors of cells undergo major modifications as a result of micro- and nanoscale topographical features. Biocomposites having structured surface designs combined with varying roughness features enable the control of cellular directionality by replicating tissue architecture. Electrospun nanofibers matching ECM structure effectively align cells in ways that prove highly beneficial to constructing nerve and muscle tissues. The modern scientific community demonstrates that designing scaffolds with hierarchical structures which duplicate natural tissue gradients enables the direction of particular cell responses.

The manner in which biocomposites behave mechanically impacts the way cells respond to them. The differentiation pathways for stem cells largely depend on both the stiffness and elasticity properties of the materials used. The matrix stiffness of materials determines whether adipose differentiation will occur on flexible surfaces and osteogenesis will occur on rigid substrates. Scaffold design becomes more effective in breaking down cells into tissue engineering usage through the comprehension of these mechanical signals.

5.6.2 Assessment of Biocompatibility

Biocomposite interaction assessment determines how these materials affect biological systems and triggers which adverse reactions might occur after implantation. Multiple established testing methods exist to evaluate biocompatibility through assessments of cytotoxicity together with histocompatibility and local responses testing.

Testing for cytotoxicity normally requires that cultured cells endure exposure to biocomposite material extracts. The laboratory results show what toxic chemicals potentially leak out from the materials through the culture medium. Scientists use MTT and live/dead testing methods to measure cell growth and survival rates during evaluations of scaffolds [33].

Research on histocompatibility determines the immune reactions experienced by patients who receive implanted biocomposites. These research studies examine how immune cells penetrate the implantation zone as well as monitor inflammatory cytokines and assess complete tissue responses. The perfect biocomposite should create a favorable healing environment by minimizing tissue reactions to foreign bodies. Biomaterial-based research using anti-inflammatory drugs with coatings displays new approaches to reduce negative immune reactions, thus improving the acceptance of transplant materials on their placement site.

Additional information about biocomposite compatibility with living tissue comes from tests conducted on real living organisms. These researches measure the decomposition rate and entire tissue repair outcomes to the tissue host connection through the evaluation tissue prepared.

5.6.3 Role of Bioactive Molecules in Tissue Engineering

Bioactive molecules are biocomocytes and they contribute to the enhancement of cellular responses that are predetermined in speed tissue repair and regeneration. These bioactive molecules have an influence on a wide variety of issues associated with cell behavior, which means that the engineer has merged and potent efficacy of tissues.

5.6.3.1 Growth Factors

Recent research on biocomostases specifically for biomedical purposes demonstrates that they can enhance biological interactions when they contain growth factors and enzymes. Biocomoth material which contains chitosan and solely natural polymer bound with bioactive ceramics will provide an environment which encourages cell division with cell expertise.

Researchers proved the significance of incorporation of the recruitment human bone morphogenetic protein (RHBMP) into biocomposite scaffolds to stimulate new bone development. Research has shown that rhBMP-2 treatment of scaffolds creates better osteoblast activity which generates higher bone regeneration rates. Research scientists have investigated enzymes as agents for biocomposite degradation while studying their remodeling potential. Studies demonstrate that lysozyme enzymes work to degrade chitosan-based composites through controlled processes which help fusion of tissues in addition to healing processes. Laboratory advances demonstrate the use of growth factor-releasing nanoparticles embedded inside biocomposites to supply VEGF and FGF and other factors to specific treatment locations. Future biomaterial research appears promising because this specific targeting approach enhances the combination of tissue healing alongside tissue regeneration capabilities [34].

5.6.3.2 Peptides

Cell and tissue interactions with biocomposites can also be promoted through the use of peptides, whether they are from the extracellular matrix or are synthesized to mimic specific bioactive motifs. Sequences of peptides, including the RGD-motif, functionally mediate cell adhesion through interaction with integrins present on the cell surface. Recent innovations have cyclic peptides or peptide-functionalized nanoparticles, providing greater target specificity and biocompatibility.

Additionally, peptides could act as signaling molecules that influence cellular behavior, as is the case of bioactive peptides released into the ECM which can induce inflammation or move healing pathways through the modulation of activities of various other cell types. Designing peptide-based hydrogel in a way that mimics the native external matrix has achieved results that promise to increase cellular performance positively to promote tissue regeneration [35].

Furthermore, emerging research has provided some insight into the use of engineered peptides that self-assemble into nanostructures to develop biomimetic environments to direct cell behavior compared to traditional scaffolds. Peptide nanofiber systems are found to be biocompatible and positively responsible for cellular activities.

Understanding biological responses to biocomposites seeks to pave the way for advancing tissue engineering strategies. The interaction of cells with biocomogens involves complexity marked by multiple influences, including surface properties, mechanical parameters, and the presence of bioactive molecules. Biocompatibility assessment proves integral to assuring the safety and efficacy of engineered tissues. Furthermore, using growth factors and peptides, scientists can strategically control cell behavior to drive tissue regeneration. Continued research into these biological interactions will prompt the design of next generation biohomeostasis that are tailored to specific medical needs in regenerative medicine.

5.7 Challenges in Cartilage and Soft Tissue Engineering

Looking at cartilage and soft tissue regeneration, there are still major challenges, although progress has been made in tissue engineering. Some important issues need attention, such as the longevity and degradation profile, proper integration with the host tissue, and appropriate mechanical properties to tolerate the physical environment. Such challenges must occur as patients continue to grapple with clinical outcomes.

5.7.1 Integration with Host Tissue

Connecting cartilage and soft tissue into host tissue is one of the major issues in tissue engineering. Effective integration reduces the likelihood of inflammation by providing effective treatment options. Surgeons treat the engineered construct on the margin scaffold to restore function and vascularity in the host tissue. Integration depends heavily on the biochemical and mechanical cues that the engineered scaffold will provide. Recent studies have highlighted the importance of mimicking the native extracellular matrix (ECM) of the tissue with respect to structure, mechanical, and architecture properties. There is a better opportunity that the mating materials with biochemical signals of native tissues will achieve a favorable cellular response. In such a case, in the example of scaffolds, to reduce integration to reduce cell spreads and migration functionally with adhesion proteins such as fibronectin. Another important part of integration is about the formation of the vascular network, which is also a unique challenge in the upliftment of auskular cartilage. On the other hand, adequate vascularization for all engineer soft tissues is a requirement to ensure cell survival and function. Advanced scaffolding construction has captured vascular development factors such as vascular endothelial growth factors (VEGF) to help attract on sites where new blood vessels have to be built. Several approaches have been researched to induce the formation of vascular networks, such as scaffolds with microchannels to promote vascularization for graft survival. Integration is hugely affected by an immune response to implanted materials. Biocompatible materials, having being demonstrated to provoke minimal inflammatory responses, could, therefore, lead to better integration with host tissue. Recent developments regarding clones of biomaterials have directed the design of scaffolds that can avoid immune detection or modulate the immune response actively. Incorporating anti-inflammatory agents or surface modifications by hydrophilic coating vastly attenuates a foreign body reaction and enhances tissue integration [36].

5.7.2 Longevity and Degradation Studies

The issue that is most studied is the degradation rate of the scaffolds. It is important to match the decline rate of scaffolding with the rate of growth of new tissue. If the scaffolding reduces very rapidly, it is not going to support any tissue formation. If it is very slowly decreases, however, it is causing encapsulation of tissues with chronic inflammatory reactions and fibrous tissues. Therefore, a good deal of research is being done in the areas of polyacrylic acid and polylactic acid as biodegradable polymers, which have focused on engineering through copolymerization and cross-linking methods to reach the more controlled decline mechanisms. The second problem that is related to this decline is the general and complete understanding of the mechanical properties of the scaffolding during the decline. Scaffolds, as they lower, lose mechanical strength. It can compromise the proper function of scaffolds to support living tissue. Thus, another thing that should be the specialty is this mechanical recession so that the scaffolding can bear a physical weight when healing. New techniques such as insect mechanical testing during the decline study have highlighted how the scaffolding over time in terms of mechanical properties and what those properties are with processes in cells. In contrast, an significant field of N study is of declining products and biohyacinability. The decline process gives rise to many sub-products that can cause inflammatory reactions. Therefore, one should study whether the fall products adversely affect the surrounding tissue or cellular activities. The work is going on in the investigation of the use of enzymes that may modify the process of decline to promote a controlled decline mechanism that reduces adverse effects when supporting regeneration [37].

5.7.3 Mechanical Properties and Stress Responses

Mechanical properties of biocomposites have to closely emulate the physiological conditions of the target tissues. Cartilage and soft tissues withstand different sets of mechanical loads and forces; hence the engineered scaffolds have to withstand better dynamic forces. Dynamic loading conditions: The response of scaffolds to dynamic mechanical loads is an active area of research. Compressive, tensile, and shear stresses represent the different mechanical loads acting on cartilage in joint movements. For prediction of *in vivo* performance, tissue-engineered constructs need to be subjected to test conditions mirroring these dynamic loading situations. The recent development of a dynamic culture system that is capable of periodically applying mechanical loads to the scaffolds appears promising in enhancing both mechanic-related and cell-related properties of engineered tissues. Biomimicking mechanical properties: In order to achieve biomimetic mechanical properties, the proper balance between stiffness and elasticity of the scaffold has to be struck. Such mechanical properties can be finely tuned by modulating the biocomposite's composition and architecture. In other words, scaffolds, in which some component is of soft material and others of hard material, would impart strength coupled with flexibility to meet the mechanical demand of the target tissue,

Stress reactions and adaptation: This is important for the functional lifetime of engineer tissues to understand their respective reactions to stress. Recent studies have shown that mechanical signals activate cellular pathways directly related to matrix discrimination and synthesis. For example, mechanical loading led to an increase of glycosaminoglycan and collagen synthesis in cartilage ECMs. Thus, by strategically controlling the specific loading regime and applying these mechanical signals, scientists can effectively use engineer tissues that mimic the tissues found in nature [38].

Translational issues observed within cartilage and soft tissue engineering are not unique to cartilage to soft tissue engineering. These translational issues include: integration with host tissues; The Longevity of engineer constructs; And mechanical properties to encourage optimal efficiencies. While ongoing research and innovation in technology may be useful in finding solutions to these issues, they can also adoptively be deducting from successful strategies to develop which will become biocomposited that has the potential to apply long term solutions within the field of tissue uplifting. Given the nature of an applied and emergent field or domain, outcomes through successfully continued clinical uses, which has changed outcomes for patients, and provide meaningful improvements associated with pain, injury, and disease by the provision of the musculoskeletal system can become conceivable.

5.8 Applications of Biocomposite-Based Methods

The advancement of methods based on biocomposites has greatly enriched the approach to the treatment of musculoskeletal disorders and can show great promise for regenerative therapy. Such material allows for the potential discovery of treatment and repair about tissue regeneration and for very complex sets of problems for novel solutions.

5.8.1 Clinical Applications in Musculoskeletal Disorders

Applications of biocomposites have become more common in traditional clinical indications for the treatment of cartilage defects related to musculoskeletal pathology, osteoarthritis, and broken bones. The properties of the solution enable the solution to meet the lost structural and biological demands for the respective injured tissues. Among them, the repair of cartilage lesions is most notable; Cell-based autologous chondrocyte implantation (ACI) is widely recognized as a successful treatment for focal cartilage lesions. In these procedures, scaffolds prepared from biocomposites serve as a temporary framework for cell growth and infiltration, helping to rejuvenate hyaline-like cartilage. Recent studies have shown promising results with 3D structures made of gelatin and polycaprolactone (PCL), which mimic the mechanical properties of natural cartilage and allow internal designs that fit specific defects. For bone grafts after surgery, biocomposite scaffolds made of synthetic ceramics such as hydroxyapatite and biodegradable polymers have demonstrated significant potential. These scaffolds not only provide the mechanical support needed in healing for load-bearing applications, but also provide sustained release of bioactive molecules such as bone morphogenetic proteins (BMPs). Recent advances include double-stage biocomposites that resemble natural bone architecture, enhancing bone regeneration in critical clinical cases with as non-fractures. Treatment of chronic osteoarthritis obtained its share from biocomose technology with the production of injectable hydrogel that allows the distribution of biological agents directly to tissues in joint space. Such hydrogel can be engineered to encounter anti-inflammatory drugs, pro-collagen development factors, or mesenchymal stem cells (MSCs) associated with local inflammatory modulation and tissue repair. Clinical trials of intra-articular injection of hyaluronic acid associated with MSc in a biocomposite have described increased functional capacity and decreased the osteoarthritis pain [39] of the knee.

5.8.2 Potential for Regenerative Medicine

Biocomozit-based methods make sufficient promises in large areas of regenerative therapy, aimed at restoring damaged tissues using biological materials and principles. Biocomposite scaffolds are being developed in regenerative medicine for the repair of tendons and all ligaments. These tissues possess unique mechanical properties, including tensile strength and elasticity. Collagen and silk fibroin biocomposites have proven effective in tendon engineering, with organized arrangements of collagen fibers formed on these substrates.

By encouraging the recruitment of mitochondria and boosting matrix synthesis, these scaffolds play a crucial role in enhancing tendon healing, especially when combined with growth factors like PDGF.

Composite scaffolds are designed to replicate the mechanical and biochemical conditions found in the dermis during skin regeneration. They create a supportive environment that fosters fibroblast growth and encourages vascularization. Hybrid scaffolds made from chitosan and gelatinous materials have shown great promise in promoting wound healing. Plus, incorporating antimicrobial agents into these scaffolds helps ward off infections, which continue to be a significant challenge in treating clinical wounds [40].

5.8.3 Future Directions in Therapy

In these procedures, scaffolds crafted from biocomposites served as a temporary framework for cell growth and infiltration, helping to rejuvenate cartilage like hyaline. Recent studies have shown promising results using 3D structures made from gelatin and polycaprolactone (PCL), which mimic the mechanical properties of native cartilage and incorporate internal designs to address defects. For post-surgery bone supplements, biocomposite scaffolds made from synthetic ceramics, such as hydroxyapatite, along with biodegradable polymers, have demonstrated great potential. Integration of stem cell technologies: An era of innovation is opening the door for integration of stem cell technologies with biocomposite scaffolds. Inspired pluripotent can achieve promising results in the cell survival and integration within the cell survival and integration within engineer tissues by combining biocomposites with iPSCs or MSCs. The design of advanced biometrics, on which these cells maintain their function within the scaffold, are likely to promote the capacity for restoration of functional tissue. Regulatory and moral views: With the advancement of such knowledge, there will be a need to carefully consider the regulatory structure and moral ideas about the applications of biocomposites in medicine. It should be noted that such treatments should be dedicated to the protection, effectiveness and long -term effects, which are medically practiced by some of the laboratory [41].

5.9 Personalized Medicine and Future Perspectives

While they provide exciting capacity for more effective and personal treatment strategies, the development of personal medicine in tissue engineering has not yet achieved their full potential. By addressing the special requirements of specific patients through optimizing biocomozites and engineering approaches, physicians can not only improve results, but can also work towards promoting more effective regeneration of specific tissues.

5.9.1 Customization of Biocomposites

Customized biocomozits are an extremely important feature of individual therapy based on the specification of each patient. Biocomposite design can apply individual mechanical performances, rate of degradation, and biological activity to achieve biocompatibility and function according to the physiological context of every patient. Patient-specific designs include advanced use of imaging technologies such as MRI and CT scans, making it possible to develop an exact 3D model of a patient's anatomy.

Patient-specific designs are taking a leap forward with the innovative use of imaging technologies like MRI and CT scans. This advancement allows us to create precise 3D models of a patient's anatomy. These models can then be utilized to craft a biological framework that aligns perfectly with biocomposites. For instance, researchers can use CAD software to design scaffolds that closely replicate the shapes and sizes of damaged tissues, enhancing the integration and performance of the engineered materials.

Moreover, fine-tuning the composition opens up new avenues for customizing biocomposites. Recent research indicates a growing trend towards hybrid materials that combine different polymers with bioactive fillers, such as hydroxyapatite or bioactive glass. These combinations not only improve bioactivity but also strengthen mechanical properties. By skillfully blending these elements in the right ratios, scaffolds can closely resemble the physiological traits of natural tissues, significantly boosting the likelihood of successful integration and functionality.

Additionally, incorporating bioactive molecules like growth factors or peptides into the biocompatible matrix can cater to the specific regenerative needs of the patient. For example, scaffolds aimed at cartilage repair can be enhanced with TGF-β or BMP-2 to promote the recruitment and differentiation of chondrocytes [42].

5.9.2 Advances in Personalized Tissue Engineering

Personalized tissue engineering is opening up a world of exciting possibilities in tissue regeneration strategies, thanks to some incredible breakthroughs in biometrics, fabrication technologies, and cellular therapy. One standout innovation in this area is 3D bioprinting—it is truly a game changer. This technology enables us to arrange cells, biomaterials, and bioactive agents in carefully designed patterns with remarkable precision. With this capability, we can create tissue constructs *in vitro* that closely resemble the intricate structures found in natural tissues.

Recent advancements have led to the integration of living cells with both synthetic and natural polymers, resulting in tissues that can support cell life and function for extended periods. Additionally, the high resolution and motion precision of bioprinting technologies allow for the creation of complex multicellular structures that incorporate gradients of various cell types and materials. This innovative approach opens up new avenues for replicating the diverse characteristics of native tissues, including vascular networks that facilitate nutrient flow and enable interactions among different cell types crucial for specific tissue functions.

In the field of regenerative therapy, personalized tissue engineering has made significant progress in medical efforts aimed at developing tissues that enhance treatment and healing for a variety of conditions. The methods for distributing mesenchymal stem cells (MSCs) or induced pluripotent stem cells (IPSCs) within biocompatible scaffolds have garnered considerable attention. These scaffolds have unique properties that influence micro-generation processes. Recent studies indicate that incorporating specific growth factors and mechanical stimuli can greatly enhance the effectiveness of cell therapies [43].

5.9.3 Ethical Considerations and Regulatory Aspects

While the world of personalized drug tissue engineering is full of exciting possibilities, it also opens up a whole can of ethical issues and regulatory challenges that we need to tackle for these technologies to really take off. Let us talk about the ethical side first: the widespread use of personalized methods in tissue engineering raises some serious moral questions about consent. It is super important that the informed consent process is solid, making sure patients truly understand what it means to use their biological material in research or treatment. Plus, we need to create guidelines that shield individuals from discrimination based on their genetic data. Now, onto the regulatory side: when it comes to the rules around individual biocomposites and tissue-engineered products, we need thorough evaluations before they hit the clinic to prove they're safe and effective. Various organizations in the US and the EMA in Europe have set up frameworks for evaluating these advanced therapy products. However, it is crucial to tweak existing regulations to keep up with new technologies, encouraging ongoing teamwork among scientists, doctors, and regulators, especially with innovations like Team Work 3D bioprinting. The approval process for biocomposite materials must tackle questions about bio-diagnosis and the decline of certain products. Additionally, it is essential to establish clear guidelines for manufacturing practices and quality control to ensure patient-specific treatments are safe as these tailored products move into rapid clinical trials [44, 45].

5.10 Conclusion

5.10.1 Summary of Key Findings

The journey into biocomposites for cartilage and soft tissue engineering has sparked some incredible breakthroughs and given us a clearer understanding of how tissue regeneration actually works. A biocomposite is a clever mix of natural and synthetic materials designed to create innovative treatments that effectively repair tissues and help regenerate damaged areas. This flexibility allows for fine-tuning of both mechanical and biological properties, making it possible to meet the specific needs of tissues, particularly in clinical environments.

Recent research has particularly spotlighted various applications, like using a biocomposite scaffold to fix cartilage defects, boost bone regeneration, or combat osteoarthritis. Adding bioactive molecules, such as different growth factors and peptides, has shown to be effective in shaping cellular responses for tissue regeneration. Plus, cutting-edge techniques like 3D bioprinting and electrospinning have broadened the design possibilities and structural complexity of scaffolds, paving the way for a more advanced approach in tissue engineering.

5.10.2 The Future of Cartilage and Soft Tissue Engineering

As we look ahead, the field of cartilage and soft tissue engineering is poised for remarkable advancements. Personalized medicine will take center stage, introducing treatments tailored to individual patients through custom scaffolds that deliver therapies designed specifically for them. Ongoing improvements in biometrics and construction technologies will enable us to create intricate structures that closely resemble the natural extracellular matrix. Moreover, the integration of smart biometrics that react to biological stimuli will enhance therapeutic outcomes. Continuous research will delve into the workings of enzymes and other bioactive molecules that can influence the modulation of decline profiles, ultimately guiding tissue maturation. It is clear that this progress will necessitate a strong commitment to ethics and regulations surrounding personalized medicine and biocompatible therapies. Overcoming these challenges is essential for the safe and effective rollout of innovative treatments. In the end, the convergence of cartilage and soft tissue engineering with biology, materials science, and engineering will shape the future of regenerative medicine, transforming patient care, and improving quality of life.

AI disclosure statement: The author acknowledges the use of AI-based tools for minor language editing or grammatical correction. However, all intellectual contributions, analyses, and conclusions presented in this chapter are entirely the author's own.

References

1. Gasser, T.C. and Grytsan, A., Biomechanical modeling the adaptation of soft biological tissue. *Curr. Opin. Biomed. Eng.*, 1, 71–77, 2017.
2. Chen, M., Jiang, Z., Zou, X., You, X., Cai, Z., Huang, J., Advancements in tissue engineering for articular cartilage regeneration. *Heliyon*, 10, 3, 2024.
3. Ding, S.L., Liu, X., Zhao, X.Y., Wang, K.T., Xiong, W., Gao, Z.L., Zhang, M.Z., Microcarriers in application for cartilage tissue engineering: recent progress and challenges. *Bioact. Mater.*, 17, 81–108, 2022.
4. Rajput, S., Malviya, R., Uniyal, P., Advances in the treatment of kidney disorders using mesenchymal stem cells. *Curr. Pharm. Des.*, 30, 11, 825–840, 2024.
5. Pramanik, S., Kharche, S., More, N., Ranglani, D., Singh, G., Kapusetti, G., Natural biopolymers for bone tissue engineering: a brief review. *Eng. Regener.*, 4, 2, 193–204, 2023.
6. Ielo, I., Calabrese, G., De Luca, G., Conoci, S., Recent advances in hydroxyapatite-based biocomposites for bone tissue regeneration in orthopedics. *Int. J. Mol. Sci.*, 23, 17, 9721, 2022.
7. Sophia Fox, A.J., Bedi, A., Rodeo, S.A., The basic science of articular cartilage: structure, composition, and function. *Sports Health*, 1, 6, 461–468, 2009.
8. Armiento, A.R., Alini, M., Stoddart, M.J., Articular fibrocartilage-Why does hyaline cartilage fail to repair? *Adv. Drug Delivery Rev.*, 146, 289–305, 2019.
9. Amirrah, I.N., Lokanathan, Y., Zulkiflee, I., Wee, M.M.R., Motta, A., Fauzi, M.B., A comprehensive review on collagen type I development of biomaterials for tissue engineering: From biosynthesis to bioscaffold. *Biomedicines*, 10, 9, 2307, 2022.
10. McCuller, C., Jessu, R., Callahan, A.L., Physiology, Skeletal Muscle, 2025.
11. Visha, P. and Sejian, V., Cardiovascular System, in: *Textbook of Veterinary Physiology*, P.K. Das, V. Sejian, J. Mukherjee, D. Banerjee, (Eds.), Springer, Singapore, 2023, https://doi.org/10.1007/978-981-19-9410-4_6.
12. Hafen, B.B., Shook, M., Burns, B., Anatomy, smooth muscle, 2018.

13. Asahara, H., Inui, M., Lotz, M.K., Tendons and ligaments: connecting developmental biology to musculoskeletal disease pathogenesis. *J. Bone Miner. Res.*, 32, 9, 1773–1782, 2017.
14. Alshammari, S.M. and Bordoni, B., Anatomy, shoulder and upper limb, arm muscles, in: *StatPearls [Internet]*, StatPearls Publishing, 2023.
15. Hemat Jouy, S., Mohan, S., Scichilone, G., Mostafa, A., Mahmoud, A.M., Adipokines in the Crosstalk between Adipose Tissues and Other Organs: Implications in Cardiometabolic Diseases. *Biomedicines*, 12, 9, 2129, 2024.
16. Richard, A.J., White, U., Elks, C.M., Stephens, J.M., Adipose tissue: physiology to metabolic dysfunction, in: *Endotext [internet]*, 2020.
17. Krakowski, P., Rejniak, A., Sobczyk, J., Karpiński, R., Cartilage integrity: A review of mechanical and frictional properties and repair approaches in osteoarthritis. *Healthcare*, 12, 16, 1648, 2024, August. MDPI.
18. Rajput, S., Malviya, R., Srivastava, S., Ahmad, I., Rab, S.O., Uniyal, P., Cardiovascular disease and thrombosis: Intersections with the immune system, inflammation, and the coagulation system. *Ann. Pharm. Fr.*, 83, 2, 228–250, 2025, March. Elsevier Masson.
19. Liu, H., Xing, F., Yu, P., Lu, R., Ma, S., Shakya, S., Liu, M., Biomimetic fabrication bioprinting strategies based on decellularized extracellular matrix for musculoskeletal tissue regeneration: Current status and future perspectives. *Mater. Des.*, 113072, 2024.
20. Andrew, J.J. and Dhakal, H.N., Sustainable biobased composites for advanced applications: recent trends and future opportunities–A critical review. *Compos. Part C Open Access*, 7, 100220, 2022.
21. Liu, S., Qin, S., He, M., Zhou, D., Qin, Q., Wang, H., Current applications of poly (lactic acid) composites in tissue engineering and drug delivery. *Compos. Part B*, 199, 108238, 2020.
22. Krishani, M., Shin, W.Y., Suhaimi, H., Sambudi, N.S., Development of Scaffolds from Bio-Based Natural Materials for Tissue Regeneration Applications: A Review. *Gels*, 9, 2, 100, 2023a, https://doi.org/10.3390/gels9020100.
23. Miri, Z., Haugen, H.J., Loca, D., Rossi, F., Perale, G., Moghanian, A., Ma, Q., Review on the strategies to improve the mechanical strength of highly porous bone bioceramic scaffolds. *J. Eur. Ceram. Soc.*, 44, 1, 23–42, 2024, https://doi.org/10.1016/j.jeurceramsoc.2023.09.003.
24. Suamte, L., Tirkey, A., Barman, J., Jayasekhar Babu, P., Various manufacturing methods and ideal properties of scaffolds for tissue engineering applications. *Smart Mater. Manuf.*, 1, 100011, 2023, https://doi.org/10.1016/j.smmf.2022.100011.
25. Bruna, T., Maldonado-Bravo, F., Jara, P., Caro, N., Silver Nanoparticles and Their Antibacterial Applications. *Int. J. Mol. Sci.*, 22, 13, 7202, 2021, https://doi.org/10.3390/ijms22137202.

26. Wu, Y., Gao, X., Wu, J., Zhou, T., Nguyen, T.T., Wang, Y., Biodegradable polylactic acid and its composites: Characteristics, processing, and sustainable applications in sports. *Polymers*, 15, 14, 3096, 2023.
27. Lutzweiler, G., Ndreu Halili, A., Engin Vrana, N., The overview of porous, bioactive scaffolds as instructive biomaterials for tissue regeneration and their clinical translation. *Pharmaceutics*, 12, 7, 602, 2020.
28. Huang, J., Ratnayake, J., Ramesh, N., Dias, G.J., Development and characterization of a biocomposite material from chitosan and New Zealand-sourced bovine-derived hydroxyapatite for bone regeneration. *ACS Omega*, 5, 27, 16537–16546, 2020.
29. Merivaara, A., Zini, J., Koivunotko, E., Valkonen, S., Korhonen, O., Fernandes, F.M., Yliperttula, M., Preservation of biomaterials and cells by freeze-drying: Change of paradigm. *J. Controlled Release*, 336, 480–498, 2021.
30. Rana, A., Mittal, A., Vashist, C., Rajput, S., Sridhar, S.B., Malviya, R., Exploration of 4D Printing and its Applications in the Biomedical Sciences. *Curr. Pharm. Des.*, 2025.
31. Chen, Y., Shafiq, M., Liu, M., Morsi, Y., Mo, X., Advanced fabrication for electrospun three-dimensional nanofiber aerogels and scaffolds. *Bioact. Mater.*, 5, 4, 963–979, 2020.
32. Xu, H., Zhang, Y., Zhang, Y., Zhao, Z., Xue, T., Wang, J., Ding, Y., 3D bioprinting advanced biomaterials for craniofacial and dental tissue engineering–A review. *Mater. Des.*, 112886, 2024.
33. Barberi, J. and Spriano, S., Titanium and protein adsorption: an overview of mechanisms and effects of surface features. *Materials*, 14, 7, 1590, 2021.
34. Vajrabhaya, L.O. and Korsuwannawong, S., Cytotoxicity evaluation of a Thai herb using tetrazolium (MTT) and sulforhodamine B (SRB) assays. *J. Anal. Sci. Technol.*, 9, 1, 1–6, 2018.
35. Paulini, M., Camal Ruggieri, I.N., Ramallo, M., Alonso, M., Rodriguez-Cabello, J.C., Esbrit, P., Feldman, S., Recombinant proteins-based strategies in bone tissue engineering. *Biomolecules*, 12, 1, 3, 2021.
36. Williams, D.F., Biocompatibility pathways and mechanisms for bioactive materials: The bioactivity zone. *Bioact. Mater.*, 10, 306–322, 2022.
37. Wei, W. and Dai, H., Articular cartilage and osteochondral tissue engineering techniques: Recent advances and challenges. *Bioact. Mater.*, 6, 12, 4830–4855, 2021.
38. Zielińska, A., Karczewski, J., Eder, P., Kolanowski, T., Szalata, M., Wielgus, K., Souto, E.B., Scaffolds for drug delivery and tissue engineering: The role of genetics. *J. Controlled Release*, 359, 207–223, 2023.
39. Hassan, N., Krieg, T., Zinser, M., Schröder, K., Kröger, N., An overview of scaffolds and biomaterials for skin expansion and soft tissue regeneration: Insights on zinc and magnesium as new potential key elements. *Polymers*, 15, 19, 3854, 2023.

40. Rajput, S., Malviya, R., Sridhar, S.B., Osteoporosis Risk Assessment and Individualized Feature Analysis Using Interpretable XAI and RAI Techniques, in: *Explainable and Responsible Artificial Intelligence in Healthcare*, pp. 89–113, 2025.
41. Golebiowska, A.A., Intravaia, J.T., Sathe, V.M., Kumbar, S.G., Nukavarapu, S.P., Decellularized extracellular matrix biomaterials for regenerative therapies: advances, challenges and clinical prospects. *Bioact. Mater.*, 32, 98–123, 2024.
42. Zahra, T., Javeria, U., Jamal, H., Baig, M.M., Akhtar, F., Kamran, U., A review of biocompatible polymer-functionalized two-dimensional materials: Emerging contenders for biosensors and bioelectronics applications. *Anal. Chim. Acta*, 1316, 342880, 2024.
43. Chong, W.J., Wright, P., Simunec, D.P., Jayashree, S., Liew, W., Heazlewood, C., Wen, C., A comprehensive study on the biodegradability, biocompatibility, and antibacterial properties of additively manufactured PLA-ZnO nanocomposites. *Smart Mater. Manuf.*, 3, 100069, 2025.
44. Yu, J., Park, S.A., Kim, W.D., Ha, T., Xin, Y.Z., Lee, J., Lee, D., Current advances in 3D bioprinting technology and its applications for tissue engineering. *Polymers*, 12, 12, 2958, 2020.
45. Cinti, C., Trivella, M.G., Joulie, M., Ayoub, H., Frenzel, M., The roadmap toward personalized medicine: Challenges and opportunities. *J. Pers. Med.*, 14, 6, 546, 2024.

6

Biocomposite Scaffolds for Bone Tissue Engineering

J. Pushpa Sweety[1], U.M. Dhanalekshmi[2], M. Arul Prakash[3], M. Kavitha[3], S. Pragatheeswaran[3] and N. Selvasudha[3]*

[1]*Department of Chemical Engineering, Indian Institute of Technology Madras, Chennai, India*
[2]*Department of Pharmacology and Biological Sciences, College of Pharmacy, National University of Science and Technology, Muscat, Oman*
[3]*Department of Pharmaceutics, Kamarajar College of Pharmacy, Tamil Nadu, India*

Abstract
Biomaterial-based scaffolds are crucial for tissue engineering, offering support for cell interaction and tissue regeneration. Ideal scaffolds should be nontoxic, biodegradable, and possess appropriate mechanical and surface properties. However, current biocomposite scaffolds face challenges, such as limited structural durability, uncontrolled degradation rates, and insufficient cellular integration, which affect long-term *in vivo* functionality. While existing biomaterials meet some physical and chemical requirements, their biomimetic properties and cell-scaffold interactions need improvement. Combining materials to offer stimuli responsiveness and employing suitable fabrication techniques like 3D and 4D printing with well-ordered micro- or nanostructures are essential to accomplish optimal scaffold design and biological responses. Additive manufacturing is utilized in 3D/4D tissue engineering for the precise fabrication of complex structures to produce the complex tissues which is unattainable by the conventional methods. Construction of scaffolds by 3D technology with specific shapes and structures by stacking biocomposite materials layer by layer offers quick and precise preparation of scaffolds tailored to individual patient needs. In case of 4D technology, it integrates specific materials that dynamically vary properties, shape or tasks with long-term relations with biological systems. Through the concept of "time" into printing, 4D printing technology enable the biocomposite materials to adapt their morphology or

**Corresponding author*: nkselvasudha@gmail.com

Rishabha Malviya and Mukesh Roy (eds.) Biocomposites in 3D and 4D Printing: Materials, Techniques, and Applications in Regenerative Medicine, (157–196) © 2026 Scrivener Publishing LLC

functionality in reaction to external stimuli (e.g., pH, heat pressure, magnetism etc.). 4D-printed biocomposite scaffolds have the potential to improve mechanical stability and functionality in the bone repair. 4D printing in tissue engineering is in its infancy. With the etiology of bone defects and repair mechanisms, in the future, it can offer advancing technology for clinical applications, can enable the design of vascularized structures to establish a bionic microenvironment, able to promote the stem cell differentiation thereby influencing cellular behavior. This chapter emphasizes the prospective of 3D and 4D printing in creating advanced, adaptable bone structures with significant clinical value. It highlights the biopolymers and smart materials used for the scaffolds fabrication, current advancements/applications, and bone materials in this advanced technology. It also discusses the limitations in the composition, properties, and performance of 3D- and 4D-printed biocomposite or polymers for bone tissue engineering.

Keywords: Conventional method, layer by layer, scaffolds, smart material, bionic environment

6.1 Introduction

Tissue engineering can be used to create functioning tissue for drug testing and regeneration. In particular, bone tissue engineering BTE is an emerging field of osteology which is a combination of biomaterials with cells. It gained much attention among the research community in the mid of 1980s, serving as an alternate treatment method for creating bone grafts. This technological advancement would prominently help in bone repair as well as bone regeneration, especially in case of bone fracture during accidents, bone-related tumors, osteoporosis, and for basic and advanced *in vivo* studies. Classical BTE involves key players for instance biocompatible scaffolds and osteogenic cells for mimicking the bone characteristic features. Due to continuous demand and reduced disease transmission rate, engineered bones are widely preferred. However, lack of vascularization at the site of injury is major limitation for clinical practice [1]. Therefore, personalized implants for patients based on the critical situation is the current need which can be made with three-dimensional (3D) printing technology. 3D printing technology has been tremendously employed in fields including latest design methodologies, rapid prototyping, and biomedical equipment construction because it makes it simple and inexpensive to transform computer 3D files into actual items [2].

3D printing technology advances the field of bone regeneration therapy by producing biological tissues or organs. However, to solve the difficulties faced by the researchers, bioink makes the process of producing 3D bone tissue simpler. This bioink comes in the form of viscous fluids and hydrogels. 3D structures that are cellularly rich are made with bioink. Thus, bone tissue can be efficiently created using 3D printing and bioink, which takes into account a number of factors like printability and mechanical integrity when creating a particular bioink [3]. Later, four-dimensional (4D) printing has been explored by incorporating the time dimension to 3D. More interestingly, 4D materials are defined as the 3D structures with incorporated properties like stimuli response and functional transformation. Stimuli like heat, humidity, pH, and light activate printed objects without the use of motors or connections. Accordingly, these smart structures functioning with respond to external stimuli are ideal examples for 4D materials, which would possibly have resulted in making stimuli-responsive as well as shape memory structures (Figure 6.1) [4].

Properties, such as structural integrity, mechanical and tensile strength, rigidity, biocompatibility, biodegradability, are under critical concern for the creation of 3D and 4D biocomposites. These composite scaffolds are often fabricated from natural or synthetic polymers, with some process specific additives for particular treatment strategy. Steriolithography, extrusion printing, selective laser sintering melting (SLS), fused deposition modelling (FDM), and bioprinting in inkjet 3D printing are some of the important techniques under practice. These methods offer affordable and industrially applicable procedures, which is briefly explained in the upcoming section. Recently, remarkable advancements in the field of 3D and 4D printing technologies are getting developed.

The advancements in printing technologies would aid in producing bioengineered scaffolds, which can imitate the extracellular matrix ECM of the native bone configuration. Accordingly, 3D and 4D printing have been recognized as the next generation manufacturing techniques not in the

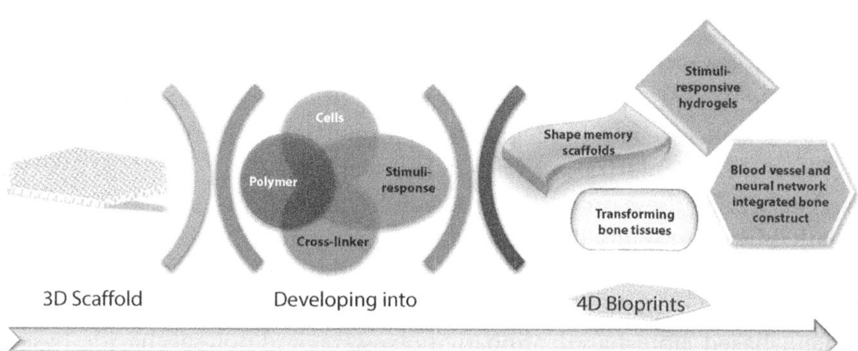

Figure 6.1 Overview on the development of 3D scaffold into 4D bioprints.

field of biomaterial for medicine, but also in computer science, chemistry, and engineering [5]. 3D and 4D printing techniques can be used to generate tissue engineering scaffolds in a variety of ways, especially for patient-specific structures. To increase the scope and speed of targeted tissue repair, printing of physiochemical guiding signals has also been suitably included. The benefits of 3D and 4D printing over traditional scaffolding methods for tissue regeneration are examined in detail in this analysis [2].

These are predicated on the concept of layer-by-layer assembly and assembling of discrete materials, have enabled the creation of bio-piezoelectric scaffolds in ever-more complex geometries. Surprisingly, 4D-printable bio-piezoelectric scaffolds with functionality-shifting characteristics can provide a programmable tissue EM for bone regeneration that is responsive to external stimuli and time-dependent [6].

First, this review organizes sample biological outcomes for bone regeneration along with the physicochemical properties of some of the frequently utilized polymers, ceramics, and their composites, along with bio-piezoelectric materials. This review covers the significant factors affecting the physical and mechanical features of the biocomposite. It mainly focuses on the incorporation of stimuli-responsive mechanism and shape memory and 3D and 4D biocomposite scaffolds. It also discusses about the recent applications and advancements in the field of bone tissue engineering mainly for treating critical size bone defects, which are found unsuccessful by other approaches. The reader can also get insight about the future trends, and some critical challenges faced by the research community along with new directions for future development.

6.2 Ideal Properties of Biocomposites in 3D and 4D Printing

While creating 3D and 4D BTE models, it is essential to consider some of the significant factors listed following, biocompatibility, mechanical qualities, structure and biodegradation properties, interconnected porous architecture, vascularization, and surface features [7].

Considering these properties enhance certain benefits, such as improved breathability, durability [8], reduced thickness, and enhanced optical characteristics. The practicality of biocompatibility, safety, and biodegradation rate influences the development of biodegradable plastics specifically protein-based, lipid-based, aliphatic polyesters, and polysaccharide-based sources from agriculture, livestock, and fish aquaculture in comparison to conventional plastics. The significance of these properties are briefly reviewed for better understanding [9, 10].

6.2.1 Structural Characteristics

Complex active structures with exceptional performance can be achieved using 3D- and 4D-printed materials. Geometric characteristics, including the form, aspect ratio, and clear edges, can influence strain gradients in a sample. For example, anisotropic fibers lead to expansion exclusively along a single axis for a specific bending curve. The printing design must be arranged in a manner that enables dispersion to create a paraboloid to achieve double curvature of the filament. This wood composite material is printed at 90° intervals between the principal axis and the angle of the specimen layer section at 10° laterally [11].

Metamaterials possess distinctive properties that are primarily characterized as either periodic or non-periodic structures. A metamaterial is a substance that is bonded not only through its molecular makeup but also through its cellular architecture, which does not exist in natural MSK materials [12].

6.2.2 Mechanical Properties

Biomaterials in three and four dimensions which are involved in BTE require substantial mechanical qualities like tensile strength, compressive, flexural, impact, and wear. These properties would probably determine how a material behaves in critical or extreme conditions. To be industry-ready, the material's mechanical properties must be thoroughly studied. For example, mechanical characteristics of natural fiber-PLA composites are analyzed widely [13].

Scaffold mechanical qualities should enable acceptable anatomical views for implantation, resulting in more robust surgical procedures with fewer clinical consequences. Although essential in all tissues, there are some unique hurdles to overcome, particularly in cardiovascular and orthopedic settings. Creating load-bearing scaffolds is the most difficult aspect of bone creation. Some implants failed despite their high load-bearing capacity. Certain materials may make it challenging to get the desired porosity structure. To ensure the success of any scaffold, two changes are necessary. Mechanical load-bearing capabilities and a porous structural volume would substantially improve strength, cell infiltration, and also vascularization, respectively [14].

Strong mechanical qualities are necessary when building 3D scaffolds to maintain structural integrity until tissue regeneration occurs. Scaffold mechanical qualities vary according to the anatomical site of application, such as soft as well as hard tissue engineering [15]. Human bone has a compressive strength of approximately 2 MPa and 100 to 230 MPa for cancellous bone and cortical bone, respectively [16].

Porous scaffolds, which are weaker than bone, require additional metal devices, such as plates, screws, or nails, to stabilize defects and external fixation to transfer load. When the scaffold has a larger elastic modulus than the host tissue, the differential results in stress shielding and bone absorption. Pore size (350–1000 μm) greatly affects mechanical properties of porous implants, resulting in lower elastic modulus and also poor stress shielding *in vivo* [7].

A scaffold that can totally replace bone must be designed as a multilayered structure with different varieties of mechanical properties across its thickness. The rigidity of the scaffold can be altered by altering its porosity [17]. The mechanical strength of scaffolds varies depending on the density and texture. The inter-pore wall thickness and the pore size to beam thickness ratio are critical components of 3D-printed porous scaffolds [18].

To maintain stability, the host tissue requires scaffolds with improved compressive strength, elasticity, and also some fatigue strength. Organs and tissues mimic scaffolds, therefore maintaining mechanical integrity *in vivo* is critical [19].

6.2.3 Biocompatibility

Biocompatibility refers to how well a polymer interacts with the body and its fluids. Therefore, biocompatible polymers are widely applied in biomedical applications to investigate, manage, improve, or replace a certain tissue or the complete organ in some cases. A biocompatible polymer improves body functions without interfering with regular operation, creating allergies, or having any other negative consequences [20].

The degradation rate should be favorable, and hazardous substances should be avoided as they may produce unpleasant reactions or immunological rejection by the host. If organic solvents are used during the manufacturing process, residual chemicals may remain in the device, potentially causing injury to newly created tissue and nearby tissue [21]. More importantly, biocompatibility along with the biological response of 3D-printed biomaterials in hard tissue. 3D printing can create surgical models and customized implants for craniomaxillofacial surgery [20].

Three-dimensional powder-printed tricalcium phosphate (brushite or monetite) powder scaffolds with phosphoric acid binder were tested for cytocompatibility with osteoblastic cells (MC3T3-E1). The researchers investigated the effect of size of the brushite 3D powder-printed scaffold and also its pore architecture on bone ingrowth in a goat model. Herein, this 12-week study, researchers implanted open and closed pore channels of different widths into goats' decorticated lumbar transverse processes. Both examples demonstrated new bone growth in pore channels, with no clear distinction between both open and closed channels [22].

PLA, PCL, PLGA, PHA, and related copolymers are biodegradable aliphatic polymers used in biomedical applications that are deemed "biocompatible" [23]. Biodegradable polymers backbones contain hydrolyzable linkages, such as carbonate, urea, urethane, amide, esters, and orthoesters. Ester-linked aliphatic polymers are biocompatible and unveil a wide variety of physical, chemical, and biological properties [24]. In addition to these, PLA and PHA are popular aliphatic polymers in biomedical applications due of their enhanced biocompatibility [25].

Long-term therapies necessitate an ongoing interaction between the living environment and the materials. Bioresorbable and biodegradable polymers are more complicated due to the possible interaction of byproducts with biological systems. Polymer biodegradability and bioresorbability should be assessed before use in medical implants. Acidic substances released during implant breakdown may cause inflammation in the body. The site of implantation can also influence the host's inflammatory response [26].

6.2.4 Biodegradability

Biodegradability is an indispensable property of any material to consider for tissue engineering-based application. Estimating the rate of biodegradation is considered as a significant factor which is primarily depending on the molecular weight of the scaffold. Presently, thermoplastics, such as PET, PE, PP, and PS, make up 31% of global plastic demand. Plastics are largely made from petrochemicals; however, there is a growing interest in exploring renewable alternatives [27]. Bioplastics may be biodegradable, biobased, or both some are discussed here [28].

Bioplastics are often classified into three categories, which are briefly explained below.

- Biobased and biodegradable bioplastics. Examples include polylactic acid (PLA), polyhydroxyalkanoates (PHAs), thermoplastic starch (TPS), and polybutylene succinate (PBS) [29].

- Second, petrochemical-based bioplastics are non-biodegradable [30] some classical examples for this kind are polycaprolactone (PCL) and polybutylene adipate terephthalate (PBAT).
- Third comes the non-biodegradable otherwise called partially biodegradable bioplastics. These kind of materials are usually derived from biobased monomers and technical polymers. BioPE, bioPET, bioPP, and bioPTT are some of the vital components playing role in this field [31].

6.2.5 Degradation Tests Aiding Medical Applications

The breakdown of PLA scaffolds under PBS buffer environment demonstrates the significance of the print's three-dimensional structure in affecting the rate of degradation. Prints with random porosity degraded more quickly than cubic and gyroid [32].

Monitoring metrics outside the print mass can aid in identifying biomaterial degradation pathways. Research into molecular weight changes during degradation has produced unexpected outcomes. Research has shown that the mechanical strength is a crucial parameter of any printed material, especially for tissue engineering applications. Natural fibers significantly affect the mechanical properties (tensile and elasticity) of PCL prints, causing deterioration. Biomass breaks down structures faster than biological substrates. For up to 2 weeks, reinforced PCL surpasses unreinforced PCL in tensile strength and modulus of elasticity [33].

External environmental variables, inherent biopolymer properties, and filler parameters in blends/composites all have an impact on bioplastic breakdown rates. Biodegradation is the process by which microorganisms generate enzymes that cause deterioration. Chemical degradation is the process of breaking down biopolymers utilizing water, oxygen, acids, alkalis, and solvents. In addition to oxidation in biological aerobic conditions, oxygen consumption accelerates physical, atmospheric, and hydrolytic deterioration. Bioplastic biodegradability methods are classified into three types: soil, compost, and aquatic systems, which are briefly explained in the followed subheadings [34].

6.2.5.1 *Biodegradation Analysis in Freshwater and Seawater*

Biodegradation measurements are conducted under controlled settings using saltwater and sediment, with the first measuring oxygen consumption and the second detecting CO_2 emission [35].

Some materials degrade quickly when exposed to water. Notably, the PCL/wool combination breaks down in seawater as early as Day 3. Adding wool to the composite raises the measured biological oxygen requirement. An analysis of seawater testing during a period of 5 months revealed changes in print patterns (black spots) that indicated changing currents over time [36].

6.2.5.2 *Biodegradation of Composites in Soil Environments*

With its varied porosity, eration, granularity, pH, water holding capacity and fraction composition, soil is a diversified ecosystem. Temperature is an important aspect that changes depending on the season and environmental conditions [37].

Soil has a diverse range of microorganisms, including bacteria and fungi, which influence how things breakdown in the environment. Standard protocols for researching plastic biodegradation in soil involve burying at the optimal conditions, as well as measuring the amount of carbon dioxide emitted as the item disintegrates. Methods for mass loss and/or deconstruction of material aspects, such as structure, tensile strength, morphology, and surface texture, have been also used. The effects of breakdown by-products on live creatures are also studied through ecotoxicity trials [35].

6.3 Factors Influencing 3D and 4D Bioprinting

6.3.1 Selection of Natural Fibers

Some of features of plant fibers like its structural characters and functional groups composition influence its mechanical strength. These features are influenced strongly by the sources from which it is obtained, soil characters, growth parameters, and the isolation process. Many plant materials like stalks, seeds, fruits, leaves and bast are used for fiber fabrication. Among all, Bast fibers exhibit good tensile strength. Bast fibers like flax, kenaf, jute, hemp, and ramie are generally used in industry. According to Mahrishi Enviro Engineering, combining multiple plant fibers can produce hybrid fibers and thermoplastic composites [38].

In addition to the above said properties, the hygroscopicity of natural fibers usually plays a prominent role in the determination of wood plastic composite's properties. To combine with the polymer matrix in the extruder, natural fibers need to be dried before injection or hot-press molding to 0% to 1% moisture content and it bypass voids.

6.3.2 Effect of Fibre Size on the Biocomposite Properties

The properties of biocomposite vary based on the fiber dimension and orientation, aspect ratio, fiber position in the matrix, and volumetric fiber percentage [39]. The aspect ratio, fiber placement, and arrangement all have a significant impact on the biocomposite's interfacial shear strength, which is responsible for supporting external loads [40].

The polymer matrix's randomly oriented fibers effectively transfer the external loads. Accordingly, hand-laid natural fiber composites perform better mechanically. Matrixes at right angles oppose tensile forces parallel to the load, producing optimum strength. By modifying flow properties of the polymer, the short fibers may be aligned during injection molding [39].

6.3.3 Effect of Filler Content

Increased proportion of wood flour to 50% improves bending strength in biocomposites. A coupling agent can boost bending modulus performance by up to 70 wt%. The flexural characteristics of WPCs improved dramatically as the fiber length and content increased and adding fiber improves energy absorption while making the composite brittle. The tidy plastic absorbs a lot of energy, which results in a high impact bending strength. These areas are typically seen near the ends of fibers or where the polymer matrix and fiber are incompatible, resulting in microvoids or gaps [41].

6.3.4 Compatibility between Fibre Matrix and Surface Modification

The natural filers and polymers possess the contrary properties like first one is hydrophilic in nature while the second one is hydrophobic in nature and together they influence the mechanical and physical properties of the biocomposites [42].

When an external load is applied, it will be transferred from the matrix to the filler due to the presence of strong interfacial bond. When no binding agent is present, the natural fiber surface reacts only minimally with the hydrophobic polymer matrix. Coupling agents boost the compatibility among polymer surfaces and fiber biocomposites [43].

6.3.5 Type of Polymer Matrix

The nature of the polymer matrix formation would significantly impact the complexity an behavior of the 3D- or 4D-printed materials. In case of natural fibres, these are embedded in the biocomposite polymer matrix, giving resilience against weathering and biological impacts. It acts as a stress disperser in composites, transferring loads to reinforcements. Moreover, in case of thermoplastics, melt flow index (MFI) has a substantial impact on the physical, mechanical, and thermal properties of biocomposite materials, whether thermoplastic or thermoset. As the MFI value increases, the matrix's viscosity decreases, allowing for greater flow and access to natural fiber channels, which resulted in smooth matrix formation [44].

6.3.6 Process Conditions in Biocomposite Manufacturing

As we know, almost all fabrication methods primarly depend on the operating conditions, biocomposite manufacturing process follows the same. Biocomposite properties are greatly influenced by changes in temperature, pressure, viscosity, and holding duration. High temperatures often cause the chemical components of natural fibers to break down. Some of the common examples are cellulose, lignin, and hemicelluloses. Thermal degradation and browning of natural fibers brought on by processing temperatures above 200°C may affect the technological performance of biocomposites [45].

6.3.7 Effect of Voids and Porosity

The mechanical and physical properties of biocomposites can be significantly impacted by voids and porosity. If we see it in detail, microvoids as a matrix form at the interface between natural fiber and polymer. More important causes to consider for controlling this void are fiber moisture content, interfacial imperfection or incompatibility in between the polymer and matrix, presence of volatile organic compounds. Processing should be performed in a degassing a sense to prevent microvoids. The porosity of biocomposites has a major impact on their mechanical characteristics. It is brought on by poor wettability of the fibres, air and gas bubbles during processing, and thermal deterioration [46].

6.3.8 Nanocellulose-Supported Biocomposites

Over the past 20 years, nanocellulose has grown significantly in popularity as a reinforcing element for a variety of biocomposites due to its unique properties [47].

The most well-known is nanocellulose, which stands out for having an extraordinarily large surface area. As a result, even small loadings of nanocellulose (1–2 wt%) in the composites are adequate. Based on the literature, it is observed that weight % of nanocellulose varies from 1 to 5. However, because the nanocellulose clumps within the matrix, the composite's properties are reduced at higher concentrations. Nanocellulose employs a number of methods, such as high shear mixers and ultrasonic procedures, due to these process influence the biocomposites used for 3D and 4D [4].

6.3.9 Type of Additive Manufacturing Process

Several additive manufacturing techniques like fused deposition modeling (FDM) and stereolithography (SLA) require digital data that provides parametric solutions to the spatial geometry in order to be utilized by the CAD/CAMP systems [48, 49]. As that of material selection step in the 3D technology, screening the suitability of smart material for the 4D manufacturing techniques is prerequisite step. The chosen additive manufacturing technique must enable the simultaneous printing of multiple, ideally uniformly distributed, materials in addition to creating a mono-printed shape in multi-material structures where the contrast of material properties drives the desired shape- or function-change [2].

6.3.10 Type of Responsive Material

Production of stimuli-responsive smart solid is the most imperative component of 4D printing since it gives printing a fourth dimension. In reaction to an external stimulus, the sensitive substance progressively changes chemically or physically. These reactions include things like folding, curling, twisting, expanding, contracting, changing color, degrading, and so forth [50].

6.3.11 Type of Stimulus

In preparing 4D material, with incorporating some stimuli-response element, certain requirements and constraints of the particular application area should be taken into account. Additionally, the appropriateness of the interaction mechanism between the application agglomerate and the application under consideration should be considered properly. The interaction mechanism is the act of exposing sensitive material to stimuli. For certain persons, exposure alone is not enough to elicit a reaction from the material and the intended consequence; they need antecedent processes (prior to the stimulus being delivered). For example, in restricted thermomechanics, a smart material shows a shape memory feature when heated. The material must be pre-stretched and bent at a high temperature with an external force in order to activate its shape memory feature. The load is released at low temp. once the material has been molded (into the required shape). Heat acts as a catalyst to return the form to its original state [51].

6.4 Polymers in Biocomposite-Based 3D and 4D Scaffolds

6.4.1 Natural Polymers

6.4.1.1 Collagen

Animal tissues have been used to make collagen, which is used in many different fields. This material has been widely used in a variety of biological applications due to its exceptional biocompatibility and biodegradability [52].

The collagen consists of more proline/glycine and exhibits 2° α-helical structure. Through non-covalent and covalent bonding triple handed helix are formed. The main structural components of collagen's self-assembled fibrils are these constituent units and companions [53]. The bulk of bone tissue comprises of inorganic hydroxyapatite (HA) and type I collagen, a biopolymeric. It is plausible that Col and HA promote osteoblast differentiation together [54].

6.4.1.2 Chitosan

Chitosan is a famous biopolymer that is produced when chitin undergoes partial deacetylation in an alkaline environment. It is chiefly linear copolymer of glucosamine and N-acetyl glucosamine, connected by the link β-1→4. It is often present in the form of chitin, a special polysaccharide, is abundant in the calcareous exoskeleton of arthropods, including insects and crustaceans [55].

Recently, researchers used chondroitin sulphate and hyaluronic acid to produce a second-generation biomimetic hybrid scaffold. The findings indicate that the nano composites contain compartments for differentiation and proliferation as well as hierarchically arranged micro/nanostructures that serve as osteogenic pros [56].

6.4.1.3 Alginate

Brown algae's cell walls include a polymer called alginate. It is made up of copolymerized mannuronic and glucuronic acids that alternate and are joined by α-1,4-glycosidic linkages [57]. Alginate has gelling property and hence used as thickening and stabilizing agents [58].

Many studies utilizing alginate for BTE were carried out. Biodegradable and biocompatible materials are used to improve mechanical qualities, promote bone regeneration, and heal wounds [59]. Chemistry to produce magnesium-doped HA scaffolding and porous scaffolding alginate with a dimensional hierarchy resembling that of bone. The inner half of the pore structures is covered with uniformly distributed magnesium-substituted HA (MgHA) particles, which encourage preosteoblastic cell adhesion and growth [60].

6.4.1.4 Hyaluronic Acid

These days, hyaluronic acid in many forms is essential to the extracellular matrix of human bodies. The fields of dentistry and craniofacial medicine have greatly benefited from the strong trend of bone regeneration in recent decades. It was found that hyaluronic acid (HyA)-soaked composite scaffolds significantly affected osteogenesis and mineralization. These HyA derivatives enabled the efficient loading of various osteoinductive or osteogenic chemicals, resulting in controlled release, by functioning as localized release vectors and counteracting the scaffolding elements. "the use of packed vectors anchored on implant surfaces can greatly improve osteointegration." Collagen, chitosan, and hyaluronic acid are lyophilized using nano-hydroxyapatite (nHA) to create scaffold structures. Evaluations of the overall biocompatibility of these inventions were then carried out [61].

Compared to micro-sized fiber scaffolds, the higher porosity of nanofibrous scaffolds favorably enhances the passage of waste products from metabolism and vital nutrients. All scaffolds exhibited higher osteoblast proliferation than their microfibrous counterparts, with some scaffolds displaying up to 53% higher levels. Additionally, the scaffolds' remarkable suitability as biocomposites for BTE applications was confirmed by the significant 63% increase in osteogenic differentiation and mineralization that followed the addition of diverse bioactive chemicals inside the scaffolds [62].

6.4.2 Synthetic Polymeric Biocomposites

6.4.2.1 Poly (Lactic-Co-Glycolic Acid) (PLGA)/HA Composite Scaffold

Extensive characterization demonstrates that PLGA is biocompatible and has the right mechanical properties for BTE applications. The copolymer PLGA comprises of PLA and PGA. It is possible to reverse the hydrolysis and breakdown of PLGA by altering the mixture of two polymers. The metabolites of this polymer are easily eliminated by urine [63].

In this regard, the system performance and tissue engineering effectiveness of the copolymer were evaluated in human skin tissue [57] and physical strength, tissue development, bone, and tissue engineering techniques and properties were briefly explored. In a different work, hydroxyapatite nanoparticles-PLGA scaffolds (3D) are analyzed and the notable improvements were observed in the mechanical properties, hydrophilicity, and osteoblast viability. use either supervised or semi-supervised neural networks [64].

6.4.2.2 HA-Based Composite/Ceramics

Ceramic materials are an excellent choice for 3D-printed scaffold construction with mechanical qualities and the requirement for biocompatibility because they effectively blend metallic and nonmetallic components [65].

Ceramic materials are recently become attractive for the progress of an artificial scaffold especially in the area of bone regeneration mainly because of their special apatite-mineralization property. Hydroxyapatite (HA) is a common ceramic material and an essential part of human teeth and bones [66, 67].

In this manner, HA or other comparable ceramics could be used to create scaffolds with mechanical properties akin to those of real bone. Enhancement in tissue compatibility is beneficial and accordingly, HA has garnered a lot of interest in the field of regenerative medicine. These days, it is widely employed as a basic building block for 3D-printed macromolecules [15].

Human osteoblasts derived from human iliac crest cancellous bone were used to seed the scaffolds. The experimental scaffold showed enhanced biocompatibility and decreased cytotoxicity. These results provide credence to the special qualities of HA materials, such as their capacity to encourage cell survival and division and their biocompatibility [68].

6.4.2.3 Polyethylene Glycol

Polyethylene glycol (PEG) which is a hydrophilic polymer that can enhance HA scaffolds' mechanical qualities and biocompatibility. In BTE, this biocompatible composite scaffold may be used [69].

6.4.2.4 Polycaprolactone

Polycaprolactone (PCL) is a petroleum-based polymeric material finding application in the field of packaging and drug delivery. This is one material that can be used to encourage the production of new bone also. The effective scaffold construction and the ensuing *in-vitro* and *in-vivo* investigations employing HA and polylactic acid [70].

PCL hydrogels have high rheological and viscoelastic capabilities, specified reabsorption, and predictable mechanical properties. The absence of bioactive groups in PCL could change its surface chemistry and improve cell adherence, leading to low biocompatibility. Consequently, it most likely acts as a scaffold, particularly for hard tissues. Some articles claim that it is utilized in bone, cartilage, and heart engineering [71].

6.4.2.5 Pluronic Acid

Pluralonic acid (olaxamer) is a tri-block thermoplastic copolymer that is created when two hydrophilic segments of poly(ethylene oxide) and a hydrophobic segment of poly(propylene oxide) mix. Whether a non-ionic surfactant gels when cooled depends on its concentration and structure. Low cytotoxicity, poor cell viability, rapid disintegration rate, rapid dissolution in aqueous solution, poor mechanical properties, and good biocompatibility are the main characteristics of the gel form. In the field of tissue engineering, polymer hydrogels are often studied for various tissue regeneration methods [72].

6.4.2.6 *Poly(N-Isopropylacrylamide)*

Poly(N-isopropylacrylamide) (PNIPAM) hydrogels, the most widely used thermoresponsive material. When heated over 32°C to 35°C, these biocompatible polymers constrict, but at lower temperatures, they expand. Additionally, they facilitate 3D printing and provide reversible folding and unfolding. The incapacity of PNIPAM to function as a bioresorbable polymer in human applications is its main disadvantage [73]. Research has explored innovative methods for developing bioresorbable PNIPAM-based hydrogels, incorporating bioresorbable crosslinkers and natural polymers such as proteins and polysaccharides, alongside synthetic jelly-like polymers including polyester, PCL, and PEG [74]. The bottom-up approach was employed to determine the complete hydrogel requirements for the drug delivery system (DDS) [75].

6.5 Overview of 3D and 4D Printing Techniques

The process of 3D printing is assisted using design optimization, which is based on the combined analysis of experimental and computational results. The main laboratory and industrial scale preparation of three printed materials involves stereolithography, extrusion printing, selective laser printing, fused deposition modelling, inkjet 3D printing techniques.

Steriolithography apparatus SL/SLA deals with a primary light source which get projected into photocurable bioink. The main approach to fabricate a 3D microstructure is by using a layer-by-layer approach, also called as laser-based modern and rapid prototyping technology. Both traditional lithography techniques and digital light processing (DLP), are being followed here. Lithography techniques builds geometries from top to bottom, and digital light processing builds geometries from bottom to top. More recently, ceramic SLA (cerSLA) utilizing ultraviolet UV-curable system has been reported for providing high yield during the fabrication of 3D systems [76]. Extrusion printing is a technique ranging from cell-free printing to cell-laden ink system. The main principle involved is the heating of a polymer-based filament in a print head and subsequently deposited out of a nozzle. Most commonly processing of nanomaterial-incorporated thermoplastics are done using this advanced methodology. The temperature of the printhead is frequently set higher than the melting point of thermoplastics [77]. It is also established that adding nanoparticles raises the formulation's melt viscosity [78]. Selective laser sintering is considered an AM technique is primary involves the production of tissue engineering scaffolds. This is also called as a rapid prototyping technique, in which binding of thin layers of powder particles with the aid of a CO_2 or Nd:YAG laser beam is employed for producing 3D objects. In this method, slicing the digital CAD image, followed by the assembly of each layer of the pattern sequentially to construct a 3D pattern more importantly in a layer-by-layer approach. In order to sinter or fuse the powder bed's polymeric or metallic particles—which are heated above the glass transition or melting temperature—the laser moves in a precise 2D pattern across the bed's surface during the production process [79]. Some of the important process factors as well as the material properties need to be considered in this case, can affect the structural and mechanical properties of fabricated parts. Mainly, laser energy density and scanning speed, particle size along with the particle bed temperature, and more significantly layer thickness, and hatch distance. SLM/ SLS has the advantage of producing structures with almost full relative density, which has been a major issue for many other AM techniques [80].

The Fused Deposition Modelling (FDM) technique and inkjet 3D printing techniques are widely being employed in 3D printing as well as 4D material fabrication. When comparing these two techniques, both are found as cost-effective manufacturing techniques; however, later one may not provide optimal results when fabricating voluminous and some complicated structures [81].

There are two distinct modes of operation in inkjet 3D printing, one is a continuous inkjet printing system and the other is drop-on-demand inkjet printing, generating singular droplets [82].

4D material fabrication involves the addition of stimuli-responsive element to the 3D-printed scaffold, either with or without cells. Accordingly, the above discussed printing processes such as steriolithography, fused deposition modelling, extrusion-based bioprinting are usually employed for creating the structures. Moreover, magnetic 3D printing, digital light processing, inkjet and 3D droplet printing are also explored for fabricating the smart structures [83].

In this regard, shape-memory polymers and shape-memory hydrogels are two significant players in creating advanced 4D biomaterials for BTE application. Firstly, SMPs are capable of changing its shape in a defined or programmable way, either bending, stretching, and twisting. This property is responsible for making the material as a stimuli-responsive one when exposed to thermal, light, or any chemical environment. Temperature-actuated, aqueous-actuated, magnetic-actuated, and electrical-actuated bioprinting are some of the important approaches for the fabrication of these 4D scaffolds.

With their shape tunability, biodegradability and biocompatibility, SMPs and SMHs are widely applicable in treating irregular bone defects. Polycaprolactone is a widely studied polymer, which is mainly known for its blending and crosslinking property with other polymeric materials. Hydroxyapatite-loaded PCL (PCL-HAp)-based porous scaffolds, polycaprolactone diacrylate, and also shape-memory polyurethanes are some of the notable examples. These kind of polymeric materials find application cell-laden scaffolds for bone and muscle repair, skeletal muscle-powered bioactuator, temperature-sensitive orthopaedic implant [84].

6.6 3D- and 4D-Printed Biocomposites in Drug Delivery

The advancement in BTE has come through several challenges. The first generation of treatments focussed on selection of materials for scaffolding as that was the most important part in bone repair. It was soon realized that additional parameters such as surface modifications and specifically inculcation of growth factors could speed up the repair. This led to second generation treatments that included growth factors in the scaffolds [85]. The third most important factor in tissue engineering is the cells that are seeded with a focus on type and age of the cells. Thus, in BTE the third generation used stem cells along with growth factors and scaffolds. The current emphasis in tissue engineering in the fourth generation is being laid on the functionality of the regenerated tissue with respect to mechanical strength. It can be achieved by inducing certain cellular signalling pathways. In this section, the types of drug delivery methods and their mechanism will be discussed [86].

To enhance the growth of tissues around bones requires scaffolding. The scaffold materials used range from ceramics, natural polymers including collagen, alginate, chitin etc. to synthetic polymers and metals such as strontium, iron, chromium etc. Composite materials and Hydrogels are also used as scaffold materials in BTE [87]. However, it was found that scaffolding alone was insufficient. In order to stimulate or accelerate the process, the scaffold was supplemented with varied range of drugs or growth regulators that can regenerate the damaged cells. For instance, platelet-rich plasma (PRP), bone morphogenetic protein (BMP), insulin like growth factor (IGF) and vascular endothelial growth factor (VEGF) are used to repair the bone tissues [88]. There are different drug delivery types used in BTE. For example, bolus injection approach in which one growth factor is injected at the site of injury. However, it fails to imitate the cascading effect and hence it has limited effect [89]. Another popular method has been to impregnate the drug with scaffold during the modelling process using temperature or solvent based processes making thermos-responsive gels. Supercritical fluid method is also used to load drugs onto scaffold materials under appropriate conditions of pressure and temperature excluding the toxic solvents [88]. Other methods such as pH-responsive gels and osmotic pumps have also been developed for drug delivery in BTE. However, all single drug delivery systems are marred by the fact that bone tissue is a complex system and its regeneration requires regulated release of multiple bioactive molecules. This led to development of dual drug delivery methods through the scaffold in order to emulate the *in vivo* tissue regeneration. In this system, one material is used to provide strength at load bearing sites while the other is used to deliver the drug/growth factor [90]. Another recent technology in controlled drug release is layer by layer technology in which the drug is impregnated between layers of sequential degradable layers of for example poly-L-lysine on a silica particle [91]. Finally, the most recent revolutionary approach is 2D and 3D printing allowing a more precise control on the dose and location of delivery of the drugs. The 3D-printed scaffold can be coated or impregnated with drugs/growth factors etc. to be released at the site of injury for faster tissue engineering [88].

In recent advances, nanoparticle technology has gained importance in BTE. Nanofibers, nanotubes, nanoparticles and hydrogels have found an eminent candidate to produce and resemble extra cellular matrix and tissue scaffolds. Nanoparticles have some unique properties that can improve blood circulation, cellular uptake, controlled drug release and improved retention and contact time of drugs in the bod with lower level of drug toxicity. For example, BMP loaded on chitosan-polyethyleneimine (CS-PEI) could effectively promote the proliferation and differentiation of MC3T3-E1 cells *in vitro* without cytotoxicity [92]. It has also been reported that Cefazolin/BMP loaded mesoporous silica nano particles significantly help in repairing of open fractures with reduced rate of inflammation *in vitro* [93]. Similar positive results were reported *in vivo* also, for instance, the BMP-CS-PEI and BMP-mesoporous silica-chitosan nanoparticles accelerated new bone formation within 12 weeks after implantation in mice [94].

The mechanism of regeneration in bone repair can happen either through intramembranous ossification or endochondral ossification. In the former one, mesenchymal progenitor cells that can typically differentiate into osteoblasts adhere to the damaged bone leading to repair of the bone. However, in the latter process, these cells attach to the damaged site and grow into chondrocytes, which forms the cartilage template that mineralizes to form the bone. In case of bone damage, inflammation is the trigger of fibrin network formation and adhesion of different progenitor cells. Various growth factors such as VEGF, TNF-α, BMPs, ILs etc. and attachment of cells such as osteoclasts, fibroblasts, endothelial and chondrocytes lead to the repair of damaged bone [90]. Calcium homeostasis plays an important role in bone health. Calcium based nanoparticles can modulate calcium metabolism and bone microenvironment. The construction of calcium-scaffolds with nanoscale structures had gained importance and has shown promising results *in vivo* in rats [95].

The mesenchymal cells differentiate to osteoclasts through a complex process involving signaling pathways as PI3K/Akt/mTOR, BMP/Smad, MAPK, Wnt/β catenin. It has been shown that 3D-printed scaffolds constituting of strontium with mesoporous bioactive glass nanoparticles (Sr-MBGNs) and gelatin methacarylate (GelMA) promoted the osteoblast differentiation from mesenchymal cells *via* the Kindlin-2/ PTH1R/OCN pathway [96]. In another study, HA-coated Fe_3O_4 magnetic nanoparticles enhanced ALP (alkaline phosphatase) gene activity. Additionally, researchers have reported the improved collagen synthesis and calcium deposition of osteoblast cells through increased amounts of fibronectin [97].

These studies shows that advanced methods of drug delivery in BTE are gaining momentum in opening avenues to accelerated bone damage repair. The *in vitro* as well as *in vivo* studies have shown that the integration of nanosized drugs would probably improve the properties of scaffolds in differentiation of cells finally leading to bone tissue repair.

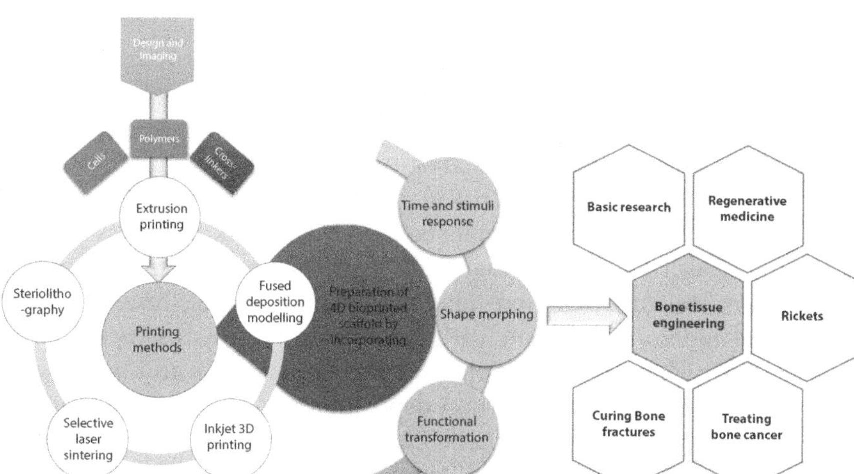

Figure 6.2 Schematic representation for the fabrication of 3D and 4D scaffolds and their application in the bone tissue engineering.

6.7 Application of 3D and 4D Biocomposites Scaffolds for BTE

In general, biocomposite-based scaffolds are beneficial in the area of bone regeneration, dental/orthopedic implants, wound dressing and more importantly BTE. More recently, researchers are comparing the application of 3D and 4D scaffolds in times of critical bone defects, and focusing on the stimuli-responsive mechanism behind it. Some of the significant recent applications of these novel biocomposites scaffolds in BTE are covered in this section (Figure 6.2).

6.7.1 Correction of Critical-Sized Bone Deformities

BTE is transforming the conventional methods of treating large bone lesions because it is based on cutting-edge 3D printing technology and basically solves these problems. By utilizing existing materials, 3D printing techniques aim to create novel scaffolds with improved mechanical qualities for load-bearing applications. Furthermore, 3D printing makes it possible to produce patient-specific bone replacements with complex configurations without the need for a mold thanks to biomaterials and precise macro- and micro-scale structure design. These replacements, when based on medical imaging data, yield the best *in vivo* results, including enriched bone regeneration through osteoconduction. Numerous tissue-engineered scaffolds with osteoinductive potential have been found in preclinical research as the 3D printing industry has expanded over the last 10 years [98].

Furthermore, heat shock protein (HSP90) expression is upregulated by NIR-based photothermal stimulation, which encourages osteogenesis and vascularization even more. By strongly encouraging osteogenic differentiation and vascularization, activation of the PI3K/AKT pathway also improves bone regeneration in critical-sized bone lesions [99].

For the first time, Liu *et al.* employed the 4D fabrication process to integrate aligned cell sheets over malleable hydrogel and preserve the bone regeneration milieu by adding adjustable morphologies, resulting in a customized bionic periosteum with an anisotropic microstructure. Du *et al.* also developed bone scaffolds with shape memory and bioactivity following the mechanism 4D fusion deposition modeling with the aid of biodegradable polyester copolymers. Eventually, complicated bone abnormalities can be repaired using this method. FDM is a feasible technique for producing scaffolds with stunning architecture using cross-linked form memory linear copolyesters that are 4D-printed. The composite scaffolds that were made can be used to mend soft tissues with little invasiveness [100].

The innovative use of 4D (bio)printing is one way to get over these limitations. Synthetic tissues with dynamic qualities are easier to construct with 4D bioprinting than with its 3D predecessor. The encapsulation of vascular or stem cells can hasten their development and ultimately their functional behavior. The capacity of 4D bioprinted materials to change dynamically makes them highly effective in treating organ-specific illnesses, such as heart issues [101].

Cardiovascular medicine specifically faces serious health hazards from aortic valve malfunction, which manifests as either regurgitation or stenosis. Blood flows unidirectionally from the heart to the aorta under the supervision of the complex tri-leaflet structure called the aortic valve. While improper closure, which causes retrograde blood flow, is known as aortic valve regurgitation, aortic valve stenosis is the thickening or stiffness of these cusps [102].

The development of reversible aortic valves or stents that can adapt to these degenerative alterations may be possible with 4D level printing technology. In the biomedical division, 4D bioprinting is not just useful for tissue engineering and cardiovascular operations. Numerous intelligent medical devices, including occlusion devices, microneedles for minimally invasive procedures, customized drug delivery systems, wound closure mechanisms, and various scaffolds and implants, might be created using this technology. Every one of these uses makes use of the dynamic, flexible qualities of 4D-printed structures to enhance treatment outcomes.

6.7.2 Scaffolds for Load-Bearing Applications

In order to produce 3D collagen scaffolds with strong collagen anisotropy, an extrusion-based 3D printing method combined with a magnetic alignment technique is used in a sophisticated 4D printing scheme. In particular, the amount of collagen to xanthan gum, the amount of streptavidin-coated magnetic particles, and the print flow rate were changed. The effects of these changes on the mixture's viscosity, printing precision, and collagen orientation were evaluated. The relationship between collagen anisotropy and the shape, orientation, metabolic activity, and ligamentous development of human mesenchymal stem cells (hMSCs) were also critically examined. The microarchitecture has an equally vital role as the composition in reproducing the anisotropic properties and the three-dimensional complexity of the natural tissue [103].

It has been already reported that the matrix anisotropy of collagen scaffolds causes tissue-specific cell responses, such as enhanced cell proliferation and differentiation, greater cellular alignment, and the formation of *de novo* oriented matrisomes. While mechanical, electrical, or magnetic stimulation can initiate collagen fiber orientation, these techniques have limitations. For example, they cannot create a custom scaffold using 3D printing architectures; some collagen alignment techniques, such as electrospinning, use inappropriate fabrication methods or harsh chemical solvents that have denatured the natural protein architecture; and they require time-consuming post-treatment procedures on 2D fibrous or sheet products in order to produce an appropriate implantable 3D scaffold [104].

Shapeshifting also called hydrogel-based smart scaffolds which is capable of display time-dependent *in situ* scaffold remodeling can be created by combining 2D printing techniques with hydrogels that react to external stimuli depending on pH/temperature, light, as well as electrical and magnetic fields. 3D complex scaffolds with the desired anisotropic properties for use in soft tissues, including tendons and ligaments, can be emulsified using the 4D printing technique, which can introduce the ability to arrive for remodeling of the collagen construct as a function of time throughout the printing process [105].

Researchers have designed and constructed 4D-printed shape memory polymer (SMP) scaffolds loaded with collagen–dexamethasone (Col–Dex) covering and bioactive fillers (hydroxyapatite and alendronate) after closely analyzing their microstructures, along with its porous and mechanical properties, and significant biological roles [106].

The SMP scaffolds fabricated in this study demonstrated the ability to cover large bone defects by temporarily assuming microscopic shapes and then significantly regaining their working sizes and shapes when subjected to an alternating magnetic field. Additionally, the biological analysis using fluorescence staining imaging, cytotoxicity tests, gene expressions associated with osteogenesis, and *in vivo* experiments confirmed that the 4D-printed SMP scaffolds had biological activities and an osteogenic effect in addition to a remarkable potential particularly for bone tissue regeneration [107].

Because of its obvious osteogenic and biocompatible properties, HAP has been extensively used extensively in traditional BTE. Although the technology is still in its early stages, 3D and 4D printing technologies are being used to construct personalized HAP scaffolds that are tailored to the specific needs of each patient, making progress toward clinical usage. The bioactivity of these printed scaffolds can be increased by including extra biomolecules, such as growth factors, peptides, pharmaceuticals, and nanomaterials [108].

Additive manufacturing (AM) techniques, like 3D printing, can be employed to efficiently generate medical implants with complex morphologies or even bespoke ones. It is finding possible application in the medical industry due to its advantages in developing production methods for a variety of specialty commodities [109]. It speeds up production processes, encourages medical model innovation, and saves time and money. Clinical imaging can be utilized in tissue engineering to develop scaffolds for defect repair and replacement of damaged tissues or organs using 3D printing. Therefore, they play a role in the intelligent repair of tissue damage which can be made in a way to react with respect to different physiological or external factors (pH, temperature, solvent, etc.). This will produce time-based physical changes and mimic the functionality of the original tissues for improved environmental adaption [110].

They perform minimally invasive surgery. More intriguingly, dynamic scaffolds that adapt to the unique needs of host implantation sites and moieties based on such bionic supremacy have great promise for use. Specifically, the 4D-printed scaffolds offer the greater benefits of fitting human organs and tissue with non-stationary structures and being appropriate for the device [105].

6.7.3 Stimuli-Responsive Regeneration Using 4D Scaffold

This stimulation allows a cellular response to be fine-tuned as needed into significant reactions in cells. They are used to activate the implanted cells and provide temporal and spatial control over stimulation. Any of these stimuli combined with a biomaterial scaffold can cause these kinds of alterations in our cells, enabling the translation of these stimuli for cells with the least amount of design complexity [111].

The most intriguing and universal will be those that helps in the conversion of an external energy source into a signal that is relevant. Examples of external signals include photostimulation or light, electrical and magnetic field-based stimulation, ultrasound activation. Certain stimuli-responsive biomaterials, chiefly photoresponsive biomaterials, have drawn a lot of attention in the field of tissue engineering, but other stimuli-responsive biomaterials have not gotten as much attention for regulating the fate of stem cells [108].

Electrical stimulation, for example, has been used to great effect in stimulating the brain and nerves, but not so much in the differentiation of multipotent stem cells. There is a growing interest in employing conductive biomaterial scaffolds to provide electrical signals to cells due to promising studies that demonstrate the profound effects of electrical stimulation on stem cells.

4D printing has gained popularity in the biomedical industry due to its capacity to produce anatomically accurate organoid models for drug analysis and disease modeling as well as new tissue engineering patterns, which involve the development of intricate functional structures that can replicate the characteristics of regular tissues and aid in tissue regeneration [112].

These 4D-printed bio-architectures must feature flexible, dynamic shapes that mimic real tissues for biomedical applications, principally drug delivery and tissue engineering. Additionally, they are able to alter their characteristics in reaction to environmental changes or outside stimuli. Additional benefits of 4D printing technology include the capacity to control printing processes in time and place and the utilization of intelligent stimuli-responsive materials to produce objects customized for specific patients [113].

Because scaffolds in TE are porous and three-dimensional, they provide the perfect conditions for tissue repair and regeneration. Scaffolds need to meet several standard requirements. An inexpensive, mechanical scaffold that is biodegradable and biocompatible that has a highly porous structure is ideal. Cell adhesion and proliferation are dependent on the biocompatibility and biodegradability of the scaffold. An immunological reaction is yet another sign of the biocompatibility of a scaffold. An immune response that is low is ideal to prevent chronic inflammatory reactions that hinder the healing rate or may even result in rejection [114].

There has been a lot of interest in AM technology for smart materials (SMs) and stimuli-responsive materials (SRMs). These materials can alter their shape and characteristics by applying physical, chemical, along with the biological stimuli. The stimuli used in the 4DP of SMMs are listed. Biomedical devices and dynamic 3DPed scaffolds for bone, heart, and cartilage regeneration were developed using different environmental alterations or cell traction forces (CTFs). The several stimuli utilized in the process of creating tissue constructs with the aid of 4DP technology are covered in this subsection [115]. This study offers a chance to learn more about the key characteristics of this new technology and could serve as a basis for further research on 4D printing and its uses in other fields. With the use of 4D printing, smart materials that react to stimuli and alter shape can be created [116].

6.8 Challenges and New Directions

The field of orthopedics still has challenges in fixing bone abnormalities. Bone defects are caused by two main components: acquired factors and congenital factors. The main factors influencing how to prevent bone defects are acquired. Aside from bone damage caused by external factors, acquired variables may include the effects of various diseases and surgeries. Managing and preventing bone stress injuries primarily aims to lower the risk of these injuries. The best course of action for bone stress injuries depends on the site of injury and probability of healing problems. In low-risk areas, bone stress injuries can usually be repaired with exercise. Then, gradually resume the activity [34].

Treatment options for high-risk areas for bone stress damage include non-weight-bearing immobilization, pharmaceutical treatment, as well as surgery. Though the precise mechanism of bone stress injury is yet unknown, it is believed that sustaining a balanced bone metabolism can help prevent the accumulation of bone damage [117].

The plasma membrane of skeletal muscle cells is damaged by these skeletal injuries, and mitochondrial function is essential for the membrane's ability to recover itself [118]. The advent of BTE technology significantly aided in the treatment of bone deformities and resulted in noteworthy breakthroughs in the study of new materials. But there are still issues with poor biocompatibility, low bionic materials, followed by lower cell migration, adhesion, and also proliferation. To accurately control cell differentiation, gene expression, growth factors, and the safety of medicinal applications, more study is still needed. Bioactive bone-like tissues can be produced either directly *in vitro* or by generating new bone repair substitutes by integrating and developing interdisciplinary concepts and technologies in the field of bone repair, such as materials science, cell biology, immunology, and manufacturing. It is projected that regenerative healing of bone defects will be applied in clinical practice soon [119].

In order to effectively treat severe infections, residual tumor tissue, or other bone-related diseases in the short term, a therapeutic biomaterial should encourage cell adhesion. Consequently, smart stimuli-responsive materials will release bioactive components or particularly, osteogenic-related elements to stimulate cell proliferation and differentiation. In the end, entire factors encourage bone regeneration. Therefore, these novel biomaterials may play a significant role in future bone tissue regeneration and precise and hopefully efficient therapy of bone disorders [69].

Despite years of research, very few responsive bioprinted materials have been developed for tissue engineering applications through 4D printing. These substances must therefore be biocompatible and ideally biodegradable (resorbable) with lower level of cytotoxicity. The capacity to perform dynamic operations in physiological conditions and a certain level of mechanical strength are also necessary. The stimulus must be safe and easy to control when it is applied to the body or in the presence of cells [120].

For example, severe pH values and high range of temperatures need to be critically considered and avoided. Therefore, due of these strict necessities, only few polymeric materials can be chosen and applicable for this kind of purpose. Additionally, most materials mentioned so far only react to one stimulus, yet tissues in nature are exposed to a wide variety of stimuli. More exertion and expertise should come into picture for developing novel and multifunctional 4D inks in order to improve the overall bioprinting process. Moreover, little is known about how cells influence the material's ability to change shape, and little is known about how printed tissues laden with cells mature. In summary, 4D printing is a powerful, innovative, and exciting technology that mimics the composition and biological functions of real tissues. However, before applying these technological advanced for clinical studies, it must be enhanced in all the mentioned parameters [71].

6.9 Conclusion

3D printing technology is become advanced and can manufacture a wide range of complex scaffolds, which mimics the internal organs of human anatomy. While taking into account of the noteworthy properties, such as mechanical, structural, biodegradable, and biocompatible characteristics, along with the incorporation of stimuli-responsive and shape memory behavior, This 3D and 4D architectures are finding a vast range of applications. Apart from its benefits in the field of BTE, 3D- and 4D-printed materials are nowadays employed in cutting-edge fields like digital medicine, programmable biomaterials, latest Janus structures for multi-step transformation, and many more on the way. Still to overcome the challenges and for continuous development for more engineering-based applications, researchers are working on biocomposite scaffolds for skin, bone as well as cartilage repair. 3D and 4D biomaterials are also found to perform a crucial role in targeted drug delivery and treating critical size bone defects in case of orthopaedic trauma. Moreover, nanoparticle-incorporated, surface-functionalized bioprinted scaffolds are applicable in stem cell-driven bone regeneration. nowadays, 3D additive manufacturing would result in printing of 3D organs, as well as cryogenic printing in fabricating nanocomposite scaffolds are the hot research areas in BTE. With this, this chapter summarizes the tremendous advancement of 3D and 4D bioprinting technology for multidisciplinary application.

AI disclosure statement: The author acknowledges the use of AI-based tools for minor language editing or grammatical correction. However, all intellectual contributions, analyses, and conclusions presented in this chapter are entirely the author's own.

References

1. Amini, A.R., Laurencin, C.T., Nukavarapu, S.P., Bone tissue engineering: recent advances and challenges. *Crit. Rev. Biomed. Eng.*, 40, 5, 363–408, 2012.
2. Tamay, D.G., Usal, T.D., Alagoz, A.S., Yucel, D., 3D and 4D Printing of Polymers for Tissue Engineering Applications, 2019;7(July).
3. Haleem, A., Javaid, M., Khan, R.H., Suman, R., 3D printing applications in bone tissue engineering. *J. Clin. Orthop. Trauma.*, 11, Suppl 1, S118–24, 2020 Feb.
4. Rana, A., Mittal, A., Vashist, C., Rajput, S., Sridhar, S.B., Malviya, R., Exploration of 4D Printing and its Applications in the Biomedical Sciences. *Curr. Pharm. Des.*, 2025.
5. Dinesh, B., Kumar, D.J.M., Kumar, S.S., Trends in 3D and 4D Printing Methods of Polymer Resources for the Applications in Tissue Engineering. *NeuroQuantology*, 20, 10, 2305–32, 2022.
6. Chen, A.N., Su, J., Li, Y., Zhang, H., Shi, Y., Yan, C., *et al.*, 3D/4D printed bio-piezoelectric smart scaffolds for next-generation bone tissue engineering. *Int. J. Extrem. Manuf.*, 5, 2023 Jun 27.
7. Kumar, A., Nune, K.C., Murr, L.E., Misra, R.D.K., Nune, K.C., Murr, L.E., *et al.*, Biocompatibility and mechanical behavior of three-dimensional scaffolds for biomedical devices: process–structure–property paradigm, 2016, 6608(March).
8. Zarna, C., Opedal, M.T., Echtermeyer, A.T., Chinga-Carrasco, G., Reinforcement ability of lignocellulosic components in biocomposites and their 3D printed applications – A review. *Compos. Part C Open Access [Internet]*, 6, 100171, 2021 Oct, Available from: http://dx.doi.org/10.1016/j.jcomc.2021.100171.
9. Daria, M., Krzysztof, L., Jakub, M., Characteristics of biodegradable textiles used in environmental engineering: A comprehensive review. *J. Clean. Prod. [Internet]*, 268, 122129, 2020 Sep, Available from: http://dx.doi.org/10.1016/j.jclepro.2020.122129.

10. Horue, M., Rivero Berti, I., Cacicedo, M.L., Castro, G.R., Microbial production and recovery of hybrid biopolymers from wastes for industrial applications- a review. *Bioresour. Technol. [Internet]*, 340, 125671, 2021 Nov, Available from: http://dx.doi.org/10.1016/j.biortech.2021.125671.
11. Le Duigou, A., Correa, D., Ueda, M., Matsuzaki, R., Castro, M., A review of 3D and 4D printing of natural fibre biocomposites. *Mater. Des. [Internet]*, 194, 108911, 2020, Available from: https://www.sciencedirect.com/science/article/pii/S0264127520304457.
12. Lincoln, R.L., Scarpa, F., Ting, V.P., Trask, R.S., Multifunctional composites: a metamaterial perspective. *Multifunct. Mater. [Internet]*, 2, 4, 43001, 2019 Dec, Available from: http://dx.doi.org/10.1088/2399-7532/ab5242.
13. Mishra, S.K., Dahiya, S., Gangil, B., Ranakoti, L., Agrawal, N., Mechanical properties of fibre/filler based polylactic Acid (PLA) composites: a brief review. *Acta Innov.*, 41, 5–18, 2021.
14. Murugan, R. and Thangamuthu, M., Mechanical Behavior of Additive Manufactured Porous Biocomposites [Internet], in: *Mechanical and Dynamic Properties of Biocomposites*, pp. 27–47, Wiley, Verlag Chemie, German, 2021, Available from: http://dx.doi.org/10.1002/9783527822331.ch2.
15. Leukers, B., Gülkan, H., Irsen, S.H., Milz, S., Tille, C., Schieker, M., *et al.*, Hydroxyapatite scaffolds for bone tissue engineering made by 3D printing. *J. Mater. Sci. Mater. Med. [Internet]*, 16, 12, 1121–1124, 2005 Dec, Available from: http://dx.doi.org/10.1007/s10856-005-4716-5.
16. Hench, L.L. and Wilson, J., An Introduction to Bioceramics [Internet], in: *World Scientific*, Intech Open, London, UK, 1993, Available from: http://dx.doi.org/10.1142/2028.
17. Bignon, A., Chouteau, J., Chevalier, J., Fantozzi, G., Carret, J.P., Chavassieux, P., *et al.*, Effect of micro- and macroporosity of bone substitutes on their mechanical properties and cellular response. *J. Mater. Sci. Mater. Med.*, 14, 12, 1089–97, 2003 Dec.
18. Furmanova, K., Gratzl, S., Stitz, H., Zichner, T., Jaresova, M., Lex, A., *et al.*, Taggle: Combining overview and details in tabular data visualizations. *Inf. Vis. [Internet]*, 19, 2, 114–36, 2020, Available from: https://doi.org/10.1177/1473871619878085.
19. Sinha, S.K., Additive manufacturing (AM) of medical devices and scaffolds for tissue engineering based on 3D and 4D printing, pp. 119–60, 2020.
20. Gbureck, U., Dembski, S., Barralet, J.E., Thull, R., Bruschit-Knochenzemente aus biphasigen b-Tricalciumphosphat/ Calciumpyrophosphat Keramiken. *BIOmaterialien [Internet]*, 5, 4, 2004 Jan, Available from: http://dx.doi.org/10.1515/biomat.2004.5.4.255.
21. Gunatillake, P., Mayadunne, R., Adhikari, R., Recent developments in biodegradable synthetic polymers. *Biotechnol. Annu. Rev.*, 12, 301–47, 2006. doi: 10.1016/S1387-2656(06)12009-8. PMID: 17045198.

22. Kanter, B., Geffers, M., Ignatius, A., Gbureck, U., Control of *in vivo* mineral bone cement degradation. *Acta Biomater. [Internet]*, 10, 7, 3279–3287, 2014 Jul, Available from: http://dx.doi.org/10.1016/j.actbio.2014.04.020.
23. Arif, U., Haider, S., Haider, A., Khan, N., Alghyamah, A.A., Jamila, N., *et al.*, Biocompatible Polymers and their Potential Biomedical Applications: A Review. *Curr. Pharm. Des. [Internet]*, 25, 34, 3608–3619, 2019 Nov, Available from: http://dx.doi.org/10.2174/1381612825999191011105148.
24. Tokiwa, Y. and Calabia, B.P., Review Degradation of microbial polyesters. *Biotechnol. Lett. [Internet]*, 26, 15, 1181–1189, 2004 Aug, Available from: http://dx.doi.org/10.1023/b:bile.0000036599.15302.e5.
25. Bano, N., Jikan, S.S., Basri, H., Adzila, S., Badarulzaman, N.A., Yunus, Z.M., *et al.*, Fabrication and Properties of Polylactic Acid/Hydroxyapatite Biocomposites for Human Bone Substitute Materials, in: *Composites in Biomedical Applications*, pp. 207–20, CRC Press, Howick Place, London, 2020.
26. Kawahara, J., Ohmori, T., Hattori, S., Kawamura, M., Contractile Myosin Gel and the Mechanism of Chemical Cross-Linking Reaction by Glutaraldehyde [II]. *Polymer Prepnnts*, Japan, 40 629, 1991.
27. Chauhan, A., Rajput, S., Malviya, R., Sridhar, S.B., Production and Multidisciplinary Applications of Bioplastic: A Biodegradable and Eco-Friendly Alternative to Plastic. *Curr. Green Chem.*, 2025.
28. Fredi, G. and Dorigato, A., Recycling of bioplastic waste: A review. *Adv. Ind. Eng. Polym. Res. [Internet]*, 4, 3, 159–177, 2021 Jul, Available from: http://dx.doi.org/10.1016/j.aiepr.2021.06.006.
29. Medeiros Garcia Alcântara, J., Distante, F., Storti, G., Moscatelli, D., Morbidelli, M., Sponchioni, M., Current trends in the production of biodegradable bioplastics: The case of polyhydroxyalkanoates. *Biotechnol. Adv.*, 42, 107582, 2020.
30. Emadian, S.M., Onay, T.T., Demirel, B., Biodegradation of bioplastics in natural environments. *Waste Manag. [Internet]*, 59, 526–536, 2017 Jan, Available from: http://dx.doi.org/10.1016/j.wasman.2016.10.006.
31. Rahman, M.H. and Bhoi, P.R., An overview of non-biodegradable bioplastics. *J. Clean. Prod. [Internet]*, 294, 126218, 2021 Apr, Available from: http://dx.doi.org/10.1016/j.jclepro.2021.126218.
32. Karimipour-Fard, P., Pop-Iliev, R., Jones-Taggart, H., Rizvi, G., Design of 3D scaffold geometries for optimal biodegradation of poly(lactic acid)-based bone tissue, in: *Proceedings of the 35th International Conference of the Polymer Processing Society (PPS-35) [Internet]*, AIP Publishing, p. 20062, 2020, Available from: http://dx.doi.org/10.1063/1.5142977.
33. Hedayati, S.K., Behravesh, A.H., Hasannia, S., Bagheri Saed, A., Akhoundi, B., 3D printed PCL scaffold reinforced with continuous biodegradable fiber yarn: A study on mechanical and cell viability properties. *Polym. Test [Internet]*, 83, 106347, 2020 Mar, Available from: http://dx.doi.org/10.1016/j.polymertesting.2020.106347.

34. Hourdé, C., Joanne, P., Medja, F., Mougenot, N., Jacquet, A., Mouisel, E., *et al.*, Voluntary Physical Activity Protects from Susceptibility to Skeletal Muscle Contraction–Induced Injury But Worsens Heart Function in mdx Mice. *Am. J. Pathol. [Internet]*, 182, 5, 1509–1518, 2013 May, Available from: http://dx.doi.org/10.1016/j.ajpath.2013.01.020.
35. Anwajler, B. and Witek-Krowiak, A., Three-Dimensional Printing of Multifunctional Composites: Fabrication, Applications, and Biodegradability Assessment. *Mater. (Basel) [Internet]*, 16, 24, 7531, 2023 Dec, Available from: http://dx.doi.org/10.3390/ma16247531.
36. Haque, A.N.M.A. and Naebe, M., Material Extrusion of Wool Waste/Polycaprolactone with Improved Tensile Strength and Biodegradation. *Polym. (Basel) [Internet]*, 15, 16, 3439, 2023 Aug, Available from: http://dx.doi.org/10.3390/polym15163439.
37. Choe, S., Kim, Y., Park, G., Lee, D.H., Park, J., Mossisa, A.T., *et al.*, Biodegradation of 3D-Printed Biodegradable/Non-biodegradable Plastic Blends. *ACS Appl. Polym. Mater. [Internet]*, 4, 7, 5077–5090, 2022 Jun, Available from: http://dx.doi.org/10.1021/acsapm.2c00600.
38. Saxena, M., Pappu, A., Sharma, A., Haque, R., Wankhede, S., Composite Materials from Natural Resources: Recent Trends and Future Potentials, in: *Advances in Composite Materials - Analysis of Natural and Man-Made Materials [Internet]*, InTech, London, United Kingdom, 2011, Available from: http://dx.doi.org/10.5772/18264.
39. Li, Y., Effect of Coupling Agent Concentration, Fiber Content, and Size on Mechanical Properties of Wood/HDPE Composites. *Int. J. Polym. Mater. [Internet]*, 61, 11, 882–890, 2012 Oct, Available from: http://dx.doi.org/10.1080/00914037.2011.617338.
40. Shesan, O.J., Stephen, A.C., Chioma, A.G., Neerish, R., Rotimi, S.E., Improving the Mechanical Properties of Natural Fiber Composites for Structural and Biomedical Applications, in: *Renewable and Sustainable Composites [Internet]*, IntechOpen, London, United Kingdom, 2019, Available from: http://dx.doi.org/10.5772/intechopen.85252.
41. Basiji, F., Safdari, V., Nourbakhsh, A., Pilla, S., The effects of fiber length and fiber loading on the mechanical properties of wood-plastic (polypropylene) composites. *Turk. J. Agric. For. [Internet]*, 2010 Jan, Available from: http://dx.doi.org/10.3906/tar-0903-16.
42. Ayrilmis, N., Jarusombuti, S., Fueangvivat, V., Bauchongkol, P., Effect of thermal-treatment of wood fibres on properties of flat-pressed wood plastic composites. *Polym. Degrad. Stab. [Internet]*, 96, 5, 818–822, 2011 May, Available from: http://dx.doi.org/10.1016/j.polymdegradstab.2011.02.005.
43. Ayrilmis, N., Kaymakci, A., Ozdemir, F., Sunflower seed cake as reinforcing filler in thermoplastic composites. *J. Appl. Polym. Sci. [Internet]*, 129, 3, 1170–1178, 2012 Nov, Available from: http://dx.doi.org/10.1002/app.38806.

44. Tufan, M. and Ayrilmis, N., Potential Use of Hazelnut Husk in Recycled High-Density Polyethylene Composites. *BioResources [Internet]*, 11, 3, 2016 Jul, Available from: http://dx.doi.org/10.15376/biores.11.3.7476-7489.
45. Al-Maharma, A.Y. and Al-Huniti, N., Critical Review of the Parameters Affecting the Effectiveness of Moisture Absorption Treatments Used for Natural Composites. *J. Compos. Sci. [Internet]*, 3, 1, 2019, Available from: https://www.mdpi.com/2504-477X/3/1/27.
46. Mehdikhani, M., Gorbatikh, L., Verpoest, I., Lomov, S.V., Voids in fiber-reinforced polymer composites: A review on their formation, characteristics, and effects on mechanical performance. *J. Compos. Mater. [Internet]*, 53, 12, 1579–1669, 2018 Nov, Available from: http://dx.doi.org/10.1177/0021998318772152.
47. Dufresne, A., Nanocellulose: a new ageless bionanomaterial. *Mater. Today [Internet]*, 16, 6, 220–227, 2013 Jun, Available from: http://dx.doi.org/10.1016/j.mattod.2013.06.004.
48. Hendrikson, W.J., Rouwkema, J., Clementi, F., van Blitterswijk, C.A., Farè, S., Moroni, L., Towards 4D printed scaffolds for tissue engineering: exploiting 3D shape memory polymers to deliver time-controlled stimulus on cultured cells. *Biofabrication [Internet]*, 9, 3, 31001, 2017 Aug, Available from: http://dx.doi.org/10.1088/1758-5090/aa8114.
49. Miao, D., Margolis, C.A., Gao, W., Voss, M.H., Li, W., Martini, D.J., *et al.*, Genomic correlates of response to immune checkpoint therapies in clear cell renal cell carcinoma. *Sci. (80-) [Internet]*, 359, 6377, 801–806, 2018 Feb, Available from: http://dx.doi.org/10.1126/science.aan5951.
50. Kumar, D., Malviya, R., Sridhar, S.B., Wadhwa, T., Talath, S., Shareef, J., Trends in 4D Printed Shape Memory Biomaterials for Tissue Engineering Applications. *Curr. Pharm. Des.*, 31, 41, 3285–302, 2025 Dec.
51. Kolesky, D.B., Truby, R.L., Gladman, A.S., Busbee, T.A., Homan, K.A., Lewis, J.A., 3D Bioprinting of Vascularized, Heterogeneous Cell-Laden Tissue Constructs. *Adv. Mater. [Internet]*, 26, 19, 3124–3130, 2014 Feb, Available from: http://dx.doi.org/10.1002/adma.201305506.
52. Gopal Shankar, K., Udhaya Kumar, S., Sowndarya, S., Suresh Babu, P., Rose, C., Isolation, characterization, and *in vitro* evaluation of bovine rumen submucosa films of collagen or chitosan-treated collagen. *J. Biomater. Appl. [Internet]*, 30, 6, 780–792, 2015 May, Available from: http://dx.doi.org/10.1177/0885328215584492.
53. Ielo, I., Calabrese, G., De Luca, G., Conoci, S., Recent Advances in Hydroxyapatite-Based Biocomposites for Bone Tissue Regeneration in Orthopedics. *Int. J. Mol. Sci. [Internet]*, 23, 17, 9721, 2022 Aug, Available from: http://dx.doi.org/10.3390/ijms23179721.
54. Calabrese, G., Giuffrida, R., Fabbi, C., Figallo, E., Lo Furno, D., Gulino, R., *et al.*, Collagen-Hydroxyapatite Scaffolds Induce Human Adipose Derived Stem Cells Osteogenic Differentiation *In Vitro*. Vanella, L. (Eds.), *PLoS*

One [Internet], 11, 3, e0151181, 2016 Mar, Available from: http://dx.doi.org/10.1371/journal.pone.0151181.
55. Teli, M.D. and Sheikh, J., Extraction of chitosan from shrimp shells waste and application in antibacterial finishing of bamboo rayon. *Int. J. Biol. Macromol. [Internet]*, 50, 5, 1195–1200, 2012 Jun, Available from: http://dx.doi.org/10.1016/j.ijbiomac.2012.04.003.
56. Soleymani, S. and Naghib, S.M., 3D and 4D printing hydroxyapatite-based scaffolds for bone tissue engineering and regeneration. *Heliyon [Internet]*, 9, 9, 2023 Sep 1, Available from: https://doi.org/10.1016/j.heliyon.2023.e19363.
57. Lee, J.B., Lee, S.H., Yu, S.M., Park, J.C., Choi, J.B., Kim, J.K., PLGA scaffold incorporated with hydroxyapatite for cartilage regeneration. *Surf. Coat. Technol. [Internet]*, 202, 22–23, 5757–5761, 2008 Aug, Available from: http://dx.doi.org/10.1016/j.surfcoat.2008.06.138.
58. Ocando, C., Dinescu, S., Samoila, I., Daniela Ghitulica, C., Cucuruz, A., Costache, M., *et al.*, Fabrication and properties of alginate-hydroxyapatite biocomposites as efficient biomaterials for bone regeneration. *Eur. Polym. J. [Internet]*, 151, 110444, 2021 May, Available from: http://dx.doi.org/10.1016/j.eurpolymj.2021.110444.
59. Sikkema, R., Keohan, B., Zhitomirsky, I., Alginic Acid Polymer-Hydroxyapatite Composites for Bone Tissue Engineering. *Polym. (Basel) [Internet]*, 13, 18, 3070, 2021 Sep, Available from: http://dx.doi.org/10.3390/polym13183070.
60. Mahmoud, E.M., Sayed, M., El-Kady, A.M., Elsayed, H., Naga, S.M., *In vitro* and *in vivo* study of naturally derived alginate/hydroxyapatite bio composite scaffolds. *Int. J. Biol. Macromol. [Internet]*, 165, 1346–1360, 2020 Dec, Available from: http://dx.doi.org/10.1016/j.ijbiomac.2020.10.014.
61. Kaczmarek, B., Sionkowska, A., Gołyńska, M., Polkowska, I., Szponder, T., Nehrbass, D., *et al.*, *In vivo* study on scaffolds based on chitosan, collagen, and hyaluronic acid with hydroxyapatite. *Int. J. Biol. Macromol. [Internet]*, 118, 938–944, 2018 Oct, Available from: http://dx.doi.org/10.1016/j.ijbiomac.2018.06.175.
62. Sujana, A., Venugopal, J.R., Velmurugan, B., Góra, A., Salla, M., Ramakrishna, S., Hydroxyapatite-intertwined hybrid nanofibres for the mineralization of osteoblasts. *J. Tissue Eng. Regen. Med.*, 11, 6, 1853–64, 2017 Jun.
63. Huang, Y., Connell, M., Park, Y., Mooney, D.J., Rice, K.G., Fabrication and *in vitro* testing of polymeric delivery system for condensed DNA. *J. Biomed. Mater. Res. Part A [Internet]*, 67A, 4, 1384–1392, 2003 Nov, Available from: http://dx.doi.org/10.1002/jbm.a.20036.
64. Norouzi, M., Shabani, I., Ahvaz, H.H., Soleimani, M., PLGA/gelatin hybrid nanofibrous scaffolds encapsulating EGF for skin regeneration: PLGA/Gelatin Hybrid Nanofibrous Scaffolds. *J. Biomed. Mater. Res. Part A [Internet]*, 103, 7, 2225–2235, 2014 Nov, Available from: http://dx.doi.org/10.1002/jbm.a.35355.

65. Seitz, H., Rieder, W., Irsen, S., Leukers, B., Tille, C., Three-dimensional printing of porous ceramic scaffolds for bone tissue engineering. *J. Biomed. Mater. Res. Part B Appl. Biomater. [Internet]*, 74B, 2, 782–788, 2005 Jun, Available from: http://dx.doi.org/10.1002/jbm.b.30291.
66. Wu, C., Fan, W., Zhou, Y., Luo, Y., Gelinsky, M., Chang, J., *et al.*, 3D-printing of highly uniform CaSiO3 ceramic scaffolds: preparation, characterization and *in vivo* osteogenesis. *J. Mater. Chem. [Internet]*, 22, 24, 12288, 2012, Available from: http://dx.doi.org/10.1039/c2jm30566f.
67. Rajput, S., Malviya, R., Sridhar, S.B., Osteoporosis Risk Assessment and Individualized Feature Analysis Using Interpretable XAI and RAI Techniques, in: *Explainable and Responsible Artificial Intelligence in Healthcare*, pp. 89–113, Elsevier, UAE, 2025 Mar 13.
68. Do, A.V., Khorsand, B., Geary, S.M., Salem, A.K., 3D Printing of Scaffolds for Tissue Regeneration Applications. *Adv. Healthc. Mater.*, 4, 12, 1742–62, 2015 Aug.
69. Fu, S., Wang, X., Guo, G., Shi, S., Liang, H., Luo, F., *et al.*, Preparation and Characterization of Nano-Hydroxyapatite/Poly(ε-caprolactone)–Poly(ethylene glycol)–Poly(ε-caprolactone) Composite Fibers for Tissue Engineering. *J. Phys. Chem. C [Internet]*, 114, 43, 18372–18378, 2010 Oct, Available from: http://dx.doi.org/10.1021/jp106488t.
70. Chuenjitkuntaworn, B., Inrung, W., Damrongsri, D., Mekaapiruk, K., Supaphol, P., Pavasant, P., Polycaprolactone/hydroxyapatite composite scaffolds: Preparation, characterization, and *in vitro* and *in vivo* biological responses of human primary bone cells. *J. Biomed. Mater. Res. Part A [Internet]*, 94A, 1, 241–251, 2010 Feb, Available from: http://dx.doi.org/10.1002/jbm.a.32657.
71. Saska, S., Pilatti, L., Blay, A., Shibli, J.A., Bioresorbable Polymers: Advanced Materials and 4D Printing for Tissue Engineering. *Polym. (Basel) [Internet]*, 13, 4, 563, 2021 Feb, Available from: http://dx.doi.org/10.3390/polym13040563.
72. Müller, M., Becher, J., Schnabelrauch, M., Zenobi-Wong, M., Nanostructured Pluronic hydrogels as bioinks for 3D bioprinting. *Biofabrication [Internet]*, 7, 3, 35006, 2015 Aug, Available from: http://dx.doi.org/10.1088/1758-5090/7/3/035006.
73. Shim, M.S., Lee, H.T., Shim, W.S., Park, I., Lee, H., Chang, T., *et al.*, Poly(D,L-lactic acid-co-glycolic acid)-b-poly(ethylene glycol)-b-poly (D,L-lactic acid-co-glycolic acid) triblock copolymer and thermoreversible phase transition in water. *J. Biomed Mater. Res. [Internet]*, 61, 2, 188–196, 2002 Apr, Available from: http://dx.doi.org/10.1002/jbm.10164.
74. Zentner, G.M., Rathi, R., Shih, C., McRea, J.C., Seo, M.H., Oh, H., *et al.*, Biodegradable block copolymers for delivery of proteins and water-insoluble drugs. *J. Control Release [Internet]*, 72, 1–3, 203–215, 2001 May, Available from: http://dx.doi.org/10.1016/s0168-3659(01)00276-0.

75. Dutta, S. and Cohn, D., Temperature and pH responsive 3D printed scaffolds. *J. Mater. Chem. B [Internet]*, 5, 48, 9514–9521, 2017, Available from: http://dx.doi.org/10.1039/c7tb02368e.
76. Truby, R.L. and Lewis, J.A., Printing soft matter in three dimensions.
77. Deng, L., Deng, Y., Xie, K., Colloids and Surfaces B: Biointerfaces AgNPs-decorated 3D printed PEEK implant for infection control and bone repair. *Colloids Surf. B Biointerfaces [Internet]*, 160, 483–92, 2017, Available from: http://dx.doi.org/10.1016/j.colsurfb.2017.09.061.
78. Yu, J., Xia, H., Ni, Q.Q., A three-dimensional porous hydroxyapatite nanocomposite scaffold with shape memory effect for bone tissue engineering. *J. Mater. Sci. [Internet]*, 53, 7, 4734–44, 2018, Available from: https://doi.org/10.1007/s10853-017-1807-x.
79. Nganga, S., Moritz, N., Kolakovic, R., Jakobsson, K., Nyman, J., Borgogna, M., *et al.*, Inkjet printing of Chitlac-nanosilver - A method to create functional coatings for non-metallic bone implants. *Biofabrication.*, 6, 41001, 2014 Oct 22.
80. Tolosa, I., Garciandía, F., Zubiri, F., Study of mechanical properties of AISI 316 stainless steel processed by "selective laser melting", following different manufacturing strategies, 639–47, 2010.
81. Domingos, M., Gloria, A., Coelho, J., Bartolo, P., Three-dimensional printed bone scaffolds: The role of nano/micro-hydroxyapatite particles on the adhesion and differentiation of human mesenchymal stem cells. *Proc. Inst. Mech. Eng. H*, 231, 6, 555–64, 2017.
82. Andersson, M., Revision History.
83. Afzali Naniz, M., Askari, M., Zolfagharian, A., Bodaghi, M., 4D bioprinting: Fabrication approaches and biomedical applications, pp. 193–229, 2022.
84. Bonetti, L. and Scalet, G., 4D fabrication of shape-changing systems for tissue engineering: state of the art and perspectives. *Prog. Addit. Manuf. [Internet]*, 10, 4, 1913–43, 2025, Available from: https://doi.org/10.1007/s40964-024-00743-5.
85. Ikada, Y., Challenges in tissue engineering. *J. R. Soc. Interface [Internet]*, 3, 10, 589–601, 2006 Apr, Available from: http://dx.doi.org/10.1098/rsif.2006.0124.
86. Matić, A., Sher, E.K., Farhat, E.K., Sher, F., Nanostructured Materials for Drug Delivery and Tissue Engineering Applications. *Mol. Biotechnol. [Internet]*, 2023 Jun, Available from: http://dx.doi.org/10.1007/s12033-023-00784-1.
87. Bauso, L.V., La Fauci, V., Longo, C., Calabrese, G., Bone Tissue Engineering and Nanotechnology: A Promising Combination for Bone Regeneration. *Biol. (Basel) [Internet]*, 13, 4, 237, 2024 Apr, Available from: http://dx.doi.org/10.3390/biology13040237.
88. Costa, P.F., Bone Tissue Engineering Drug Delivery. *Curr. Mol. Biol. Rep. [Internet]*, 1, 2, 87–93, 2015 Apr, Available from: http://dx.doi.org/10.1007/s40610-015-0016-0.

89. Nguyen, L.H., Annabi, N., Nikkhah, M., Bae, H., Binan, L., Park, S., *et al.*, Vascularized Bone Tissue Engineering: Approaches for Potential Improvement. *Tissue Eng. Part B Rev. [Internet]*, 18, 5, 363–382, 2012 Oct, Available from: http://dx.doi.org/10.1089/ten.teb.2012.0012.
90. AD, V.K., Ray, S., Arora, U., Mitra, S., Sionkowska, A., Jaiswal, A.K., Dual drug delivery platforms for bone tissue engineering. *Front. Bioeng. Biotechnol. [Internet]*, 10, 2022 Sep, Available from: http://dx.doi.org/10.3389/fbioe.2022.969843.
91. Keeney, M., Waters, H., Barcay, K., Jiang, X., Yao, Z., Pajarinen, J., *et al.*, Mutant MCP-1 protein delivery from layer-by-layer coatings on orthopedic implants to modulate inflammatory response. *Biomater. [Internet]*, 34, 38, 10287–10295, 2013 Dec, Available from: http://dx.doi.org/10.1016/j.biomaterials.2013.09.028.
92. Zhao, L., Zhang, K., Bu, W., Xu, X., Jin, H., Chang, B., *et al.*, Effective delivery of bone morphogenetic protein 2 gene using chitosan–polyethylenimine nanoparticle to promote bone formation. *RSC Adv. [Internet]*, 6, 41, 34081–34089, 2016, Available from: http://dx.doi.org/10.1039/c5ra24891d.
93. Shen, M., Wang, L., Feng, L., Xu, C., Gao, Y., Li, S., *et al.*, Cefazolin/BMP-2-Loaded Mesoporous Silica Nanoparticles for the Repair of Open Fractures with Bone Defects. Liang, S. (Eds.), *Oxid. Med. Cell. Longev [Internet]*, 2022, 1, 2022 Jan, Available from: http://dx.doi.org/10.1155/2022/8385456.
94. Zhang, Q., Zhou, J., Zhi, P., Liu, L., Liu, C., Fang, A., 3D printing method for bone tissue engineering scaffold. *Med. Nov. Technol. Devices [Internet]*, 17, August 2022, 100205, 2023, Available from: https://doi.org/10.1016/j.medntd.2022.100205.
95. Wu, Z., Zheng, K., Zhang, G., Huang, L., Zhou, S., Preparation of polysulfone-based nano fiber Janus membrane for membrane distillation containing organic pollutants, pp. 1–15, 2024.
96. Xu, Z., Qi, X., Bao, M., Zhou, T., Shi, J., Xu, Z., *et al.*, Biomineralization inspired 3D printed bioactive glass nanocomposite scaffolds orchestrate diabetic bone regeneration by remodeling micromilieu. *Bioact. Mater. [Internet]*, 25, 239–255, 2023 Jul, Available from: http://dx.doi.org/10.1016/j.bioactmat.2023.01.024.
97. Tran, N. and Webster, T.J., Increased osteoblast functions in the presence of hydroxyapatite-coated iron oxide nanoparticles. *Acta Biomater. [Internet]*, 7, 3, 1298–1306, 2011 Mar, Available from: http://dx.doi.org/10.1016/j.actbio.2010.10.004.
98. Zhang, L., Yang, G., Johnson, B.N., Jia, X., Three-dimensional (3D) printed scaffold and material selection for bone repair. *Acta Biomater. [Internet]*, 84, 16–33, 2019 Jan, Available from: http://dx.doi.org/10.1016/j.actbio.2018.11.039.
99. Li, Y., You, J., Lv, H., Wang, C., Zhai, S., Ren, S., *et al.*, 4D-printed dual-responsive bioscaffolds for treating critical-sized irregular bone defects.

Chem. Eng. J. [Internet], 489, 151205, 2024 Jun, Available from: http://dx.doi.org/10.1016/j.cej.2024.151205.

100. Shen, C. and Shen, A., 4D printing: innovative solutions and technological advances in orthopedic repair and reconstruction, personalized treatment and drug delivery. *Biomed. Eng. Online [Internet]*, 24, 1, 2025 Jan, Available from: http://dx.doi.org/10.1186/s12938-025-01334-3.
101. Sahafnejad-Mohammadi, I., Karamimoghadam, M., Zolfagharian, A., Akrami, M., Bodaghi, M., 4D printing technology in medical engineering: a narrative review. *J. Braz. Soc. Mech. Sci. Eng. [Internet]*, 44, 6, 2022 May, Available from: http://dx.doi.org/10.1007/s40430-022-03514-x.
102. Lai, J., Liu, Y., Lu, G., Yung, P., Wang, X., Tuan, R.S., *et al.*, 4D bioprinting of programmed dynamic tissues. *Bioact. Mater. [Internet]*, 37, 348–377, 2024 Jul, Available from: http://dx.doi.org/10.1016/j.bioactmat.2024.03.033.
103. Mohammadi, M., Zolfagharian, A., Bodaghi, M., Xiang, Y., Kouzani, A.Z., 4D printing of soft orthoses for tremor suppression. *Bio-Des. Manuf. [Internet]*, 5, 4, 786–807, 2022 Jun, Available from: http://dx.doi.org/10.1007/s42242-022-00199-y.
104. Han, X., Saiding, Q., Cai, X., Xiao, Y., Wang, P., Cai, Z., *et al.*, Intelligent Vascularized 3D/4D/5D/6D-Printed Tissue Scaffolds. *Nano-Micro Lett. [Internet]*, 15, 1, 2023 Oct, Available from: http://dx.doi.org/10.1007/s40820-023-01187-2.
105. Patrawalla, N.Y., Liebendorfer, K., Kishore, V., An innovative 4D printing approach for fabrication of anisotropic collagen scaffolds. *Biofabrication [Internet]*, 17, 1, 15002, 2024 Oct, Available from: http://dx.doi.org/10.1088/1758-5090/ad7f8f.
106. Wang, H. and Guo, J., Recent advances in 4D printing hydrogel for biological interfaces. *Int. J. Mater. Form. [Internet]*, 16, 5, 2023 Aug, Available from: http://dx.doi.org/10.1007/s12289-023-01778-9.
107. Zhou, W., Dong, X., He, Y., Zheng, W., Leng, J., *In-vitro* and *in-vivo* studies of 4D printed shape memory scaffolds with bioactive fillers and coating for enhanced bone tissue regeneration. *Smart Mater. Struct. [Internet]*, 31, 10, 105002, 2022 Aug, Available from: http://dx.doi.org/10.1088/1361-665x/ac884a.
108. Chopra, V., Fuentes-Velasco, V., Nacif-Lopez, S.R., Melendez-Malpicca, J., Mendez-Hernandez, A.S., Ramos-Mendez-Iris, L.F., *et al.*, Advancements in 3D-4D printing of hydroxyapatite composites for bone tissue engineering. *Ceram. Int. [Internet]*, 50, 20, 38819–38840, 2024 Oct, Available from: http://dx.doi.org/10.1016/j.ceramint.2024.07.266.
109. Noroozi, R., Arif, Z.U., Taghvaei, H., Khalid, M.Y., Sahbafar, H., Hadi, A., *et al.*, 3D and 4D Bioprinting Technologies: A Game Changer for the Biomedical Sector? *Ann. Biomed. Eng. [Internet]*, 51, 8, 1683–1712, 2023 Jun, Available from: http://dx.doi.org/10.1007/s10439-023-03243-9.

110. Lui, Y.S., Sow, W.T., Tan, L.P., Wu, Y., Lai, Y., Li, H., 4D printing and stimuli-responsive materials in biomedical aspects. *Acta Biomater. [Internet]*, 92, 19–36, 2019 Jul, Available from: http://dx.doi.org/10.1016/j.actbio.2019.05.005.
111. Lai, J., Wang, C., Wang, M., 3D printing in biomedical engineering: Processes, materials, and applications. *Appl. Phys. Rev. [Internet]*, 8, 2, 2021 Jun, Available from: http://dx.doi.org/10.1063/5.0024177.
112. Gelmi, A. and Schutt, C.E., Stimuli-Responsive Biomaterials: Scaffolds for Stem Cell Control. *Adv. Healthc. Mater. [Internet]*, 10, 1, 2020 Sep, Available from: http://dx.doi.org/10.1002/adhm.202001125.
113. Antezana, P.E., Municoy, S., Ostapchuk, G., Catalano, P.N., Hardy, J.G., Evelson, P.A., *et al.*, 4D Printing: The Development of Responsive Materials Using 3D-Printing Technology. *Pharm. [Internet]*, 15, 12, 2743, 2023 Dec, Available from: http://dx.doi.org/10.3390/pharmaceutics15122743.
114. Sultana, N., Cole, A., Strachan, F., Biocomposite Scaffolds for Tissue Engineering: Materials, Fabrication Techniques and Future Directions. *Mater. (Basel) [Internet]*, 17, 22, 5577, 2024 Nov, Available from: http://dx.doi.org/10.3390/ma17225577.
115. Arif, Z.U., Khalid, M.Y., Noroozi, R., Sadeghianmaryan, A., Jalalvand, M., Hossain, M., Recent advances in 3D-printed polylactide and polycaprolactone-based biomaterials for tissue engineering applications. *Int. J. Biol. Macromol. [Internet]*, 218, 930–968, 2022 Oct, Available from: http://dx.doi.org/10.1016/j.ijbiomac.2022.07.140.
116. Ding, A., Lee, S.J., Tang, R., Gasvoda, K.L., He, F., Alsberg, E., 4D Cell-Condensate Bioprinting. *Small [Internet]*, 18, 36, 2022 Aug, Available from: http://dx.doi.org/10.1002/smll.202202196.
117. Kiuru, M.J., Pihlajamäki, H.K., Ahovuo, J.A., Bone stress injuries. *Acta Radiol. [Internet]*, 45, 3, 2004 May, Available from: http://dx.doi.org/10.1080/02841850410004724.
118. Horn, A., der Meulen, J.H., Defour, A., Hogarth, M., Sreetama, S.C., Reed, A., *et al.*, Mitochondrial redox signaling enables repair of injured skeletal muscle cells. *Sci. Signal [Internet]*, 10, 495, 2017 Sep, Available from: http://dx.doi.org/10.1126/scisignal.aaj1978.
119. Xue, N., Ding, X., Huang, R., Jiang, R., Huang, H., Pan, X., *et al.*, Bone Tissue Engineering in the Treatment of Bone Defects. *Pharmaceuticals [Internet]*, 15, 7, 879, 2022 Jul, Available from: http://dx.doi.org/10.3390/ph15070879.
120. Moud, A.A., Kamkar, M., Sanati-Nezhad, A., Hejazi, S.H., Sundararaj, U., Viscoelastic properties of poly (vinyl alcohol) hydrogels with cellulose nanocrystals fabricated through sodium chloride addition: Rheological evidence of double network formation. *Colloids Surf. A Physicochem. Eng. Asp. [Internet]*, 609, 125577, 2021 Jan, Available from: http://dx.doi.org/10.1016/j.colsurfa.2020.125577.

7

Biocomposites in 4D Printing: Enhancing Regenerative Medicine

Shruti Khare[1]*, Swatantra K. S. Kushwaha[2], Amit Mishra[1] and Neelottma Kushwaha[2]

[1]*Department of Pharmaceutics, Maharana Pratap College of Pharmaceutical Sciences, Kanpur, Uttar Pradesh, India*
Department of Pharmaceutics, Krishna Institute of Pharmacy and Sciences, Kanpur, Uttar Pradesh, India

Abstract
A new type of material in the stem cell therapy field has been a breakthrough, with the introduction of biocomposites, which are a combination of both biopolymers and reinforcing materials. The unique properties of the expression indicate that tissues can be used interchangeably, are friendly to the environment, and firm, making they quite competent to be used in medical practices. The possibility of using 3D-printed products in generating new models of regenerative medicine has been further increased by the development of four-dimensional printing techniques, which have added the new dimension of time to the products made with biocomposites. It is possible to create dynamic, trigger-responsive materials based on this innovation that can respond readily to alterations in the physiology of the system, such as shape changes, self-assembly, or functions. Examples of such bioactive additives include hydroxyapatite, different forms of graphene, various forms of metallic nanoparticles that can be incorporated into biopolymer matrices, e.g. collagen, chitosan or alginate, to control their mechanical, thermal and biological characteristics to perform specific regenerative capabilities. One of the possible areas of implementation is the tissue engineering of 4D-printed biocomposites. This kind of material may be used in implants and scaffolds that may be personalized to meet the various environments in the body of a person. As an example, the 4D-printed vascular grafts or stents could adjust the shape following the implantation in a manner that would further support their functionality and integration. Similarly, 4D-printed biocomposites can be programmed in a controlled release

**Corresponding author*: shrutiaibs@gmail.com

Rishabha Malviya and Mukesh Roy (eds.) Biocomposites in 3D and 4D Printing: Materials, Techniques, and Applications in Regenerative Medicine, (197–218) © 2026 Scrivener Publishing LLC

formulation to release bioactive agents in a controlled fashion, under physiological conditions such as inflammation or pH changes. The present studies in the area are trying to find solutions to such problems by creating highly developed biopolymer composites with numerous functions. Material science advancements have already shown promise through the inclusion of metallic nanoparticles to increase mechanical and structural properties of biocomposites. Thus, this chapter emphasises that 4D-printed biocomposites are a major step towards regenerative medicine that offers dynamic, fluidic responses to biofabrication, drug delivery, etc. With the use of various unique properties of biocomposites and life-altering potential of the Self-assembling structure printing, researchers will open the door to personalised and low-invasive procedures, which could have a tremendous effect on patient outcomes. Therefore, this chapter fills the gap between the material science and biomedical engineering, and it implies that 4D-printed biocomposites can change regenerative therapy and provide better outcomes to patients significantly.

Keywords: Biocomposites, 4D printing, vascular grafts, regeneration, specific implants to patients

7.1 Introduction

The phenomena of biomedical engineering involve cellular repair and tissue hemostasis, and such a novel technique provides stimulation and repair of the damaged muscles and other biological tissues. The significance of frameworks of 4D printing as a valuable body of research can be illustrated by their critical roles in the new regenerative medical approaches. In the current days, three-dimensional printing is commonly applied in single uproot, various implants and orthosis as well as in living biological tissues. The development of four-dimensional printing has also introduced certain opportunities in development of flexible structures in biomedical engineering. The method combines 3D printing with responsive substances that can undergo change of shape or attributes with time concerning a sequence of external influences [1]. The 4D printers uses stimulation-responsive materials which maintain hydrogels, shape memory polymers (SMPs) and bio-ink to form structures that can be converted into a certain variable like temperature, pH or moisture. Such adaptive resources help in developing new medical equipment that could individuals of the environment as it adjusts to the environment, enhancing clinical results and, consequently, allowing clinical management strategies [2, 3]. The four dimensional printing process has been shown to have a bright future in terms of nature healing systems of the body or the regeneration of the damaged tissue using innovative technologies. Thus, the four-dimensional printing method technology utilizes smart materials to form scaffolds that are used in the treatment of the damaged tissues and their healing [3, 4]. 4D printing refers to the additive manufacturing process that allows objects to autonomously alter their shape or functionality in response to external stimuli, such as variations in colloidal osmotic pressure, electrical input, ultraviolet radiation, and more. A widely accepted emerging method that has shown significant potential for developing biostructures in tissue engineering is 3D printing, which utilizes a three-dimensional framework and is commonly known as rapid prototyping [5].

The fundamental techniques for 4D printing involve various 3D printing technologies. Among these methods are direct ink writing, fused deposition modeling, vat photopolymerization (including digital light projection and stereolithography), direct laser writing, selective laser sintering, and selective laser melting, among others [6]. Additionally, hybrid methods, such as molding-integrated and electroless metallic plating-integrated 4D printing, have been developed that combine 3D printing with traditional manufacturing processes. The intended use, the length scale, the mechanical and biological characteristics of the pertinent materials, and the production technique (e.g., batch or mass manufacturing) all influence the choice of a particular 4D printing process [7, 8].

Uniqueness and interesting biofabrication protocol, which can bring innovation in the sphere of tissue engineering and promise positive outcomes, is four-dimensional (4D) bioprinting [9, 10]. There are two major ways that facilitate the process of bioengineering, which include the generation of a scaffold that has the ability to respond to external stimuli both macrosystemically and on small scales of microscopes [11]. The first approach consists of 3D printing smart self-regulating material that can change shape or structure in response to stimuli that can affect it and may include light, local electric or magnetic forces, pH, temperature or even humidity [12]. Secondly, biocompatible responsive hydrogels are successful choices of material to use in 4D bioprinting procedures. As a matter of fact, these 4D printing parts react to external forces by swelling or deswelling, altering their framework, etc. [13]. Additionally, many hydrogels can be printed at physiological temperature (37°C), which supports living cells when the 3D printing process is underway, and they mirror the physical characteristics of natural extracellular matrix [14, 15].

The domain of tissue engineering and regenerative medicine is undergoing groundbreaking advancements thanks to 4D printing technology. Scientists have created adaptive structures that greatly improve the outcomes of organ transplants and accelerate the healing of injured tissues. The best example of the utilization of four-dimensional features involves the incorporation of self-sustaining biological materials into specially designed bioinks, which is the most promising technique. The development of bioink matrices using a four-dimensional technique with stimuli-responsive material allows researchers to develop different layered structures that resemble biological tissues. The method causes the preparation of biologically active compounds that not only resemble the physical properties of human cellular organs [16, 17]. The above technique leads to the development of biomolecular assemblies that exhibit similar properties to biological tissues, so they are utilized for therapeutic applications [18, 19]. As an illustration, it is possible to employ a material that contains live cells, such as human mesenchymal stromal cells, in conjunction with a hydrogel that is reliant on swelling in order to print cartilage tissue using a four-dimensional technology [20]. If we consider an example of printing cartilage tissues by the use of four-dimensional technology, a component with living tissue, such as human mesenchymal stem cells, is combined with hydrogel.

The printed structure alters its form when it expands, closely mimicking the characteristics of real cartilage [21]. With this method, customized tissue architectures that are responsive to particular environmental cues can be produced. Highly developed approaches are utilized in the field of cartilage replacement therapies and repair which are due to tissue engineering and regenerative medicines approach [22].

Additionally, functioning blood arteries have been produced *via* 4D printing. Scientists have accomplished the 4D printing of intricate structures that could be utilized in vascularized tissues by employing a biomaterials and living systems that includes vascular endothelial cells. One illustration is the 4D bio fabrication of a vital part of blood artery networks: the T-shaped vascular bifurcation [23, 24]. The swelling of the polymer causes a regulated shape alteration in the 3D-printed T-junction structure. A complex vascular branching structure suitable for integration into engineered tissues is formed when the printed design responds to specific environmental stimuli. Vascularized tissues, which are necessary for more complex organs to survive and function properly, might be produced utilizing a cutting-edge 4D printing process [25, 26].

7.2 Composition of Biocomposites

A composite is an extensively recognized substance made up of one or more unique materials or polymers that differ considerably in their physical and chemical features. By adding or improving certain qualities that specific homogeneous materials cannot exhibit, composites are used to create tailored attributes [27]. Biocomposites are composite materials made from polymers and specific biological elements. The concept of 4D bioprinting incorporates time as a fourth dimension, aiming to enhance the adaptability of the construct for better functional responses. This technology has come in with the ability to cause deliberate changes in the constructs to be able to bridge the lab inventions with human tissues in nature [28]. This is made possible by a three-dimensional construct which is capable of changing its properties (e.g., shape) depending on a preset stimulus thus forming biologically active structures. Biocomposites, reinforcing components mixed with biodegradable polymers, have an important position in the biomedical sector, in various heart related procedures they are utilized in the production of the biodegradable stents which slowly dissolve, removing the need of having permanent foreign objects in the blood vessels-based [29]. Biocomposites decrease chances of implant rejection and increase compatibility in orthopedic implants. Regenerative medicine has utilized their biocompatibility to help in the development of new tissues and organs [30–33]. In the area of tissue engineering, biocomposites also offer necessary scaffold to cell growth. Moreover, the bio composites improve the drug delivery carriers by offering controlled release of the drug thereby optimize beneficial effects besides reducing side effects. It is

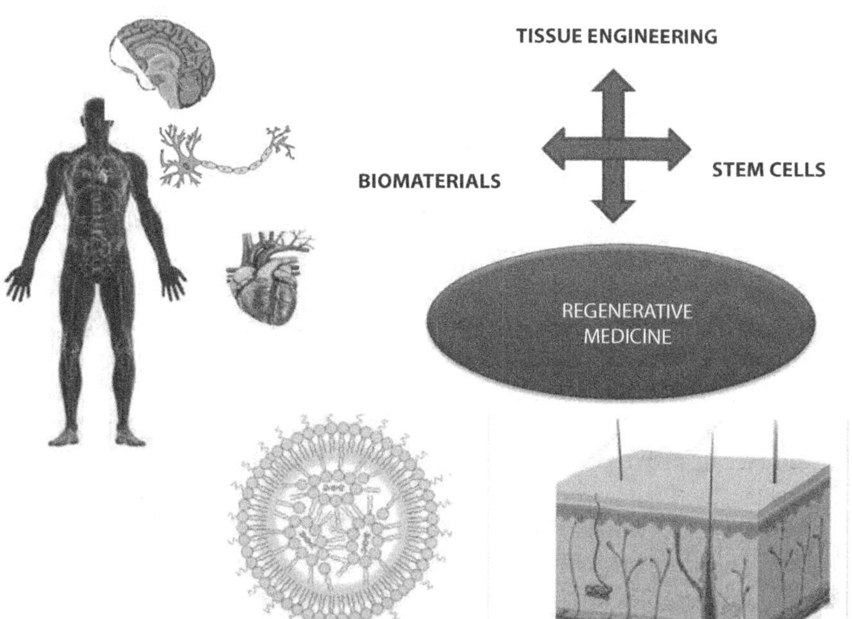

Figure 7.1 Figure summarizing the biomaterial scaffolds for regenerative medicines [34].

one of the adaptable and biocompatible materials that are transforming healthcares as the use of biomaterials scaffolds in regenerative medicines have been summarized in Figure 7.1 [34]. Different research works have revealed that the bio-composites are made of biodegradable polymers, such as polylactic acid, polycaprolactones that form natural fibers matrices, to make biocomposites useful in medical sector. These materials are also proxies of the nature of the biological tissues in the sense that they integrate biodegradable polymers with bioactive components, which facilitates the mucoadhesive nature of the cells and cell development. Biomaterials out of composite scaffolds offer mechanical stability, which facilitates tissue regeneration. Biocomposites with growth factors and stem cells aid parallel regeneration of tissues and are degraded in a controlled and safe manner [35, 36].

Biomineral composites are now developing into potential applications in an assortment of orthopedic biomedical applications, including bone regeneration, tendons and ligament repair, wound repair, dental applications, and cartilage and bone prosthetics. These materials are encouraging classes of materials in regenerative medicine. By combining inorganic and organic structure they parody the mining nature tissues of the natural world and the human body that are mineralized. Nevertheless, biomineral-based composite as a regenerative field still has different issues to usage. It is among the major issues in order to attain the demanded mechanical properties. The practicality of these composites in the practice of regenerative medicine is dependent on them in terms of power and robustness. However, sometimes it is quite hard to establish the best proportions of power and flexibility. Another new technology, which is emerging in most areas of orthopedics like bone regeneration, tendons and ligaments, wound healing, dental and cartilage and bone prostheses is biomineral based composites. Such composites have a good potential as a type of composites in regenerative medicine [37, 38]. They mimic the naturally found mineralized tissues in the environment together with the human body as they use the inorganic and the organic matter. The practicality of these composites in the practice of regenerative medicine is dependent on them in terms of power and robustness. However, the difficulty in getting the best outcome between power and flexibility can prove difficult [39]. The biomineral-based composites have already attained advanced solutions to various biomedical solutions such as orthopedics, bone and cartilage repair, tendons and ligaments repair, wound healing, and dentistry. These hybrid materials- are organic and inorganic based and resembles the structure of natural calcified tissue and that is why they are very suitable in biomaterials and scaffolds in tissue engineering [40, 41]. Regardless of their enormous potentials, there are certain challenges confronting the use of bio composites scaffolds in regenerative medicine. One of the biggest difficulties is the attainment of the desired physicochemical properties. These biocomposites rely on their strength and durability which is required in their performance. One of the most important aspects to be considered when using regenerative medicines using biocomposites biocompatibility also relates to the possible choice of the material to be used, the technique of fabrication, and the microstructure of the composite [42, 43] that directly influence the nature of mechano-chemical properties of biocomposites. The materials manufactured must be relatable to the human body and must not cause any negative effects on the immune responses [44]. Despite the consideration that there are biomineral based composites, which are biocompatible, others can be inflammatory or other host defense reactions [45, 46]. Research on new biodegradable polymer biocomposites is a developing business with massive potential of creating a more granted product as it seeks to include the entire value chain of the product like supply, processing and end-of-life of the product. It is most interesting to note that the existing problems associated with the blend of biofibers and biodegradable polymer composites, such as insufficient adhesion between the fiber and matrix, absorption by moisture, inability to control fiber orientation, lack of toughness to be improved, and improved fire resistance are all being actively addressed. The ability to biodegrade is a crucial factor to consider. The degradation of a biomineral-based composite utilized in regenerative medicine should align with the pace of tissue regeneration. Due to its biodegradable nature, it can gradually release ions or drugs such as growth factors and antibiotics, aiding in the regeneration of bone tissue [47, 48].

These are composites composed of petroleum-derived nonbiodegradable or biodegradable polymers (PLA, PHA) with natural or biofiber. Biocomposites made of crop/bioderived plastic (biopolymer) and plant-derived fiber (natural/biofibre) are referred to as "green composites" since they are probably more environmentally benign [49, 50]. Starch and cellulose are the most well-known renewable materials that can be used to create biodegradable polymers. One of the cheapest biodegradable materials on the market right now is starch. It is a polymer that can be used in a wide range of non-food sectors. For the production of cellulose plastics, cellulose derived from cotton and tree plants is used instead of petroleum feedstocks. The development of biodegradable plastics from vegetable oils such as sunflower, sesame and soybean oil is another topic that has drawn attention from all around the world [51].

Emerging from 3D printing, four dimensional printing technologies causes the structures that can change their structures in response to external stimuli. This innovative 4D printing technique incorporates intelligent materials and deformation design within the materials themselves, making it easier to convert design ideas into physical items. It enables the automatic construction of structures and ensures a smooth fusion of product design, manufacturing, and assembly by integrating time-sensitive changes into the 3D printing process [52]. The different external environmental factors like light, electric current, magnetic fields and different pH conditions causes the modification of shapes, different physiochemical properties on the printed structures unit. Four dimensional printing based on Magnetically controlled stimulus is a specialized technique that can be classified as magnetic thermal deformation, which depends on the SMP materials. This method takes advantage of the ability of SMPs to heat and conform, and cool and reform into shape using magnetic fields using power loss and thermal activation, and hence returned to its initial shape. This technique is based on the capabilities of the SMP to warm and shape, and then cool and reshape with the help of magnetic fields as a result of temperature-induced activation, in which an original shape can be put into a temporary one and back again [53, 54]. Further, such a drive system is lightweight, economical, and in addition to the exclusive memory shape and deformation, it possesses important potential and applications in different fields. Nonetheless, the traditional manufacturing of SMP has finite constraints on the usage and development of SMP deformable structures [55, 56].

7.3 Materials Employed as Biomaterials

7.3.1 Thermo-Responsive Materials

These materials operate through the Shape Memory Effect (SME) mechanism.

They change shape and deform when exposed to thermal energy or heat as a trigger. Shape memory polymers are preferred by many researchers due to their ease of printing [55].

7.3.2 Stimuli-Responsive Materials (Moisture)

Such materials are specific in nature and they response only in the presence of moisture. Water is often chosen by researchers because it is easily accessible and has numerous applications. Hydrogel is considered a smart material because of its high reactivity to water. SMPs can change their structures with change in moisture conditions. This is beneficial in environments such as greenhouses and grain silos, where it's essential to control humidity for optimal plant development and food preservation. For example, SMPs can be utilized to generate seals in response to humidity in silos, preventing moisture from entering and compromising the grains [56].

7.3.3 Photo/Electro/Magneto-Responsive Materials

These materials are influenced by light, electricity, and magnetic fields. For example, when exposed to natural light, photoresponsive chromophores are precisely integrated into polymer gels that expand and absorb light [57]. Similarly, when ethanol is present in an object, it evaporates, causing an increase in the object's volume and matrix. Additionally, magnetic nanosized particles are incorporated so that the system can be magnetically controlled [58, 59].

7.4 Integration of Biocomposites in 4D Printing Technology

4D printing represents an innovative development in different fields of additives manufacturing techniques extending beyond capabilities of traditional 3D printing. While conventional 3D printing allows users to control only the shape of the product, the three dimensional structures incorporates the time dimension to form 4D printing technology. There is targeted drug delivery approach of three dimensional structures as per their properties and functions. It enables the final shape or attributes of the object to change under different conditions and over time. While there are various interpretations of 4D printing, it is defined here as the use of a 3D printer to create objects that can alter their shape after they have been produced. Additionally, 4D printing is categorized into two types: single-material and multi-material 4D printing [60, 61].

In the field of shape-changing materials, two primary types are often studied: those that alter their shape and those with shape-memory capabilities. Shape memory describes the capability of materials to "recall" and revert to their initial form. Such characteristic allows the material's initial form to be modified and maintained in a new shape in response to external stimuli. The material remains in its altered form unless until a stimulus prompts it to revert to its original form [62, 63]. Shape memory polymers are products that have a two-part structure that is elastic and a transition part. On the other hand, shape-changing materials would need some external stimulus in order to initiate a shift in shape away from their original form and be reversed upon the removal of the stimulus [64]. The weakness can be overcome with the use of a composite method that strengthens the soft gel or polymer. This method includes composites with localized regions of two different polymers (one of these an SMP and the other non-reactive), using shear-aligned cellulose fibers or magnetically oriented rigid particles to control anisotropic swelling of hydrogels, and has developed thermally responsive shape-changing bilayers with continuous carbon fibers [65]. Natural fibers possess an anisotropic property and are sensitive to moisture, and this may restrict their applicability in structural works. However, the stimulation response through innovative activation by structured hierarchies pre-programmed, such as nastic fibrous motors, is pioneering a novel framework of bio-inspiration. This technique allows the natural fibers functional evolution to be used, which is beneficial to the development of hygromorphing (or hygroscopically powered morphing). An extraordinary hygroscopic actuation of nature, as applied in pine cones, wheat awns and other seed dispersal systems, offers opportunities of biomimicry and opens up new material capabilities. Moreover, the introduction of four dimensional techniques means that the material science backdrop is going through a metamorphosis to yield the bio composite on the basis of which the reactivity to the stimulus and adaptation shape through. Not only these 4D-printed bioarchitectures can modify its properties in relation to a specific environment, but it is also capable of competing with the versatility of biomaterials. Such feature of adapting has depicted, innovative biomedical applications, therapeutic and tissue based engineering, utilization. The benefits associated with the technology are also immense and they include, the capability to critically regulate space and time printing processes, the capability to produce intuitively stimulating materials to custom demands in line with the requirement of the patient. It is a new technology that holds the forefront to constitute the new generation of flexible, life-sustaining materials [66]. The 4D-printing technology has multiple challenges that impede its application in some fields of biomedicine. As an example, the resources used are not likely to be robust, tractable and stable, otherwise they may not hold the adverse conditions at the printing level. Furthermore, identifying materials that exhibit both shape-changing behavior and biocompatibility, along with suitable degradation rates to ensure patient safety, remains challenging [67]. Although 4D-printing technology can produce assemblies with varied characteristics by integrating multiple structure into a single printed form; it is frequently challenging to identify materials that can be joined without compromising the proposed properties of the final structure. Many current 4D printing methods either do not provide the level of accuracy and precision necessary for creating materials that perfectly fit within the human body, or they are excessively costly and time-consuming for large-scale production. This renders them unfeasible when addressing the requirements of a substantial number of patients [68].

7.5 4D Printing Utilizing Smart Polymeric Materials

In recent years, there has been an increasing curiosity in investigating new materials that can display particular reactions to external stimuli. Thus increase in importance of four dimensional technologies by expansion of sophisticated devices like sensors, and actuators used in both biomedical and electronic areas. Smart materials are also being explored (composite materials can change form, color or size in response to external parameter variations, such as light, heat, moisture and electrical or magnetic fields). The programmable materials in this array is 4D printing and this category. The process is much the same as traditional 3D printing, but the items produced with this new method can metamorphose into different shapes or subsume other features over time under the influence of exterior forces [68]. A variety of intelligent polymers, such as smart polymeric materials, hydrogels, are developed; nevertheless, these materials display specific drawbacks. Some intelligent polymers show little responsiveness for prolonged durations or have restricted reversibility in their changes. Despite the fact that additional developments are still required to achieve quick and precise modification of four-dimensional printed objects, 4D printing finds exclusive opportunities in different fields, such as textiles, aerospace, medicine, electronics and robotics [69].

7.5.1 Shape Memory Polymers

Smart materials like SMPs can remember their original shape even when placed in a temporary-shape which will be regained when of being exposed to external stimuli. SMPs then return to their original form after being deactivated and form a shape memory cycle. It is also possible to make shape memory polymer composites using SMPs, and adding different functional fillers. SMP has been adjusted to react more elements to the broader set of stimuli including temperature, electric current, light, magnetic field or chemical solutions. SMPs and shape memory polymer composites exhibit unique shape memory properties such as ability to memorize more than one shape, and irreversible deformations. As opposed to 3D printing technology, 4D can not only give shapeshifts but also hundreds of other functions such as; self-transformation, self-assembly and even the capability to self-heal in any programmed fashion (that includes shapes, properties and features of tools) with the support of its quality and precision and performance. It, however, is also a promising technology in various applications in life construction, agriculture, soft robotics (soft actuators), electronics (5D printing), smart furniture/architecture/aerospace/military industries, etc., and future is bright since SMPs have attractive characteristics of fast processing ability and a perfect fit to flexible additive manufacturing processes in comparison to other polymer type. A recent finding due to their reconfigurable processing properties that align easily with the needs of most additive manufacturing processes is the use of shape memory polymers as an outstanding choice in applications related to 4D printing [70].

7.6 Advantages of Biocomposites in 4D Printing

The development of four dimensional printing technology and the stimuli sensitive polymers has loudly become handy since the technology is bringing forth the age of innovations. The 4D printing technique is specifically relevant to the deliberate integration of Self-regulating materials and precise organized pattern of forms. This review highlights the outstanding development in the development of stimulus reactive polymers showing how they have been successfully used in manufacturing to create biomimetic structures with valuable implications in many aspects of biomedical use [71]. The impact of 4D printing has been explained and summarized in Figure 7.2. The development of the three dimensional printing technologies has changed the face of the personalized drug delivery systems. Certainly, the instances of 3D and 4D printing and their implementation in the given sphere of biology are numerous and ever evolving. The future studies in this field are set to come up with innovative solutions to tailored drug delivery and therapy.

There is the close study on smart polymers and it can be seen that they are soft materials, under the influence of heat and are capable of adapting to the alterations in their environment. In their work, the authors explain the formation, characterization and design of such polymers, paying attention to their use in the field of precision medicine: 3D printing, cell treatment and medical devices. In the article by Pineda-Castillo *et al.*, the authors discuss modern endovascular devices and the possibilities of shape memory polymers to produce solutions to the issue of total occlusion, especially

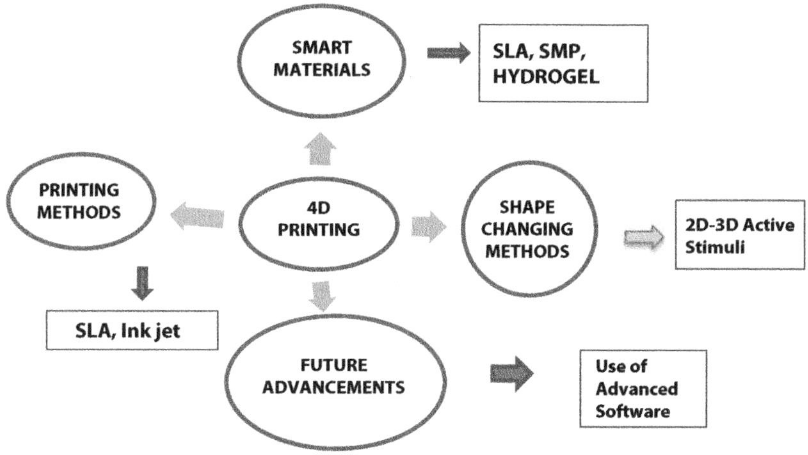

Figure 7.2 Figure summarizing the impact of four dimensional printing [71].

the application in embolzing intracranial aneurysms. Uboldi *et al.* present a self-expanding rod-shaped device that can be fabricated by using a film-coating technique on a lab scale using Polyvinyl Alcohol (PVA) that involves Hot Melt Extrusion, followed by fused filament fabrication, the technique that allows tightly controlling the diffusion and transportation processes. Afzali Naniz *et al.* dwell on the relevant developments in smart materials and prospects of their application in the production of biomimetic structure and actuation in 4D bioprinting that has been utilized therein to date both in biomedical research and pharmaceutical therapy [72]. Finally, Sirawit describes a novel approach to the hydrogel actuators 4D-printing because it is the way to introduce a new mechanism of adjusting the swelling of the hydrogel material to the demands of a certain task and the same time, create the structure similar to one that is demanded by a certain task quite closely [73]. Composite materials: Interaction with biological systems Because of existing components of the biocompatible elements in the composite materials, the interaction of the media with the biological systems can take place where each component can replace the drawbacks of the other component. The synergistic nature increases the consistency of tissues that might slow down adverse reactions and inflammation [74]. The production of composite materials may be richer and costlier than the production of single-materials. In addition, the components of the composite are incompatible as well not possible and it can influence the final performance of the materials at large. The presence of the composite materials has implied an enormous impact on the regenerative medicine, even though in some sense the utilization of the separate materials has been present since the dawn of time. The use of composite materials has led to dramatic changes in the quality of life, life span and outcome of recovery of the patients that have serious defects in their skin or bones [75]. Composite allows the choice of materials that are applicable to various purposes which allows to design the structures according to the requirements of various tissues. Composite materials are manufactured using biocompatible elements and therefore their interplay with biological systems produces a positive result with all elements completing failures of the other.

7.7 The Different Obstacles Faced During the Utilization of Biomineral Based Composite Materials

The process for creating biomineral-based composites is more intricate and pricier than that of single polymer materials. Occasionally, interactions between various components of the composites can directly impact their final performance. Composite materials, renowned for their strength, might be more vulnerable to impact damage than metals, which can expose their structural integrity [76]. Furthermore, forecasting the long-term behavior of composites in biological applications can be difficult due to potential variances arising from interactions among components. The main challenge during the fabrication of biocomposites for regenerative medicines is to represent the exact structure and biochemical properties of the existing tissues and also to get ensured about the incorporation of novel cellular organ with already existing tissue [77]. Another important parameter to be kept in mind that the biocomposites prepared must be ecofriendly and economical in nature so that they can favor pilot plant scale up also. During the fabrication of biocomposites based on regenerative medicines there must be utilization of standard and high grade quality raw materials as it have direct influence on good quality proceedings of complex processes thus to achieve such objective all the processes involved are verified on regular basis. Different studies revealed that the biomineral based materials must be biologically compatible unless they will be more susceptible to trigger adverse immunological reactions. The development of biocomposite-based regenerative medicines enhances the release of antibiotics and other growth factors, thus enhancing the biodegradability [78]. Another characteristic that is important to be considered is to fabricate biocomposites, which are composed of materials with appropriate elastic modulus to avoid collapse and stress shielding, thus providing proper vascularization [79]. For better and economic development of biocomposites, it is essential to manage all the regulatory criteria to ensure quality and safety measures. To maintain the sustainable regeneration for the environment, the fabrication process employed must meet the essential requirement.

7.8 Conclusion

The biocomposites are made ecofriendly and economical by considering important issues such as the use of standardized materials and removal of complex barriers. Because the regulations for these materials are so intricate, they often slow down widespread adoption. Lastly, it has been concluded that the use of biomaterials is significant to accomplish the sustainable future and such a new method develops the innovative clarifications that will be beneficial in the future [80, 81]. The chapter has concluded all the issues that are connected with the biocomposites which is supposed to be used in future.

AI disclosure statement: The author acknowledges the use of AI-based tools for minor language editing or grammatical correction. However, all intellectual contributions, analyses, and conclusions presented in this chapter are entirely the author's own.

References

1. Al Mahmud, M.Z., Exploring the versatile applications of biocomposites in the medical field. *Bioprinting*, 36, e00319, 2023 Dec 1.
2. Kim, S.H., Ki, M.R., Han, Y., Pack, S.P., Biomineral-Based Composite Materials in Regenerative Medicine. *Int. J. Mol. Sci.*, 25, 11, 6147, 2024 Jun 2.
3. Wang, Y., Cui, H., Esworthy, T., Mei, D., Wang, Y., Zhang, L.G., Emerging 4D printing strategies for next-generation tissue regeneration and medical devices. *Adv. Mater.*, 34, 20, 2109198, 2022 May.
4. Planell, J.A., Navarro, M., Engel, E., Developing targeted biocomposites in tissue engineering and regenerative medicine, in: *Biomedical Composites*, pp. 569–587, Woodhead Publishing, 2017 Jan 1.
5. Sahafnejad-Mohammadi, I., Karamimoghadam, M., Zolfagharian, A., Akrami, M., Bodaghi, M., 4D printing technology in medical engineering: a narrative review. *J. Braz. Soc. Mech. Sci. Eng.*, 44, 6, 233, 2022 Jun.

6. Campilho, R.D., Recent innovations in biocomposite products, in: *Biocomposites for High-Performance Applications*, pp. 275–306, Woodhead Publishing, 2017 Jan 1.
7. John, M.J., Thomas, S., Biofibres and biocomposites. *Carbohydr. Polym.*, 71, 3, 343–64, 2008 Feb 8.
8. Wang, H., Zhao, J., Luo, Z., Li, Z., Recent research developments of 4D printing technology for magnetically controlled smart materials: a review. *Magnetochemistry*, 9, 8, 204, 2023 Aug 14.
9. Hart, L.R., He, Y., Ruiz-Cantu, L., Zhou, Z., Irvine, D., Wildman, R., Hayes, W., 3D and 4D printing of biomaterials and biocomposites, bioinspired composites, and related transformers, in: *3D and 4D Printing of Polymer Nanocomposite Materials*, pp. 467–504, Elsevier, 2020 Jan 1.
10. Vatanparast, S., Boschetto, A., Bottini, L., Gaudenzi, P., New trends in 4D printing: a critical review. *Appl. Sci.*, 13, 13, 7744, 2023 Jun 30.
11. Oladapo, B.I., Adebiyi, A.V., Elemure, E.I., Microstructural 4D printing investigation of ultra-sonication biocomposite polymer. *J. King Saud Univ. Eng. Sci.*, 33, 1, 54–60, 2021 Jan 1.
12. Bodaghi, M., Wang, L., Zhang, F., Liu, Y., Leng, J., Xing, R., Dickey, M.D., Vanaei, S., Elahinia, M., Hoa, S.V., Zhang, D., 4D printing roadmap. *Smart Mater. Struct.*, 33, 11, 113501, 2024 Nov 1.
13. Ghosh, S., Chaudhuri, S., Roy, P., Lahiri, D., 4D Printing in biomedical engineering: A state-of-the-art review of technologies, biomaterials, and application. *Regener. Eng. Transl. Med.*, 9, 3, 339–65, 2023 Sep.
14. Mahmood, A., Perveen, F., Chen, S., Akram, T., Irfan, A., Polymer composites in 3D/4D printing: materials, advances, and prospects. *Molecules*, 29, 2, 319, 2024 Jan 9.
15. Rajput, S., Sharma, P.K., Malviya, R., Biomarkers and treatment strategies for breast cancer recurrence. *Curr. Drug Targets*, 24, 15, 1209–20, 2023 Nov 1.
16. Joshi, S., Rawat, K., Rajamohan, V., Mathew, A.T., Koziol, K., Thakur, V.K., Ass, B., 4D printing of materials for the future: Opportunities and challenges. *Appl. Mater. Today*, 18, 100490, 2020 Mar 1.
17. Osouli-Bostanabad, K., Masalehdan, T., Kapsa, R.M., Quigley, A., Lalatsa, A., Bruggeman, K.F., Franks, S.J., Williams, R.J., Nisbet, D.R., Traction of 3D and 4D printing in the healthcare industry: From drug delivery and analysis to regenerative medicine. *ACS Biomater. Sci. Eng.*, 8, 7, 2764–97, 2022 Jun 13.
18. Pourmasoumi, P., Moghaddam, A., Nemati Mahand, S., Heidari, F., Salehi Moghaddam, Z., Arjmand, M., Kühnert, I., Kruppke, B., Wiesmann, H.P., Khonakdar, H.A., A review on the recent progress, opportunities, and challenges of 4D printing and bioprinting in regenerative medicine. *J. Biomater. Sci., Polym. Ed.*, 34, 1, 108–46, 2023 Jan 2.
19. Mandal, A. and Chatterjee, K., 4D printing for biomedical applications. *J. Mater. Chem. B*, 12, 12, 2985–3005, 2024.

20. Bajpai, A., Baigent, A., Raghav, S., Brádaigh, CO., Koutsos, V., Radacsi, N., 4D printing: materials, technologies, and future applications in the biomedical field. *Sustainability*, 12, 24, 10628, 2020 Dec 19.
21. Soleymani, S. and Naghib, S.M., 3D and 4D printing hydroxyapatite-based scaffolds for bone tissue engineering and regeneration. *Heliyon*, 9, 9, 2023 Sep 1.
22. Van Osch, G.J., Brittberg, M., Dennis, J.E., Bastiaansen-Jenniskens, Y.M., Erben, R.G., Konttinen, Y.T., Luyten, F.P., Cartilage repair: past and future–lessons for regenerative medicine. *J. Cell Mol. Med. (JCMM)*, 13, 5, 732–810, 2009 May.
23. Thomas, S., Joseph, K., Malhotra, S.K., Goda, K., Sreekala, M.S. (Eds.), *Polymer composites, Biocomposites*, John Wiley & Sons, 2013 Nov 11.
24. Akampumuza, O., Wambua, P.M., Ahmed, A., Li, W., Qin, X.H., Review of the applications of biocomposites in the automotive industry. *Polym. Compos.*, 38, 11, 2553–69, 2017 Nov.
25. Arsuffi, B., Magrini, T., Champeau, M., Siqueira, G., Titotto, S., 4D printing of natural materials: A review. *Sustainable Mater.Technol.*, 12, e01346, 2025 Mar.
26. Damiati, L.A., Alsudir, S.A., Mohammed, R.Y., Majrashi, M.A., Albrahim, S.H., Alghamdi, F.O., Alamari, H.A., Alzaydi, M.M., 4D Printing in Skin Tissue Engineering: A Revolutionary Approach to Enhance Wound Healing and Combat Infections. *Bioprinting*, e00386, 2025 Jan 3.
27. Alghauli, M.A., Aljohani, R., Aljohani, W., Almutairi, S., Alqutaibi, A.Y., Evolution of Medical 3D Printing, Printable Biomaterials, Prosthetic and Regenerative Dental Applications. *Bioprinting*, 2025 Jan 23e00395.
28. Li, X., Yang, Y., Fan, Y., Feng, Q., Cui, F.Z., Watari, F., Biocomposites reinforced by fibers or tubes as scaffolds for tissue engineering or regenerative medicine. *J. Biomed. Mater. Res. Part A*, An Official Journal of The Society for Biomaterials, The Japanese Society for Biomaterials, and The Australian Society for Biomaterials and the Korean Society for Biomaterials, 102, 5, 1580–94, 2014 May.
29. Pina, S., Oliveira, J.M., Reis, R.L., Natural-based nanocomposites for bone tissue engineering and regenerative medicine: A review. *Adv. Mater.*, 27, 7, 1143–69, 2015 Feb.
30. Stupp, S.I., Biomaterials for regenerative medicine. *MRS Bull.*, 30, 7, 546–53, 2005 Jul.
31. Chun, H.J., Park, K., Kim, C.H., Khang, G. (Eds.), *Novel biomaterials for regenerative medicine*, Springer Singapore, 2018 Oct 24.
32. Bhat, S. and Kumar, A., Biomaterials in regenerative medicine. *J. Postgrad. Med. Educ. Res.*, 46, 2, 81–9, 2012 Jun 1.
33. Farag, M.M., Recent trends on biomaterials for tissue regeneration applications. *J. Mater. Sci.*, 58, 2, 527–58, 2023 Jan.
34. Abdulghani, S. and Mitchell, G.R., Biomaterials for *in situ* tissue regeneration: a review. *Biomolecules*, 9, 11, 750, 2019 Nov 19.

35. Seal, B.L., Otero, T.C., Panitch, A.J., Polymeric biomaterials for tissue and organ regeneration. *Mater. Sci. Eng.: R: Rep.*, 34, 4-5, 147–230, 2001 Oct 10.
36. Prakash, A., Malviya, R., Sridhar, S.B., Shareef, J., 4D printing in dynamic and adaptive bone implants: Progress in bone tissue engineering. *Bioprinting*, 28, e00373, 2024 Nov.
37. Amukarimi, S., Rezvani, Z., Eghtesadi, N., Mozafari, M., Smart biomaterials: From 3D printing to 4D bioprinting. *Methods*, 205, 191–9, 2022 Sep 1.
38. Place, E.S., Evans, N.D., Stevens, M.M., Complexity in biomaterials for tissue engineering. *Nat. Mater.*, 8, 6, 457–70, 2009 Jun.
39. Christensen, B., Alginates as biomaterials in tissue engineering, in: *Carbohydrate Chemistry: Chemical and Biological Approaches*, vol. 37, pp. 227–58, 2011 Dec 23.
40. Parenteau-Bareil, R., Gauvin, R., Berthod, F., Collagen-based biomaterials for tissue engineering applications. *Materials*, 3, 3, 1863–87, 2010 Mar 16.
41. Atala, A., Future perspectives in reconstructive surgery using tissue engineering. *Urol. Clin. North Am.*, 26, 1, 157–65, 1999 Feb 1.
42. Balint, R., Cassidy, N.J., Cartmell, S.H., Conductive polymers: Towards a smart biomaterial for tissue engineering. *Acta Biomater.*, 10, 6, 2341–53, 2014 Jun 1.
43. Khan, F. and Tanaka, M., Designing smart biomaterials for tissue engineering. *Int. J. Mol. Sci.*, 19, 1, 17, 2017 Dec 21.
44. Zhang, K., Wang, S., Zhou, C., Cheng, L., Gao, X., Xie, X., Sun, J., Wang, H., Weir, M.D., Reynolds, M.A., Zhang, N., Advanced smart biomaterials and constructs for hard tissue engineering and regeneration. *Bone Res.*, 6, 1, 31, 2018 Oct 22.
45. Ribeiro, C., Sencadas, V., Correia, D.M., Lanceros-Méndez, S., Piezoelectric polymers as biomaterials for tissue engineering applications. *Colloids Surf., B*, 136, 46–55, 2015 Dec 1.
46. Berthiaume, F., Maguire, T.J., Yarmush, M.L., Tissue engineering and regenerative medicine: history, progress, and challenges. *Annu. Rev. Chem. Biomol. Eng.*, 2, 1, 403–30, 2011 Jul 15.
47. Dzobo, K., Thomford, N.E., Senthebane, D.A., Shipanga, H., Rowe, A., Dandara, C., Pillay, M., Motaung, K.S., Advances in regenerative medicine and tissue engineering: innovation and transformation of medicine. *Stem Cells Int.*, 2018, 1, 2495848, 2018.
48. Koh, C.J. and Atala, A., Tissue engineering, stem cells, and cloning: opportunities for regenerative medicine. *J. Am. Soc. Nephrol.*, 15, 5, 1113–25, 2004 May 1.
49. Gomes, M.E., Rodrigues, M.T., Domingues, R.M., Reis, R.L., Tissue engineering and regenerative medicine: new trends and directions—a year in review. *Tissue Eng. Part B Rev.*, 23, 3, 211–24, 2017 Jun 1.
50. Pina, S., Ribeiro, V.P., Marques, C.F., Maia, F.R., Silva, T.H., Reis, R.L., Oliveira, J.M., Scaffolding strategies for tissue engineering and regenerative medicine applications. *Materials*, 12, 11, 1824, 2019 Jun 5.

51. Knopf-Marques, H., Pravda, M., Wolfova, L., Velebny, V., Schaaf, P., Vrana, N.E., Lavalle, P., Hyaluronic acid and its derivatives in coating and delivery systems: applications in tissue engineering, regenerative medicine and immunomodulation. *Adv. Healthcare Mater.*, 5, 22, 2841–55, 2016 Nov.
52. Chauhan, A., Rajput, S., Malviya, R., Sridhar, S.B., Production and Multidisciplinary Applications of Bioplastic: A Biodegradable and Eco-Friendly Alternative to Plastic. *Curr. Green Chem.*, 2025.
53. Li, Z., Xie, M.B., Li, Y., Ma, Y., Li, J.S., Dai, F.Y., Recent progress in tissue engineering and regenerative medicine. *J. Biomater. Tissue Eng.*, 6, 10, 755–66, 2016 Oct 1.
54. Wobma, H. and Vunjak-Novakovic, G., Tissue engineering and regenerative medicine 2015: a year in review. *Tissue Eng. Part B Rev.*, 22, 2, 101–13, 2016 Apr 1.
55. Harrison, R.H., St-Pierre, J.P., Stevens, M.M., Tissue engineering and regenerative medicine: a year in review. *Tissue Eng. Part B Rev.*, 20, 1, 1–6, 2014 Feb 1.
56. Esdaille, C.J., Washington, K.S., Laurencin, C.T., Regenerative engineering: a review of recent advances and future directions. *Regen. Med.*, 16, 5, 495–512, 2021 May 1.
57. Veeman, D., Sai, M.S., Sureshkumar, P., Jagadeesha, T., Natrayan, L., Ravichandran, M., Mammo, W.D., Additive manufacturing of biopolymers for tissue engineering and regenerative medicine: an overview, potential applications, advancements, and trends. *Int. J. Polym. Sci.*, 2021, 1, 4907027, 2021.
58. Chen, C., Dubin, R., Kim, M.C., Emerging trends and new developments in regenerative medicine: a scientometric update (2000–2014). *Expert Opin. Biol. Ther.*, 14, 9, 1295–317, 2014 Sep 1.
59. Gurtner, G.C., Callaghan, M.J., Longaker, M.T., Progress and potential for regenerative medicine. *Annu. Rev. Med.*, 58, 1, 299–312, 2007 Feb 18.
60. Şeker, S., Elçin, A.E., Elçin, Y.M., Advances in regenerative medicine and biomaterials, in: *Gene, Drug, and Tissue Engineering*, pp. 127–152, Springer US, New York, NY, 2022 Oct 28.
61. Zhang, J., Zhao, X., Liang, L., Li, J., Demirci, U., Wang, S., A decade of progress in liver regenerative medicine. *Biomaterials*, 157, 161–76, 2018 Mar 1.
62. Paschos, N.K., Brown, W.E., Eswaramoorthy, R., Hu, J.C., Athanasiou, K.A., Advances in tissue engineering through stem cell-based co-culture. *J. Tissue Eng. Regener. Med.*, 9, 5, 488–503, 2015 May.
63. Mariano, M., Pilate, F., de-Oliveira, F.B., Khelifa, F., Dubois, P., Raquez, J.M., Dufresne, A., Preparation of cellulose nanocrystal reinforced poly (lactic acid) nanocomposites through noncovalent modification with PLLA-based surfactants. *ACS Omega*, 2, 2678e88, 2017.
64. Nagalakshmaiah, M., Nechyporchuk, O., El-Kissi, N., Dufresne, A., Melt extrusion of polystyrene reinforced with cellulose nanocrystals modified

using poly [(styrene)-co-(2-ethylhexyl acrylate)] latex particles. *Eur. Polym. J.*, 2017, 91, 297e306, 2017.
65. Cui, H., Miao, S., Esworthy, T., Zhou, X., Lee, S.J., Liu, C., Yu, Z.X., Fisher, J.P., Mohiuddin, M., Zhang, L.G., *Adv. Drug Delivery Rev.*, 132, 252, 2018.
66. Langer, R. and Vacanti, J., Advances in tissue engineering. *J. Pediatr. Surg.*, 51, 1, 8–12, 2016 Jan 1.
67. Cheng, B., Lu, S.L., Fu, X.B., Regenerative medicine in China: main progress in different fields. *Mil. Med. Res.*, 3, 1–3, 2016 Dec.
68. Sultankulov, B., Berillo, D., Sultankulova, K., Tokay, T., Saparov, A., Progress in the development of chitosan-based biomaterials for tissue engineering and regenerative medicine. *Biomolecules.*, 9, 9, 470, 2019 Sep 10.
69. Horch, R.E., Kneser, U., Polykandriotis, E., Schmidt, V.J., Sun, J., Arkudas, A., Tissue engineering and regenerative medicine–where do we stand? *J. Cell. Mol. Med.*, 16, 6, 1157–65, 2012 Jun.
70. Ahmed, S. and Tomer, A. (Eds.), *Bionanocomposites in Tissue Engineering and Regenerative Medicine*, Woodhead Publishing, 2021 Jun 3.
71. Mishra, S., Sharma, S., Javed, M.N., Pottoo, F.H., Barkat, M.A., Alam, M.S., Amir, M., Sarafroz, M., Bioinspired nanocomposites: applications in disease diagnosis and treatment. *Pharm. Nanotechnol.*, 7, 3, 206–19, 2019 Jun 1.
72. Nakashima, M. and Akamine, A., The application of tissue engineering to regeneration of pulp and dentin in endodontics. *J. Endod.*, 31, 10, 711–8, 2005 Oct 1.
73. Rau, J.V., Antoniac, I., Cama, G., Komlev, V.S., Ravaglioli, A., Bioactive materials for bone tissue engineering. *Biomed Res. Int.*, 2016, 3741428, 2016 Jan 28.
74. Bai, Y., Wang, Z., He, X., Zhu, Y., Xu, X., Yang, H., Mei, G., Chen, S., Ma, B., Zhu, R., Application of bioactive materials for osteogenic function in bone tissue engineering. *Small Methods*, 8, 8, 2301283, 2024 Aug.
75. Stratton, S., Shelke, N.B., Hoshino, K., Rudraiah, S., Kumbar, S.G., Bioactive polymeric scaffolds for tissue engineering. *Bioact. Mater.*, 1, 2, 93–108, 2016 Dec 1.
76. Chen, Q., Zhu, C., Thouas, G.A., Progress and challenges in biomaterials used for bone tissue engineering: bioactive glasses and elastomeric composites. *Prog. Biomater.*, 1, 1–22, 2012 Dec.
77. Carrow, J.K. and Gaharwar, A.K., Bioinspired polymeric nanocomposites for regenerative medicine. *Macromol. Chem. Phys.*, 216, 3, 248–64, 2015 Feb.
78. Omidian, H., Wilson, R.L., Babanejad, N., Bioinspired polymers: transformative applications in biomedicine and regenerative medicine. *Life*, 13, 8, 1673, 2023 Aug 1.
79. Rana, A., Mittal, A., Vashist, C., Rajput, S., Sridhar, S.B., Malviya, R., Exploration of 4D Printing and its Applications in the Biomedical Sciences. *Curr. Pharm. Des.*, 2025.

80. Muthe, L.P., Pickering, K., Gauss, C., A review of 3D/4D printing of polylactic acid composites with bio-derived reinforcements. *Composites, Part C: Open Access*, 8, 100271, 2022 Jul 1.
81. Aftab, M., Ikram, S., Ullah, M., Khan, S.U., Wahab, A., Naeem, M., Advancement of 3D btowards 4D bioprinting for sustained drug delivery and tissue engineering from biopolymers. *J. Manuf. Mater. Process*, 9, 9, 2025 Aug 21.

8

Scalability and Commercialization of 3D/4D-Printed Biocomposites

Anjali Rana[1], Shivam Rajput[2], Ajesh Chauhan[1,2] and Rishabha Malviya[2*]

[1]Department of Pharmacy, Hindu College of Pharmacy, Sonipat, Haryana, India
[2]Department of Pharmacy, School of Medical and Allied Sciences, Galgotias University, Greater Noida, Uttar Pradesh, India

Abstract

Researchers are looking for alternatives to typical building materials to promote circular economy in construction. Biocomposites are in high demand for a variety of uses since they are an environmentally conscious substitute to synthetic composites. An eco-friendly substitute for synthetic fibers in biocomposite production, natural fibers, such as flax, hemp, and sisal, are seeing a rise in usage. In the dynamic field of additive manufacturing (AM), 3D printing is quickly rising to the top as a game-changing answer to conventional production limitations. The revolutionary process of 4D printing, also known as additive manufacturing, has recently attracted a lot of attention because of the drastic changes it could bring to the manufacturing business. By combining 4D printing with composite materials, the capacity to create composite things that can effortlessly transform into specific shapes in response to outside forces has been enhanced. Aerospace, military, medicine, architecture, consumer products, and textiles are just a few of the many industries that could benefit greatly from 4D-printed composites due to their unique properties. Biocomposites are the focus of this chapter, which also discusses their creation, the challenges of making them scalable for 3D/4D printing, their commercialization, and the future of 3D/4D printing.

Keywords: Biocomposites, additive manufacturing, 3D Printing, 4D printing, scalability, commercialization

**Corresponding author*: rishabhamalviya19@gmail.com; rishabha.malviya@galgotiasuniversity.edu.in

Rishabha Malviya and Mukesh Roy (eds.) Biocomposites in 3D and 4D Printing: Materials, Techniques, and Applications in Regenerative Medicine, (219–240) © 2026 Scrivener Publishing LLC

8.1 Introduction

Biocomposites have been a driving force in the development of sustainable materials, which have been accelerated in response to growing environmental concerns [1–3]. Composites are multiphase systems with a clearly defined interface that purposely combine at least two chemically different elements to provide them superior performance compared to their individual components [4]. Composites are substances that combine the best features of multiple different types of materials in order to achieve combined strengths that are greater than those of the individual components. One modern and environmentally friendly option is biocomposites, a composite material. Biodegradable reinforcements, including biofibers, and matrices make up biocomposites, which are highly recyclable and have a low environmental impact when disposed of in landfills. A new analysis [5] indicates that the worldwide biocomposite market is expected to grow from 2016 to 2024 at a CAGR of 11.8%. Biocomposite materials have a long history of practical application. The utilization of natural fibers and biodegradable matrixes enables the production of robust materials that address contemporary worries regarding sustainability and the reduction of carbon footprints. The mechanical strength and rigidity of biocomposites are derived from natural fibers, such as flax, hemp, and bamboo. A biodegradable matrix, made of either natural or synthetic polymers, improves the material's cohesiveness and ecological properties, on the other hand. Biocomposites, which are a combination of bio-based polymers with natural fibers or nanoparticles, have aroused attention in 3D and 4D printing. Several factors contribute to this, including the fact that biocomposites can substitute polymers derived from petroleum, their low environmental impact, and their low weight [6]. A biodegradable polymer matrix and fillers are the main ingredients in biocomposites.

A plethora of new manufacturing processes, including 3D and 4D printing, have found uses in fields as diverse as building materials, healthcare, transportation, and textiles. Additive manufacturing (AM), often called 3D printing, allows for the accurate fabrication of complicated structures by adding different elements in a certain sequence [7]. In contrast, "smart" materials can be created through 4D printing and manipulated by various environmental elements such as radiation, heat, solvents, electric current, pressure, pH, and magnetic fields to alter their structure, properties, and functions. 3D printing, also known as additive manufacturing, allows for the creation of intricate devices from biomaterials and bioinspired materials in accordance with designs that are unique to each patient [8]. With its many benefits over more conventional methods, 3D printing has emerged as a leading platform for additive manufacturing and an innovative perspective on composite development. Biocomposites produced by 3D printing technology demonstrate significant potential to substitute conventional non-biodegradable materials [9]. An innovative method for creating complex functional materials for passive adaptable structures, four-dimensional printing offers a fresh perspective on creating autonomous transducers that can sense their surroundings, respond to them, and transform the energy they receive into mechanical motion [10, 11]. Since they are composed of renewable, sustainable resources like natural fibers and act as smart materials that react to variations in humidity levels, hygromorph biocomposites (HBC) have been the subject of numerous 4D printing investigations [12]. Composites benefit from the mechanical properties enhanced by the natural fibers' dual role as sensors and actuators.

The combined use of 4D printing and composite materials has opened up new avenues for the production of composite objects with the ability to effortlessly assume preset shapes in response to external stimuli. This unique feature of 4D-printed composites makes them very promising for usage in many other industries, such as the consumer goods, fashion, medicine, aerospace, and military sectors, among many others. Novel 3D printing materials, machines, and processes have the potential to revolutionise a number of sectors and raise living standards around the world. The correct design of multi-material structures derived from mathematical modelling and innovative material interaction techniques allow 4D-printed objects to adapt and exhibit intelligent behaviour in reaction to certain stimuli [13]. The qualities of 3D- and 4D-printed composites are determined by the type of fiber used, the stiffness of the material, its tensile strength, and the interface that is connected to the polymer matrix. A multidisciplinary strategy known as translational research seeks to close the gap between laboratory-based scientific findings and their real-world applications [14]. It entails the conversion of knowledge acquired from fundamental scientific research into practical applications that can benefit individuals, communities, and society at large. Translational research fundamentally attempts to go beyond theoretical knowledge and actively implement scientific discoveries to enhance human health, well-being, and quality of life [15]. The integration of several disciplines, such as fundamental science, epidemiological research, health education, and implementation science, is essential for the effective translation of scientific discoveries into significant interventions and solutions. Translational research is essential for connecting scientific discovery with practical application. Translating knowledge acquired from fundamental science into practical applications fosters the advancement of novel therapies, diagnostics, and preventive strategies that enhance human health and well-being. Biocomposite materials encounter a number of difficulties, including insufficient fire and moisture resistance, incompatibility between fibers and matrix, logistical supply issues, undesirable fabrication, and a great deal of characteristic variability due to fiber inconsistency [16, 17]. Among the many topics covered in this analysis of commercial 4D bioprinters are their price, scalability, and accessibility, as well as potential solutions like the use of AI and better materials. It lays the groundwork for commercializing 4D bioprinting, which has so far only existed in research labs.

8.2 Composition of Biocomposites

Biocomposites consist of biopolymers reinforced with natural fibers. Researchers have been creating these materials as substitutes for traditional materials that may be non-renewable resistant to degradation, or produced through environmentally harmful techniques. Although industrial-scale manufacture of biocomposites is increasingly feasible, the longevity of these natural materials restricts their application in many situations. Two main components make up a biocomposites: a matrix ingredient (such as a resin or biopolymer) and reinforcing fibers (which can be either manmade or natural). Attaching fibers, transferring loads to them, and protecting them from mechanical injury and environmental deterioration all require a matrix. Reinforcing components, which might take the form of fibers, whiskers, particles, or flakes, are the main load-bearing elements [18]. The matrix serves as a mechanical support system in addition to connecting the reinforcing components [19]. Many common matrixes are starch-based polymers, polyhydroxyalkanoates, polylactic acid, and polybutylene succinate (PBS). This is because these materials are simple to process and have little impact on the environment. The material's strength and rigidity are improved by adding natural fibers like hemp, jute, flax, bamboo, or sisal. Nanomaterials, such as cellulose nanocrystals or nano clay, can further enhance the material's mechanical and thermal qualities. Biocomposites incorporate reinforcing fibers derived from environmentally friendly materials. Materials such as these can be found in nature, such as flax, hemp, and bamboo, or as agricultural waste, such as straw and husks, or as wood. The material's versatility is enhanced by the addition of these fibers, which increase its strength, stiffness, and endurance. Much progress has been made in the search for biodegradable substitutes for conventional petroleum-based materials thanks to the creation and application of biocomposites. Alginate, chitosan, and collagen are examples of natural polymers; hydroxyapatite, nanofibers, and cellulose are examples of synthetic polymers that can be combined to form biocomposites. Biocomposites rely on these naturally occurring polymers for their structural integrity. The combination of these two characteristics results in biocomposites that are ideal for numerous medical uses due to their adaptability, mechanical strength, and biocompatibility [20]. Biocomposites are materials that play a major role in medicine; they are formed when renewable elements such as starch, cellulose, or proteins are mixed with natural fibers to form a matrix [21]. It is particularly noteworthy that among these materials, fibers formed from hemp, flax, and jute are particularly significant. The reduction in fiber weight, improvement in mechanical strength, and insignificant environmental impact are all benefits. These are only some of the numerous advantages that these fibers provide. Biocomposites can be categorized in several ways based on the kind of natural fibers and matrices used to create them and how well they perform biologically and environmentally. A sustainable material with excellent mechanical performance and no environmental impact can be made by combining various natural fibers (like flax, hemp, or sisal) with biodegradable matrices (like PLA or PHA). Bioceramics and biocomposites, in particular, can be synthesized by interacting with living tissues. Durability, flexibility, and stability are all improved when biocomposites incorporate natural fibers [6]. When compared to synthetic fibers, natural ones are better because they are more accessible, cheaper, and better for the environment. They could have an animal, plant, or mineral origin. Biodegradable matrices are crucial for preserving biocomposites and improving natural fibers. Matrixes protect the fibers from outside influences and allow stress to transfer between them more easily. There will be less of an effect on the environment after biocomposites are used. The use of a biodegradable matrix allows it to dissolve in nature, which is how this is accomplished. Biodegradable polymers, either naturally occurring or synthesized, may make up these matrices. Biocomposites of good quality can only be made when the matrix polymers have low viscosities at processing temperatures, when the reinforcing fiber and matrix both have good mechanical properties, and when there is a firm attachment between the two.

8.3 Scalability: From Prototypes to Mass Production

The ability of a product to increase production volume without sacrificing quality or performance is known as scalability. Although this research demonstrates "dual" printing, it is possible to further boost scalability without significantly affecting either the cost or the amount of time required. When it originates to the essence of it, scalability encompasses a number of different characteristics, such as the flexibility of the design, the efficiency of the production process, and the cost-effectiveness. The technology that underpins 3D printing has undergone significant growth over the course of the past several years, shifting from the stage of prototyping on a small size to the stage of manufacturing and production being carried out on a large scale. One of the factors that has contributed to this progression is the scalability of filaments that are manufactured *via* 3D printing. Historically, the primary application of 3D printing was for prototyping, which enabled engineers and designers to rapidly prototype and test new concepts [22]. On the other hand, developments in filament technology and hardware for 3D printing have made it possible to move from prototyping on a small scale to manufacturing on a big scale. Scalability in four-dimensional printing is a considerable challenge to the widespread use of the technology in industry [23]. Specifically, this is driven by constraints imposed by the qualities of the material, the efficiency of the process, and the capabilities of the machine. Nylon, high-performance PLA, ABS, PETG, and carbon fiber-reinforced composites are just a few examples of the modern 3D printing filaments that have made scaling possible. A combination of these filaments' exceptional strength, heat resistance, and long-term durability makes them perfect for making finished products and functioning components. Industries, such as healthcare, aerospace, and automotive, have been revolutionized by the potential to enhance production while preserving critical material qualities for individual applications [24]. Advancements in 3D printing technology are propelling the scalability of filaments that can be 3D printed. Thanks to advancements like larger print beds, improved printing rates, and multi-extrusion systems, manufacturers can now print larger and more complicated structures or several components at once. Fused Deposition Modelling (FDM) and Selective Laser Sintering (SLS) are two examples of high-volume printing processes that enable batch manufacture of components. This, in turn, decreases manufacturing time and expenses.

Among the many benefits of 3D printing is the ability to raise manufacturing quantities without sacrificing customization. Traditional manufacturing technologies, such as injection moulding, necessitate a large initial investment, which makes it impossible to make adjustments or build low-volume, bespoke components. Now that scalable filaments can be 3D-printed, manufacturers may offer on-demand customization at any scale, allowing for personalized goods and limited production runs without costly retooling. The scalability of 3D printed filaments is creating new opportunities in sectors that require large production as well as personalization. Due to large-scale 3D printing and modern filament materials, manufacturers can now mass-produce high-quality, functioning items, surpassing the capabilities of prototypes altogether [25]. It is essential for the transformation of production methods and the expansion of additive manufacturing into new industries that 3D-printed filaments can be scaled up, especially with the rapid improvement of technology.

The incorporation of dynamic biomaterials and environmental inputs into 4D bioprinting structures allows them to evolve over time, presenting a groundbreaking approach to bio manufacturing. The use of stimuli-responsive biomaterials, such as shape-memory polymers, hydrogels, and composites, is central to this approach [26]. For manufacturing to be efficient and cost-effective, standardized procedures for material synthesis, printer operation, and product validation must be created. Modular printer designs that can handle various materials might be a better fit for a variety of industrial applications. Scaling up the use of self-healing hydrogels as wound dressings is a good example of a case study that shows how important it is for businesses and universities to work together. Using automated workflows and forming partnerships with material suppliers are two ways to make 4D bioprinting more appealing to the industry. In addition, if the government offered incentives for R&D in this area, its acceptance could be accelerated, making the leap from lab to market much easier. To sum up, 4D bioprinting has the potential to generate structures that are dynamic and adaptive, which could lead to a revolution in bio manufacturing [27]. This method can help bioprinters work better with existing standards, solve problems with materials and scalability, inspire new developments in commercial bioprinters, and bridge the gap between lab dreams and real-world production. The scalability of 4D printing is hindered by limitations in material properties, process efficiency, and machine performance, which is a big deal for its widespread industrial adoption. Scalability is an issue with Stratasys' Poly Jet technology because their patented smart materials are expensive. Materials, such as shape-memory polymers (SMPs) and hydrogels, are specifically engineered to function inside their systems.

Massive implementations would be financially unfeasible because to the high expense of producing these materials, despite their great degree of programmability [28]. Even if it is accurate, the layer-by-layer jetting method is too slow and energy intensive for industrial output. Progress in materials science and technology are essential for overcoming this challenge. Develop smart materials that are open-source or compatible with all devices to reduce dependence on proprietary options; use advanced deposition techniques like hybrid extrusion or continuous liquid interface production (CLIP) to speed up production without compromising resolution. The Multi Jet Fusion (MJF) technique developed by HP is highly scalable and voxel-based, enabling efficient fusion of powder beds with minimal waste [29]. Automating powder recovery and material management can improve production and make it more scalable for large production. Improving the scalability of printed components through increased lifetime and superior resolution could be achieved through the development of multi-curing systems. These systems combine UV curing with thermal or chemical curing processes. To lessen the impact of the scalability issue in industrial applications, printer designs can be modularized to facilitate large-scale production (Figure 8.1). This technique can bridge the gap between laboratory promise and industrial reality by fixing scalability and material concerns, boosting industry compatibility, and promoting innovation in commercial bioprinters.

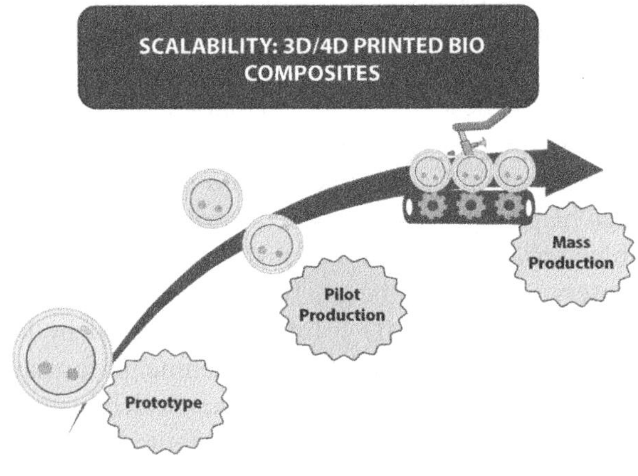

Figure 8.1 Industrial realization through innovation using 3D/4D-printed biocomposites.

8.4 Challenges in Scaling Up

Transitioning 3D- and 4D-printed biocomposites from laboratory research to industrial manufacturing encounters numerous substantial obstacles. The scalability of 3D/4D-printed biocomposites encounters numerous challenges that extend beyond conventional laboratory achievements, each substantiated by practical constraints in material, technological, economic, and regulatory domains. In order to make 4D-printed biomedical equipment more accessible, the manufacturing process needs to be both cost-effective and time-efficient [30]. However, a number of contemporary 4D printing methods are laborious and have restricted scalability, rendering them unsuitable for large-scale manufacturing. To solve the problem of 4D printing not being able to handle large orders, new printing technologies are needed that can make structures faster while using less material [31]. Automating 4D printing and integrating it with other manufacturing methods, such as injection moulding or additive manufacturing, could further reduce costs and improve production efficiency, thus researchers should investigate this possibility [32]. When it comes to moving 4D bioprinting from labs to factories, scalability is still a major hurdle. Material synthesis, printer operation, and product validation must all adhere to established norms for production to be both economical and efficient. Modular printer designs that can handle different kinds of materials could make them more adaptable to different kinds of industrial demands [33].

Biocomposites face a significant obstacle in the combustibility of biofibers. Ceramics, silicone, phenolics, glass matting, and ablatives are among the additives and coatings that can be utilized to decrease the flammability of biocomposites [34]. Many 3D and 4D printing techniques are inherently time-intensive and optimized for accuracy rather than rapidity, making them appropriate for prototyping but unsuitable for mass production. Issues like as nozzle obstruction, inconsistent fiber distribution, warping during fabrication, and inadequate fiber–matrix adhesion become increasingly evident with higher production numbers, diminishing product reliability. The effectiveness of biomedical devices and constructions that are 4D-printed relies heavily on the materials that are used. Hydrogels, shape-memory polymers, and stimuli-responsive materials are examples of smart materials that, when exposed to environmental signals, must display the intended behaviour [35]. Nonetheless, locating materials with the requisite qualities might be arduous. There is a possibility that the mechanical properties of a material, which include its strength, flexibility, and durability, might not always meet the requirements for a particular biomedical application. Ongoing research is being conducted on biocompatible smart materials, and it can be difficult to identify materials that not only meet high biocompatibility standards but also demonstrate the requisite shape-changing characteristics. The final product's biocompatibility may also be affected by the processing steps used in 4D printing. A significant difficulty is the printing speed and production efficiency, as most laboratory printers' function slowly and fail to satisfy the requirements of large-scale manufacturing [36]. 4D printing poses a significant challenge in biomedical engineering when it comes to scalability, since printed structures often need to be mass-produced or created at different sizes to meet the needs of different patients. Additionally, in order to support the widespread deployment of 4D-printed biomedical equipment, the production technique must be both economically viable and time-efficient [37].

Envision TEC's DLP technology has unique scalability problems because it uses UV-curable photopolymers, which are costly and not very durable when it comes to mechanical use. These materials are great for applications that need a lot of precision, but they cannot be used for big or heavy-duty jobs since they are sensitive to things like UV light and humidity. Research into multi-curing systems that combine UV curing with thermal or chemical curing procedures could improve the strength and resolution of printed items. This would make them more scalable. In order to produce a consistent light intensity across larger construction areas, scaling up the DLP process necessitates further advancements in light projection technology, like laser-array systems or complicated optics. The outstanding degree of detail for which DLP is renowned could be maintained as larger quantities of material cured simultaneously [33]. By improving materials, optimizing processes, and making machines more scalable, 4D printing can overcome these technical obstacles and achieve manufacturing on an industrial scale. This might have a big impact on industries including aerospace, healthcare, and infrastructure.

8.5 Commercialization of Biocomposites

The commercialization of 3D- and 4D-printed biocomposites is a new but complicated process that links research in the lab with real-world uses. The development of increasingly effective methods of 3D printing has paved up new paths for the production of biocompatible polymers that have the potential to be used in medical applications, particularly in the domains of tissue engineering and regenerative medicine. Biocomposites are made up of a biodegradable polymer matrix and one or more additives. The combination of these parts makes the mechanical properties better. Biocomposites made from natural fibers, such as flax, sisal, hemp, and others, are gaining popularity as an eco-friendly substitute for synthetic fibers [31]. The main markets for biocomposites are the building and automotive industries, where they are mostly used. The major obstacle to using biocomposites instead of traditional composites is ensuring that they are just as structurally and functionally stable [38, 39], despite the growing interest in these materials. Biocomposites are expected to see an uptick in commercialization and market potential as a result of rising consumer and producer awareness of environmental and social issues, the identification of new applications, and advancements in processing technology. The underappreciation of R&D efforts in underdeveloped nations, despite the abundance of bio-based materials, has been a key barrier to the economic viability of biocomposites. Numerous industrialized countries, particularly in Europe, have effectively overcome this challenge and have established themselves as frontrunners in this domain [40–42]. Researchers in materials science still have trouble setting standards for biocomposites, especially when it comes to cost and performance. Setting product standards that guarantee consistent performance is also very important [43–45].

In recent years, industries like healthcare, automotive, aerospace, construction, and packaging have been more interested in 3D-printed biocomposites for making parts that are light, good for the environment, and cheap. 4D printing opens up new possibilities in biomedical implants, self-healing structures, and adaptive packaging solutions by enabling smart biocomposites that react to environmental stimuli like moisture, temperature, or pH, greatly increasing economic value [46]. Due to this special feature, the composites made with 4D printing could be employed in many fields, including aerospace, military, biomedical, architecture, consumer products, and fashion. High production costs, slower printing rates, and differences in materials still make it hard to use on a broad scale [47, 48]. Although commercially accessible 4D bioprinters, such those made by RegenHU and Cellink, have shown great potential, they also come with a number of drawbacks. High costs, limited compatibility with different materials, and the difficulty of programming dynamic reactions are some of the biggest problems [49, 50]. If scientists keep studying materials, businesses keep looking for ways to cut costs, and governments keep making rules that help, these technologies will probably migrate from specialist markets to mainstream ones soon.

8.6 Emerging Trends in the 3D/4D Printing of Biocomposites Made from Natural Fibers

3D/4D printing techniques will advance due to ongoing research aimed at integrating innovative printing technologies, resulting in expedited and cost-effective product production, portable apparatus, and improved printing resolution [51]. Despite the revolutionary promise of 4D bioprinting, achieving high precision and biocompatibility—two factors that are crucial for sophisticated medicinal applications—is a huge challenge. Improving commercial 4D printers to address accuracy issues should centre on integrating stimuli-responsive material functions with high-resolution techniques, similar to those used in SLA-based systems [52]. Biocompatibility can be improved by utilizing materials like thermos-responsive polymers, which are less prone to eliciting unpleasant reactions [53, 54]. This section expands the scope to include hydrogel-based materials and cellulose-based composites so that we may look at expected trends in a wider range of applications. With a projected compound annual growth rate (CAGR) of more than 39% from 2019 to 2022, the global 4D printing market is anticipated to reach over USD 162 million, primarily driven by the aerospace and military, healthcare, automotive, construction, apparel, and packaging sectors. Researchers have shown that 4D printing could provide computer-generated materials the ability to incorporate sensors and actuators directly into them [55]. Therefore, the system does not require any external electromechanical systems. By incorporating this feature, modern electromechanical systems can drastically cut down on the number of components needed for a building, the amount of time and energy spent on assembly, and the frequency of devices that fail. A novel approach to using local resources to create individualized, self-sufficient, shape-morphing structures is offered by 4D-printed biocomposites. Solar tracking systems for architecture [58, 59], adaptive underwater systems for marine engineering [56], shading systems, architectural skin systems, and solar tracking systems for artificial reefs [57], all of which can be used in a variety of settings where independence is crucial, even in the most hostile of settings like war zones [60, 61].

Hygromorph biocomposites, a new class of smart materials made of natural fibers, can be used in 4D printing to make structures and systems with unique properties that can change shape. The bilayer architecture of these biocomposites mimics that of biological hydraulic actuators (like pine cones) and allows them to respond to changes in humidity or a mix of temperature and moisture [62, 63]. The ability to precisely control the orientation and distribution of fibers opens up a world of possibilities for

Table 8.1 4D printing techniques with their industrial applications, MOA, scalability issues and its solution.

S. no.	Techniques	Mechanism of action	Scalability issues	Solution	Applications	References
1.	Stereolithography (SLA)	It uses photopolymerization to transform liquid photopolymer resins into solid three-dimensional objects, incrementally, utilizing ultraviolet (UV) light.	Expensive tools and supplies, restricted material compatibility	By presenting economical alternatives for photo polymeric resins and enhancing material diversity for SLA procedures.	Tissue Engineering, dental and orthopaedic implants, Automotive and aerospace, scaffolds	[67, 68]
2.	Fused Deposition Modelling (FDM)	Intelligent polymers offer shape memory or dynamic reactivity when melting and extruding thermoplastic filaments to produce layers.	Limited resolution for complex designs, anisotropic mechanical properties	Improve resolution through nozzle design optimization, multi-nozzle system integration, and composite filament development.	Rapid prototyping, aerospace, adaptive implants, biomedical, drug delivery systems	[69, 70]

(Continued)

Table 8.1 4D printing techniques with their industrial applications, MOA, scalability issues and its solution. (*Continued*)

S. no.	Techniques	Mechanism of action	Scalability issues	Solution	Applications	References
3.	Direct Ink Writing (DIW)	Layer-by-layer extrusion-based deposition of bioinks facilitates high-resolution printing with dynamic characteristics.	Restricted structural integrity, reduced printing velocity, challenges in managing viscous bioinks	Improve ink compositions for increased printability and stability, and create extrusion systems with faster speeds.	Bio-actuators, soft tissue engineering, flexible electronics	[71, 72]
4.	Selective Laser Sintering (SLS)	Uses a laser to combine powdered materials, allowing for the creation of things with complex forms and properties that respond to stimuli.	Reduced production velocity, elevated equipment expenses, and restricted biocompatibility of powdered substances	Investigate hybrid printing methodologies, emphasizing material innovation to enhance biocompatibility and decrease expenses.	Agriculture sector, aerospace and automotive industries, smart materials and load-bearing structures	[73, 74]

hygromorph biocomposites made possible by additive manufacturing through 4D printing. According to the literature, there is a lack of systematic mechanical characterization of 3D-printed biocomposites. Natural fiber composites can be printed continuously using in-nozzle impregnation or customized printers with pre-impregnated filaments, thanks to a new technology [64]. Biocomposites, such as those made from continuous strands of flax or jute, have far superior mechanical characteristics to composites made from natural discontinuous fibers [65]. Natural fibers that were pre-impregnated showed a considerable improvement in performance. The composite made of PLA with 6% jute by volume had a strength of 57.1 ± 5.33 MPa and a stiffness of 5.11 ± 0.41 GPa, whereas the composite made of PLA with 30% flax by content had a strength of 253.7 ± 15 MPa and a stiffness of 23.3 ± 1.8 GPa. In addition, thermos compressed biocomposites with the same fiber content as 3D-printed continuous flax/PLA biocomposites are strong and have tensile stiffness that is equivalent. These technologies could be part of future plans to enhance the performance of biocomposites made from continuous natural fibers. The improved production of filaments using low-twist yarn with diminished linear weight presents considerable possibilities at the material level [66]. The development of processing equipment that is both individualized and economical possesses the potential to bring about an improvement in the performance of natural fiber composites that have been manufactured using 3D printing (Table 8.1).

8.7 Conclusion

Technological advancements and expanded industrial applications are poised to facilitate significant growth in 3D/4D printing utilizing composite materials. The adaptability and distinctive properties of 3D-printed composite materials have attracted considerable interest across multiple industries, each possessing specific performance criteria and design limitations. 4D bioprinting presents a revolutionary method for bio fabrication by incorporating dynamic biomaterials and environmental inputs to create structures that evolve over time. Scalability continues to be a major obstacle to the industrial adoption of 4D bioprinting from experimental settings. The creation of standardized procedures for the synthesis of materials, the operation of printers, and the validation of products are essential for efficient and cost-effective manufacturing. A more versatile solution for a wide range of industrial needs could be to adopt printer designs that are modular and can accommodate different types of materials. To reduce biocomposites' degradation, future studies should look into scalable and cost-effective pretreatment strategies for biofibers and fillers. The primary objective of this study is to analyze the difficulties encountered during the last ten years in the process of expanding prototype modelling from a small to a large scale. Additionally, sought to formulate innovative design solutions for optimizing the printing process, thereby minimizing time and enhancing production without sacrificing efficiency.

AI disclosure statement: The author acknowledges the use of AI-based tools for minor language editing or grammatical correction. However, all intellectual contributions, analyses, and conclusions presented in this chapter are entirely the author's own.

References

1. Gurunathan, T., Mohanty, S., Nayak, S.K., A review of the recent developments in biocomposites based on natural fibres and their application perspectives. *Composites, Part A*, 77, 1–25, 2015 Oct 1.
2. Ryłko-Polak, I., Komala, W., Białowiec, A., The reuse of biomass and industrial waste in biocomposite construction materials for decreasing natural resource use and mitigating the environmental impact of the construction industry: a review. *Materials.*, 15, 12, 4078, 2022 Jun 8.
3. Saini, K., Matsagar, V.A., Kodur, V.R., Recent advances in the use of natural fibers in civil engineering structures. *Constr. Build. Mater.*, 411, 134364, 2024 Jan 12.
4. Agarwal, B.D., Broutman, L.J., Chandrashekhara, K., *Analysis and performance of fiber composites*, John Wiley & Sons, New Jersey, USA, 2017 Oct 30.
5. Ahmad, W., Jaiswal, A.K., McCormack, S.J., Byrne, A., An overview of the resource availability in the European region to develop eco-efficient biocomposites for construction applications.
6. Bălțatu, M.S., Vizureanu, P., Sandu, A.V., Achitei, D.C., Nergis, D.D., Perju, M.C., Nabiałek, M., Biocomposites: Materials, Properties, and Applications.
7. Appuhamillage, G.A., Ambagaspitiya, S.S., Dassanayake, R.S., Wijenayake, A., 3D and 4D printing of biomedical materials: Current trends, challenges, and future outlook. *Explor. Med.*, 5, 1, 17–47, 2024 Feb 6.
8. Hart, L.R., He, Y., Ruiz-Cantu, L., Zhou, Z., Irvine, D., Wildman, R., Hayes, W., 3D and 4D printing of biomaterials and biocomposites, bioinspired composites, and related transformers, in: *3D and 4D Printing of Polymer Nanocomposite Materials*, pp. 467–504, Elsevier, Amsterdam, Netherlands, 2020 Jan 1.

9. Shoeb, M., Kumar, L., Haleem, A., Biocomposites based 3D printing, processes and applications: a brief review. *Int. J. Sustain. Mater. Struct. Syst.*, 5, 4, 326–41, 2021.
10. Tibbits, S., 4D printing: multi-material shape change. *Archit. Des.*, 84, 1, 116–21, 2014 Jan.
11. Bodaghi, M., Van Hoa, S., Gries, T., Le Duigou, A., Tadesse, Y., Yao, L., Zolfagharian, A., Focus on 4D materials design and additive manufacturing. *Smart Mater. Struct.*, 32, 110401, 4pp, 2023 Jan 1.
12. De Kergariou, C., Demoly, F., Perriman, A., Le Duigou, A., Scarpa, F., The design of 4D-printed hygromorphs: state-of-the-art and future challenges. *Adv. Funct. Mater.*, 33, 6, 2210353, 2023 Feb.
13. Rana, A., Mittal, A., Vashist, C., Rajput, S., Sridhar, S.B., Malviya, R., Exploration of 4D Printing and its Applications in the Biomedical Sciences. *Curr. Pharm. Des.*, 2025.
14. Fenby-Hulse, K., Situating early career researchers within a dynamic research and innovation ecosystem, in: *Research Impact and the Early Career Researcher*, pp. 179–198, Routledge, New York, 2019 May 24.
15. Frese, M. and Gielnik, M.M., The psychology of entrepreneurship: Action and process. *Annu. Rev. Organ. Psychol. Organ. Behav.*, 10, 1, 137–64, 2023 Jan 23.
16. Jawaid, M.H. and Khalil, H.A., Cellulosic/synthetic fibre reinforced polymer hybrid composites: A review. *Carbohydr. Polym.*, 86, 1, 1–8, 2011 Aug 1.
17. Yang, Y., Dang, B., Wang, C., Chen, Y., Chen, K., Chen, X., Li, Y., Sun, Q., Ultrastrong lightweight nanocellulose-based composite aerogels with robust superhydrophobicity and durable thermal insulation under extremely environment. *Carbohydr. Polym.*, 323, 121392, 2024 Jan 1.
18. Rinaudo, M., Main properties and current applications of some polysaccharides as biomaterials. *Polym. Int.*, 57, 3, 397–430, 2008 Mar.
19. Kim, Y.Y., Hore, K., Hall, S.R., Walsh, D., Controlled Nanoparticle Formation by Enzymatic Deshelling of Biopolymer-Stabilized Nanosuspensions. *Small.*, 5, 8, 913–8, 2009 Apr 20.
20. Al Mahmud, M.Z., Exploring the versatile applications of biocomposites in the medical field. *Bioprinting.*, 36, e00319, 2023 Dec 1.
21. Swetha, M., Sahithi, K., Moorthi, A., Srinivasan, N., Ramasamy, K., Selvamurugan, N., Biocomposites containing natural polymers and hydroxyapatite for bone tissue engineering. *Int. J. Biol. Macromol.*, 47, 1, 1–4, 2010 Jul 1.
22. Solanki, A., Ranganath, M.S., Singholi, A.K., Optimizing 3D/4D Printing for Large-Scale Prototyping: A Comprehensive Review on Speed, Cost, and Quality Challenges. *J. Inst. Eng. India Ser. C*, 3, 1–4, 2025.
23. Parida, S.K., Processing Techniques and Challenges of 3D-, 4D-, and 5D-Printed Polymer Composites, in: *Polymer Nanocomposites for 3D, 4D and 5D Printing: Fundamental to Applications*, pp. 87–108, Springer Nature Singapore, Singapore, 2025 May 4.

24. Burela, R.G., Kamineni, J.N., Harursampath, D., Multifunctional polymer composites for 3D and 4D printing, in: *3D and 4D Printing of Polymer Nanocomposite Materials*, vol. 1, pp. 231–57, Elsevier, Amsterdam, Netherlands, 2020 Jan.
25. Stansbury, J.W. and Idacavage, M.J., 3D printing with polymers: Challenges among expanding options and opportunities. *Dent. Mater.*, 32, 1, 54–64, 2016 Jan 1.
26. Kumar, P., Suryavanshi, P., Dwivedy, S.K., Banerjee, S., Stimuli-responsive materials for 4D printing: mechanical, manufacturing, and biomedical applications. *J. Mol. Liq.*, 410, 125553, 2024 Sep 15.
27. Deng, D., Kwok, T.H., Chen, Y., Four-dimensional printing: design and fabrication of smooth curved surface using controlled self-folding. *J. Mech. Des.*, 139, 8, 081702, 2017 Aug 1.
28. Khare, V., Sonkaria, S., Lee, G.Y., Ahn, S.H., Chu, W.S., From 3D to 4D printing–design, material and fabrication for multi-functional multi-materials. *Int. J. Precis. Eng. Manuf. Green Technol.*, 4, 3, 291–9, 2017 Jul.
29. Hamza, H., Dental 4D printing: an innovative approach. *Innovation.*, 1, 9, e17, 2018 Oct.
30. Silva, L.P., Current trends and challenges in biofabrication using biomaterials and nanomaterials: Future perspectives for 3D/4D bioprinting, in: *3D and 4D Printing in Biomedical Applications: Process Engineering and Additive Manufacturing*, pp. 373–421, Wiley, Hoboken, New Jersey, USA, 2019 Feb 19.
31. Siddiqui, V.U., Sapuan, S.M., Ahmad, S., Noorazlan, L.H., Selvan, N.M., Ilengo, S., Mailvaganan, P., Ravindran, V.K., Sustainable Biocomposites: Characterization and Applications in 3D and 4D Printing, in: *Additive Manufacturing for Biocomposites and Synthetic Composites*, pp. 1–19, CRC Press, Florida, USA, 2023 Dec 27.
32. Ramezani, M. and Mohd Ripin, Z., 4D printing in biomedical engineering: advancements, challenges, and future directions. *J. Funct. Biomater.*, 14, 7, 347, 2023 Jun 29.
33. Mathur, V., Agarwal, P., Kasturi, M., Varadharajan, S., Devi, E.S., Vasanthan, K.S., Transformative bioprinting: 4D printing and its role in the evolution of engineering and personalized medicine. *Discover Nano.*, 20, 1, 1–57, 2025 Dec.
34. Chauhan, A., Rajput, S., Malviya, R., Sridhar, S.B., Production and Multidisciplinary Applications of Bioplastic: A Biodegradable and Eco-Friendly Alternative to Plastic. *Curr. Green Chem.*, 2025.
35. Amukarimi, S., Rezvani, Z., Eghtesadi, N., Mozafari, M., Smart biomaterials: From 3D printing to 4D bioprinting. *Methods.*, 205, 191–9, 2022 Sep 1.
36. Demoly, F., Dunn, M.L., Wood, K.L., Qi, H.J., André, J.C., The status, barriers, challenges, and future in design for 4D printing. *Mater. Des.*, 212, 110193, 2021 Dec 15.

37. Alpár, T., Hasan, K.F., Horváth, P.G., Introduction to biomass and biocomposites, in: *Value-Added Biocomposites*, pp. 1–33, CRC Press, Florida, USA, 2021 Sep 6.
38. Satyanarayana, K.G., Arizaga, G.G., Wypych, F., Biodegradable composites based on lignocellulosic fibers—An overview. *Prog. Polym. Sci.*, 34, 9, 982–1021, 2009 Sep 1.
39. Andrew, J.J. and Dhakal, H.N., Sustainable biobased composites for advanced applications: recent trends and future opportunities–A critical review. *Composites, Part C: Open Access*, 7, 100220, 2022 Mar 1.
40. Faruk, O., Bledzki, A.K., Fink, H.P., Sain, M., Biocomposites reinforced with natural fibers: 2000–2010. *Prog. Polym. Sci.*, 37, 11, 1552–96, 2012 Nov 1.
41. Cardon, L.K., Ragaert, K.J., De Santis, R., Gloria, A., Design and fabrication methods for biocomposites, in: *Biomedical Composites*, pp. 17–36, Woodhead Publishing, Cambridge, United Kingdom, 2017 Jan 1.
42. Dahal, R.K., Acharya, B., Dutta, A., Mechanical, thermal, and acoustic properties of hemp and biocomposite materials: a review. *J. Compos. Sci.*, 6, 12, 373, 2022 Dec 6.
43. Aiduang, W., Jinanukul, P., Thamjaree, W., Kiatsiriroat, T., Waroonkun, T., Lumyong, S., A comprehensive review on studying and developing guidelines to standardize the inspection of properties and production methods for mycelium-bound composites in bio-based building material applications. *Biomimetics.*, 9, 9, 549, 2024 Sep 11.
44. Vitola, L., Gendelis, S., Sinka, M., Pundiene, I., Bajare, D., Assessment of plant origin by-products as lightweight aggregates for bio-composite bounded by starch binder. *Energies.*, 15, 15, 5330, 2022 Jul 22.
45. Özdemir, E., Saeidi, N., Javadian, A., Rossi, A., Nolte, N., Ren, S., Dwan, A., Acosta, I., Hebel, D.E., Wurm, J., Eversmann, P., Wood-veneer-reinforced mycelium composites for sustainable building components. *Biomimetics.*, 7, 2, 39, 2022 Mar 31.
46. Komal, U.K., Yadav, M., Siddiqui, A.H., Dewangan, M.K., 4D printing of composites: Challenges, opportunities, and paths to commercialization, in: *4D Printing of Composites*, vol. 1, pp. 111–38, Woodhead Publishing, Sawston, Cambridge, 2024 Jan.
47. Howard, S., The challenges and opportunities of commercialisation. *New Z. Sci. Rev.*, 68, 1, 33–6, 2011.
48. Ballardini, R.M., Lindman, J., Ituarte, I.F., Co-creation, commercialization and intellectual property-challenges with 3D printing. *Eur. J. Law Technol.*, 7, 3, 1–37, 2016.
49. Yang, Q., Gao, B., Xu, F., Recent advances in 4D bioprinting. *Biotechnol. J.*, 15, 1, 1900086, 2020 Jan.
50. Noroozi, R., Arif, Z.U., Taghvaei, H., Khalid, M.Y., Sahbafar, H., Hadi, A., Sadeghianmaryan, A., Chen, X., 3D and 4D bioprinting technologies: a game changer for the biomedical sector? *Ann. Biomed. Eng.*, 51, 8, 1683–712, 2023 Aug.

51. Sadasivuni, K.K., Deshmukh, K., AlMaadeed, M.A. (Eds.), *3D and 4D printing of polymer nanocomposite materials: processes, applications, and challenges*. Elsevier, Amsterdam, Netherlands, 2019.
52. Le Duigou, A., Correa, D., Ueda, M., Matsuzaki, R., Castro, M., 3D and 4D printing of natural fibre biocomposites, in: *FabAddComp JST Fabrication additive des Composites 2020*, 2020.
53. Correa, D., Poppinga, S., Mylo, M.D., Westermeier, A.S., Bruchmann, B., Menges, A., Speck, T., 4D pine scale: biomimetic 4D printed autonomous scale and flap structures capable of multi-phase movement. *Philos. Trans. R. Soc. A.*, 378, 2167, 20190445, 2020 Mar 20.
54. Sullivan, F., Advances in 4D printing—next paradigm in manufacturing, Rep D545-TI, 2014.
55. Ngo, T.D., Kashani, A., Imbalzano, G., Nguyen, K.T., Hui, D., Additive manufacturing (3D printing): A review of materials, methods, applications and challenges. *Composites, Part B*, 143, 172–96, 2018 Jun 15.
56. Le Duigou, A., Requile, S., Beaugrand, J., Scarpa, F., Castro, M., Natural fibres actuators for smart bio-inspired hygromorph biocomposites. *Smart Mater. Struct.*, 26, 12, 125009, 2017 Nov 2.
57. Rana, A., Mittal, A., Vashist, C., Rajput, S., Sridhar, S.B., Sharma, S., Malviya, R., Advancements in Herbal Drug Loaded Nanoparticulate System for Antimicrobial Effect. *Curr. Top. Med. Chem.*, 25, 25, 2918–36, 2025 Oct.
58. Lovras, M., Rajput, S., Sridhar, S.B., Shareef, J., Malviya, R., Cancer Management Using Photodynamic Therapy: Fundamentals, Mechanism and Advances. *Curr. Pharm. Des.*, 2025.
59. Vazquez, E., Gürsoy, B., Duarte, J.P., Formalizing shape-change: Three-dimensional printed shapes and hygroscopic material transformations. *Int. J. Archit. Comput.*, 18, 1, 67–83, 2020 Mar.
60. Lee, A.Y., An, J., Chua, C.K., Two-way 4D printing: a review on the reversibility of 3D-printed shape memory materials. *Engineering.*, 3, 5, 663–74, 2017 Oct 1.
61. Momeni, F., Liu, X., Ni, J., A review of 4D printing. *Mater. Des.*, 122, 42–79, 2017 May 15.
62. Ge, Q., Qi, H.J., Dunn, M.L., Active materials by four-dimension printing. *Appl. Phys. Lett.*, 103, 13, 2013 Sep 23.
63. Pei, E., 4D Printing: dawn of an emerging technology cycle. *Assem. Autom.*, 34, 4, 310–4, 2014 Sep 9.
64. Matsuzaki, R., Ueda, M., Namiki, M., Jeong, T.K., Asahara, H., Horiguchi, K., Nakamura, T., Todoroki, A., Hirano, Y., Three-dimensional printing of continuous-fiber composites by in-nozzle impregnation. *Sci. Rep.*, 6, 1, 23058, 2016 Mar 11.
65. Le Duigou, A., Barbé, A., Guillou, E., Castro, M., 3D printing of continuous flax fibre reinforced biocomposites for structural applications. *Mater. Des.*, 180, 107884, 2019 Oct 15.

66. Le Duigou, A., Correa, D., Ueda, M., Matsuzaki, R., Castro, M., A review of 3D and 4D printing of natural fibre biocomposites. *Mater. Des.*, 194, 108911, 2020 Sep 1.
67. Wang, X., Jiang, M., Zhou, Z., Gou, J., Hui, D., 3D printing of polymer matrix composites: A review and prospective. *Composites, Part B*, 110, 442–58, 2017 Feb 1.
68. Melchels, F.P., Feijen, J., Grijpma, D.W., A review on stereolithography and its applications in biomedical engineering. *Biomaterials*, 31, 24, 6121–30, 2010 Aug 1.
69. Mwema, F.M. and Akinlabi, E.T., Basics of fused deposition modelling (FDM), in: *Fused deposition modeling: strategies for quality enhancement*, pp. 1–15, Springer International Publishing, Cham, 2020 May 30.
70. Vyavahare, S., Teraiya, S., Panghal, D., Kumar, S., Fused deposition modelling: a review. *Rapid Prototyping J.*, 26, 1, 176–201, 2020 Jan 8.
71. Saadi, M.A., Maguire, A., Pottackal, N.T., Thakur, M.S., Ikram, M.M., Hart, A.J., Ajayan, P.M., Rahman, M.M., Direct ink writing: a 3D printing technology for diverse materials. *Adv. Mater.*, 34, 28, 2108855, 2022 Jul.
72. Cheng, M., Ramasubramanian, A., Rasul, M.G., Jiang, Y., Yuan, Y., Foroozan, T., Deivanayagam, R., Tamadoni Saray, M., Rojaee, R., Song, B., Yurkiv, V.R., Direct ink writing of polymer composite electrolytes with enhanced thermal conductivities. *Adv. Funct. Mater.*, 31, 4, 2006683, 2021 Jan.
73. Kumar, M.B., Sathiya, P., Varatharajulu, M., Selective laser sintering, in: *Advances in additive manufacturing processes*, pp. 28–47, Bentham Science Publishers, United Arab Emirates, 2021 Nov 29.
74. Fina, F., Goyanes, A., Gaisford, S., Basit, A.W., Selective laser sintering (SLS) 3D printing of medicines. *Int. J. Pharm.*, 529, 1–2, 285–93, 2017 Aug 30.

9

4D Printing Biocomposites: Fabricating Dynamic Tissues

Atifa Khan[1], Abdullah Khan[2], Syed Salman Ali[3]*, Ashwin Kumar Saxena[4], Aditya Sharma[5] and Priyanka Saroj[6]

[1]Department of Pharmacy, Lloyd School of Pharmacy, Knowledge Park-II, Greater Noida, Uttar Pradesh, India
[2]School of Pharmacy, Mangalayatan University, Beswan, Aligarh, Uttar Pradesh, India
[3]Lloyd Institute of Management and Technology, Knowledge Park-II, Greater Noida, Uttar Pradesh, India
[4]School of Pharmacy, Shri Venkateshwara University, Amroha, Uttar Pradesh, India
[5]School of Pharmaceutical Sciences, Shri Venkateshwara University, Gajraula, Uttar Pradesh, India
[6]Department of Pharmacology, Amity Institute of Pharmacy, Amity University, Noida, Uttar Pradesh, India

Abstract

Four-dimensional (4D) printing technology has revolutionized the field of biomedical engineering by enabling the creation of dynamic structures that can change shape or function in response to environmental stimuli. This chapter explores the application of 4D printing in fabricating dynamic tissues using biocomposites, which are engineered materials composed of biodegradable polymers and reinforcing materials. The chapter discusses the fundamentals of 4D printing, including smart materials and stimuli responsiveness, and highlights the potential of biocomposites in biomedical applications. It also examines the various fabrication techniques, such as direct ink writing and fused deposition modeling, and their applications in tissue engineering, regenerative medicine, and controlled drug delivery. The chapter concludes by discussing the challenges and limitations of 4D bioprinting, including material and structural constraints, cellular compatibility, and regulatory complexities, and provides future perspectives on the potential of this technology to transform the field of medicine.

**Corresponding author*: salmanali.ali32@gmail.com

Rishabha Malviya and Mukesh Roy (eds.) Biocomposites in 3D and 4D Printing: Materials, Techniques, and Applications in Regenerative Medicine, (241–270) © 2026 Scrivener Publishing LLC

Keywords: 4D printing, biocomposites, dynamic tissues, biomedical engineering, tissue engineering

9.1 Introduction

9.1.1 Definition and Evolution of 4D Printing

4D printing is a revolutionary evolution of the concept of additive manufacturing where the new revolutionary method of 4D printing takes the main idea of 3D printing, adding time as a new dimension, and turns it into a multi-dimensional, transformative concept. In contrast to the 3D-printed objects that are left in their original form after printing process, the 4D-printed objects can evolve in shape, functionality or property over time due to a particular external influence. Examples of such stimuli are changes in temperature, exposure to light, pH, moisture, magnetic field or electric field. These materials often called smart materials or stimuli-responsive materials are targeted to behave in a programmable manner and the materials respond to certain environmental conditions by morphing or reconfiguring. The visualization of 4D printing came into renown in 2013 when Skylar Tibbits of Massachusetts Institute of Technology (MIT) released a new era of smart manufacturing. The term also embodies the concept of manufacturing the objects that would be able to adapt or evolve after manufacturing without any human interaction to fulfil the pattern of dynamic processes similar to those manifested by biological systems [1].

The 4D printing has presented new avenues of creating next-generation medical devices and treatment in the biomedical application context. Its functions allow it to create self-assembling or self-folding structures, deployable implants, which are activated once inside the body, and drug delivery systems, which dispense medications in response to a particular physiological signal. Such dynamic systems are especially useful to generate patient-tailored therapies which are able to respond irrespective of time to the varying situations in the human body. For example, shape-memory polymers and hydrogels used in 4D printing can be engineered to expand or contract when exposed to body temperature or moisture, making them ideal for minimally invasive surgeries or controlled therapeutic delivery. Scaffolds created through 4D printing can provide a dynamic environment for tissue regeneration, adapting to the growth and mechanical needs of developing tissues something traditional scaffolds cannot achieve [2].

9.1.2 Difference between 3D and 4D Printing

Functioning of the final product printed is where the main difference between 3D and 4D printing exists wherein final product printed will have the ability to change over time in reaction to an outside stimulation. Objects in standard 3D printing are created by adding material layer to layer in order to create a pre-determined, fixed structure. Thermoplastics, metals, ceramics and resin materials are typically used to produce components of precise geometric shape and mechanical properties. But as soon as one is done with the 3D print object, the structure will not be capable of alteration under any kinds of environment. Contrastingly, 4D printing contains

Table 9.1 Comparative overview of 3D and 4D printing in biomedical applications.

Parameter	3D printing	4D printing	References
Dimensionality	Constructs are static in X, Y, Z dimensions	Incorporates time (4th dimension) for dynamic transformation	[17]
Material Type	Thermoplastics, ceramics, metals, resins	Stimuli-responsive/ smart materials	[18]
Post-Fabrication Behavior	Fixed geometry and functionality	Morphological or functional changes in response to stimuli	[19]
Stimulus Responsiveness	Not responsive to environmental changes	Responsive to pH, temperature, humidity, light, magnetic field, etc.	[19]
Biological Mimicry	Limited ability to mimic natural tissue dynamics	High biomimicry with adaptive, programmable behaviors	[20]
Representative Applications	Rigid implants, prosthetic components, static scaffolds	Self-expanding stents, dynamic drug delivery systems, morphing tissue scaffolds	[20]

a revolutionary aspect, whereby stimuli-responsive or intelligent materials are incorporated into printing. The materials used have the inbuilt characteristics of reacting to the environmental stimuli like temperature, humidity, pH, light or magnetic field. The programmed response of the material during the designing and manufacture process dictates that it responds to certain stimuli, like bending, swelling, shrinking, coiling, or folding among others. This fundamental advancement enables 4D-printed constructs to perform functions that would be impossible for static 3D-printed structures [3]. Basic comparative study has been depicted in Table 9.1.

As an example, a biodegradable stent manufactured using 4D-printing can be shaped in such a manner that, in order to be inserted, it is small in size, and that, when in contact with human body heat, it should open by itself. This avoids the use of mechanical expansion which requires invasive procedures thereby reducing the discomfort experienced by the patient and also making the surgery complicated [3].

This adaptability in programming can greatly expand the use of the biomedical applications. A dynamic and permissive environment during the healing process may be achieved in regenerative medicine through the changing of the structure of 4D-printed scaffolds in response to the growth and orientation of the new tissue. Shape-shifting also can be used with implantable devices, where it may be necessary to ensure that the device adheres to a particular anatomical structure, or to respond to body changes as the body is used or over time. In addition, it is possible to greatly exaggerate the release of therapeutic agents toward physiological stimuli in targeted drugs delivery, making them more precise and effective [4].

9.1.3 Relevance of Biocomposites in Biomedical Applications

Biocomposites are engineered materials composed of a biodegradable polymer matrix reinforced in combination with different fillers, e.g., natural or synthetic fibers, nanomaterials, bioactive ceramics, or other functional agents. Such materials are specifically designed to fulfill the needs of biomedical usage by blending vital qualities of mechanical strength, biocompatibility, bioactivity, and regulated degradation characteristic. The polymer matrix is considered as a supportive phase and the reinforcing phase improves structural stability, exposes a certain type of biological interaction, and provides any other functionality [5].

Biocomposites in biomedical engineering, biocomposites find use in preparing scaffolds, implants, prosthetics and drug delivery systems allowing controlling of drug release. They are especially useful in the application of tissue engineering in which the scaffolds are required to hold the tissue developing in place temporarily, help the cells grow and also breakdown at a rate that is sensible to the replacement of natural tissue. Strengthening the polymer system by incorporation of nanoparticles or bioceramics (hydroxyapatite, zinc oxide or silica) enhances not only the mechanical stability of the construct but also the osteoconductivity and surface reaction capability of the construct with the cells. Such modifications enable better cell adhesion, proliferation, and differentiation, which are crucial in effective tissue regeneration [6].

Biocomposites will add two properties once incorporated into 4D printing platforms: physical durability and stimuli responsiveness. 4D printing leverages the fourth dimension, time, to allow printed materials to transform after printing in response to stimuli in the environment, including temperature, pH, magnetic fields, light, or moisture. The intrinsic properties that govern this transformation are the ability of the smart materials used to change reversibly or irreversibly under certain conditions of physiology. Biocomposites that are intended to be used as 4D printing materials therefore enable the fabrication of dynamic implants and scaffolds that change their shape, porosity, or mechanical properties after being implanted within the human body [7].

As an example, biocomposites based on shape-memory polymer and reinforced with nanofillers can be printed in a compact form and programmed to expand, fold or self-assemble when body temperature is reached. This is especially useful in minimally invasive surgery where, in the least invasive surgical procedures, the devices may be implanted in a minimally invasive form and triggered on site. In the same way, the drug delivery systems created with 4D-printed biocomposites can be programmed to respond to environmental signals to release active substances with a specificity in space and time, making it more effective than other delivery forms and eliminating systemic effects [8].

It is possible to accurately control the degradation of biocomposites, by changing the composition of the polymer matrix, crosslinking density and kind of reinforcement. This is crucial in applications whereby the material is required to break down in tandem with tissue regeneration and to only act as a mechanical support as needed and not to cause chronic inflammation or when a secondary surgical removal is necessary [9].

The materials science side of the biocomposites also takes care of the environmental issues that are attached to the traditional synthetic polymers. Many biocomposites are derived from renewable resources such as starch, chitosan, alginate, polylactic acid (PLA), and polycaprolactone (PCL), and degrade into non-toxic by-products. Their biodecomposable nature is coupled with the rising attention to green biomaterials in healthcare and manufacturing of biomedical devices [10].

The integration of biocomposites into 4D printing technologies represents a convergence of responsive material science and personalized medicine. It enables the fabrication of multifunctional, patient-specific constructs that are not only biologically and mechanically compatible but also capable of responding to dynamic physiological environments. These features are essential for advancing regenerative therapies, improving clinical outcomes, and reducing the invasiveness of medical interventions [11].

9.1.4 The Need for Dynamic Tissue Constructs

The physiological conditions of a human body are naturally dynamic, which is accompanied by constant biochemical signaling, mechanical loading, tissue remodeling, and cell turnover. Conventional biomaterials and medical implants tend to be developed as stiff systems, which provide little or no means of interacting with the biological environment. Although these traditional structures can be useful in the short term, they cannot always be as adaptable as necessary in the long term to fit and integrate into physiological balance. Therefore, such issues as poor tissue integration, persistent inflammation, immune rejection, and device failure of the device are often observed [12].

Native tissues have complex behaviors that incorporate self-healing, remodeling, contractility, and responsiveness to mechanical and biochemical stimulations. That is why artificial structures that replace or supplement biological tissues should have the capacity to simulate such dynamic functions. Static scaffolds and implants lack the ability to be biomechanically and biochemically responsive to living tissues, especially when the microenvironment of a tissue is changing with time in the healing or growth process [13].

The latter is being addressed by creating dynamic tissue constructs to integrate materials and structures, which have the potential to actively adapt to physiological stimuli with the help of 4D bioprinting. These are developed with the usage of stimuli-responsive biocomposites or smart materials which could be modified during the course of time both in shape, stiffness, porosity, or functional behavior. The constructs can be induced to re-organize, in a controlled way, morphologically or functionally, when subjected to endogenous conditions in the form of temperature, pH and moisture, or biochemical cues.

The ability to have dynamic behavior in tissue construct results in the ability to better integrate with surrounding biological systems. As an example, a 4D-printed vascular graft can be grown or shaped by the changes in blood pressure, thus keeping it hemodynamically compatible and limiting the chance of stenosis or thrombosis. Otherwise, in skin regeneration, a scaffold that will accommodate and wound contraction and enhance a guided cellular migration may also quicken the healing process and enhance the tissue quality. In soft tissue engineering, the constructs that alter their mechanical behavior in relation to cellular activity or due to the deposition of extracellular matrix can promote tissue maturation and cell differentiation [14].

The integration of dynamic capabilities into tissue engineering platforms represents a significant advancement in regenerative medicine. It enables the development of personalized, adaptable, and functional constructs that align with the complex and evolving requirements of human tissues, thereby improving the clinical performance and longevity of biomedical implants and scaffolds.

9.2 Fundamentals of 4D Printing

9.2.1 Concept of Time as the Fourth Dimension

Traditional 3D printing involves layer-by-layer fabrication in three spatial dimensions (X, Y, and Z), resulting in static structures with fixed geometry and function. After being made, these objects are not able to react to changes in the environment. The 4D printing adds time as a functional dimension to printed objects that can change their shape or behavior after being printed in response to external stimuli like temperature, pH, moisture, or light. The design is done in such a way that this transformation is programmed by applying stimulus responsive materials and engineered structures. With the addition of time, the constructs of print can become dynamic systems, and have the ability to respond to dynamic situations. Such flexibility is necessary in biomedical uses, where tissues must be exposed to dynamic physiological conditions. 4D printing can be used to print such devices as self-morphing scaffolds and adaptive implants, as well as responsive drug delivery systems that behave more like natural tissues and can lead to better treatment [14].

9.2.2 Smart Materials and Stimuli Responsiveness

These technologies of 4D printing have become closely connected to the development and usage of stimuli-responsive materials, which are sometimes called smart or intelligent materials. Such materials have the intrinsic capability of being physically or chemically transformed in a controlled way when they are exposed to certain environmental stimuli. These transformations can be shape, volume, stiffness, porosity, or functional activity changes based on the chemical structure and design Smart materials used in 4D printing are typically classified according to the type of external stimulus they respond to:

The property of thermoresponsive materials is the thermal induced structural/phase transitions upon temperature variation. Among them, shape-memory polymers (SMPs) are extensively utilized due to their capacity to be temporarily deformed and then return to a pre-defined shape upon heating above a critical transition temperature. These are preferable materials particularly in the less invasive and self-deployable implants [15].

The PH-sensitive materials are either ionic in their character or swellable based on the local pH. Targeted drug delivery systems, particularly in the gastrointestinal tract, can be of great interest to this type of material, since there are specific pH gradients. They enable delivery of therapeutic agents at the local levels and limit the exposure to the systemic administration.

Photo-responsive materials undergo a conformational (or chemically) induced change in response to certain wavelengths of light. Photo-crosslinkable hydrogels are a demonstrative example, since their spatial patterning and tailoring of mechanical properties and structural integrity of tissue engineering constructs with a very high degree of precision is achievable by the impact of ultraviolet or visible light [16].

Some natural polymers, and hydrogels, are hygroscopic and their volumes increase or decrease volumetrically in response to changes in moisture/hydration. These materials can be used in applications where dermal tissue, wound healing, and mucosal environments are concerned as moisture dynamics is a key factor in such applications.

The dynamic and patient adaptive constructs can be made using these smart materials that are used in biocomposites matrices. These types of systems are able to be responsive to the physiological microenvironment, thereby making better cellular interactions, tissue integration and controlled therapeutic effect. In addition, the capacity to engineer material functionality by composition and structure design and fabrication is aiding in the development of the next generation biomedical devices that are increasingly aligned with the requirements of human tissues [21].

The stimuli-responsive materials coupled to the bioactive composites have the potential to produce self-morphing scaffolds and responsive implants and the programmable drug delivery systems when addressed in the framework of 4D bioprinting. Such innovations do not only promote the sphere of regenerative medicine but also suggest the basis of the creation of highly personalized and minimally invasive treatment strategies [22].

9.2.3 Mechanisms of Shape Transformation

The ability to programmatically respond to certain stimuli through changes in the functional shape of the 4D-printed biomedical construct is a major determinant of the functional capacity of this type of structure. Such transformations are not random but rather come as a consequence of interaction between material setup, geometric organization and the surrounding cues. To take advantage of this behavior, many different varieties of actuation have been developed to apply them to specific biomedical uses [23]. Mechanism of 4D printing biocomposites for tissue morphing is shown in Figure 9.1.

Hydro-swelling or deswelling is one of the most well-known mechanisms that can be often seen in hydrophilic polymers like hydrogels. These materials undergo changes in their dimensions when they absorb or release

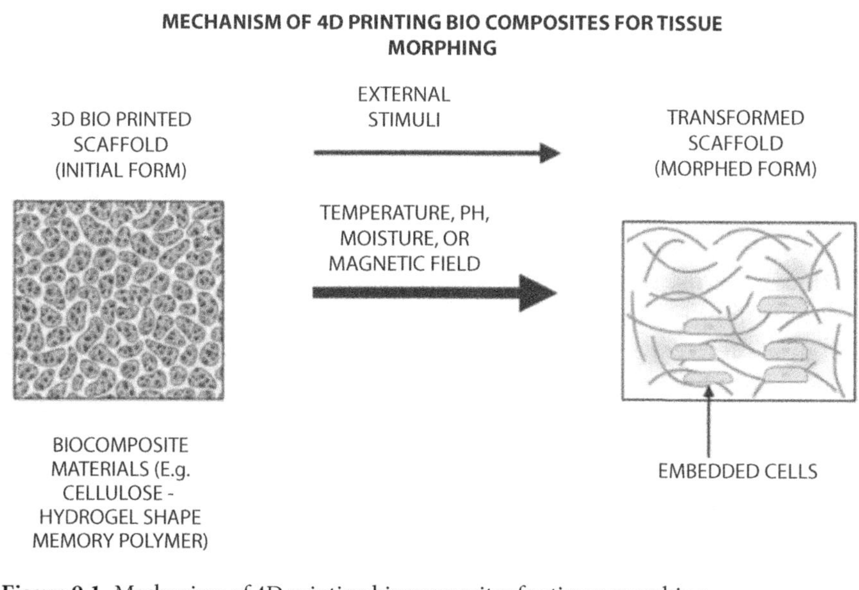

Figure 9.1 Mechanism of 4D printing biocomposites for tissue morphing.

aqueous fluids, and this effect occurs due to environmental stimulus (environmental pH, temperature or ionic concentration). This property makes them good dynamic scaffold in tissue engineering whereby interaction with the physiological milieu becomes very essential [23].

Thermoresponsive shape-memory polymers (SMPs) represent another widely used class of materials. Such polymers may be stretched out, only to restore their initial shape when they encounter a particular temperature level. They are particularly useful in the case of implants that have a minimal invasive nature in which they can be implanted and compressed and then expanded or activated internally.

The difference in the response of two affixed materials with different swelling characteristics or different thermal characteristics is used in bilayer actuation. When the construct is exposed to a stimulus, the dimensional mismatch produces a bending, curling or twisting of the construct. The principle finds its use in applications where curvature is needed, e.g., gastrointestinal or vascular implants [24].

The crease patterns of programmable origami or stimuli-responsive folding are pre-programmed, by printing the crease patterns onto a material substrate. The flat structure folds up into three-dimensional shapes in a controlled manner when external stimuli, e.g., in form of light, heat, or humidity is applied to it. Such strategy is beneficial in those devices where spatial transformation is paramount after implantation [25].

9.2.4 Design and Modeling Tools

The predictive design strategy that should be employed to generate functional 4D-printed biomedical constructs is computational modeling grounded in material science. Advanced tools like finite element analysis (FEA), computational fluid dynamics (CFD), and 3D computer-aided design (CAD) software are used to simulate and refine the behavior of smart biocomposite structures. They are instruments that predict deformation patterns, simulate interactions with biological environment and optimise internal architecture and distributions of materials. Important design parameters are the choice of smart materials, orientation of printing and layout of geometry. The best modeling is the secret to any valuable performance in biological systems and the computational and experimental validation are the one of the factors to bring the 4D bioprinting to the clinical practice [26].

9.3 Biocomposites in 4D Printing

9.3.1 Definition and Classification of Biocomposites

Another form of developed material is biocomposites, which are made by introducing the biodegradable polymer matrices and reinforcing materials into the structure, such as natural fibers, inorganic nanoparticles, or bioceramics. Such hybrid materials are expected to fill the gap between mechanical functionality and biological compatibility, so they are especially applicable to biomedical use such as load bearing implants, tissue engineered scaffolds, and intelligent drug delivery systems. The classification of biocomposites usually falls under three main criteria, namely the nature of the polymer matrix, nature of the reinforcing component, and functional use of the material. In terms of the matrix, both natural polymers (such as alginate, chitosan, gelatin, or cellulose) and synthetic biodegradable polymers (including polylactic acid [PLA], polycaprolactone [PCL], and polyglycolic acid [PGA]) are widely employed. The compounds that are used as reinforcement agents may be organic (e.g. collagen, silk fibroin or keratin) or inorganic (e.g. hydroxiapatite, silica nanoparticle or calcium phosphates), and this enables them to provide mechanical strength and osteoconductivity to the composite. Biocomposites can be designed to meet many possible functions in the biomedical field: They may be used structurally to increase strength in load-bearing conditions, be bioactive with better tissue integration, or have stimuli-responsive elements to biodegrade or release drugs in a controlled manner in response to physiological stimuli. It is the combination of the reinforcement materials with the matrix that enables the user to adjust to meet the requirements of the targeted tissue or application because of its capability to customize physical, chemical, and biological properties. Therefore, biocomposites have become one of the pillars of developing 4D bioprinting and regenerative medicine and provide a strong foundation in generating dynamic and multifunctional biomedical devices [27].

9.3.2 Natural vs Synthetic Biopolymers

Natural and synthetic biopolymers are critical components in the composition of bio-based biocomposites that are intended to be used in the biomedical field. Natural biopolymers, such as gelatin, hyaluronic acid, and collagen, closely resemble the structural and biochemical properties of the native extracellular matrix (ECM). The materials are excellent in terms of biocompatibility and natural bioactivity which supports cell attachment, cell growth, and differentiation. Nonetheless, their usefulness is in most cases constrained by the limitations associated with them, which are their low mechanical strength, high solubility in water and short degradation half-lives in physiological conditions. On the other hand, synthetic biopolymers like polylactic acid (PLA), polycaprolactone (PCL), and polyethylene glycol (PEG) offer superior control over mechanical integrity, degradation kinetics, and reproducibility. In spite of these merits, synthetic polymers in general do not possess the biointeractive properties that would ensure optimal cell interaction and might need further surface functionalization or addition of bioactive molecules to improve the biological performance. To overcome the weakness of individual polymers classes, even hybrid biocomposites are getting popular which are the combination of both naturally occurring as well as synthetic polymer. These combinations allow tuning the mechanical, biochemical and degradation characteristics, so that it is possible to obtain multifunctional materials that suit specific needs, in various biomedical applications, such as tissue engineering, regenerative medicine and drug delivery system [28].

9.3.3 Nanomaterials in Biocomposite Fabrication

Nanotechnology has a remarkable contribution to the improvement of the performance of biocomposites through the incorporation of nanoscale materials that possess functional properties. Carbon nanotubes work in further boosting the mechanical strength and electric conductivity and nanoclays work in increasing thermal stability and controls rates of biodegradation. Nano-hydroxyapatite is similar to bone mineral and can therefore be used in bone tissue engineering and the silver nanoparticles have an innate antimicrobial property. In addition to reinforcement, such nanomaterials can act as stimuli-responsive components, which can shape the materials in a controlled manner and deliver drugs to specific locations in bioapplications [29].

9.3.4 Biocompatibility and Biodegradability

Biocompatibility guarantees the absence of the immune reaction or toxicosis. Biodegradability is the properties of the ingredient to decompose without harming the body, over time. The main details that are to be taken into account are:

- The rate of change that is proportional to time of healing the tissue

- Attachment, migration and differentiation of the cells
- Attachment, migration, and differentiation of the cells

The generic one will involve a heavy duty testing of the cytotoxicity, genotoxicity and *in vivo* biocompatibility as part of regulatory approval [30].

9.4 Smart Biomaterials for Dynamic Tissues

9.4.1 Hydrogels and Shape Memory Polymers

Hydrogels are 3-dimensional, water loving, polymer networks capable of taking in a lot of water without disintegrating. Due to their high water content and soft, elastic nature, hydrogels closely resemble natural extracellular matrix (ECM), making them ideal for applications in soft tissue engineering. Materials such as alginate, gelatin-methacrylate (GelMA), and polyethylene glycol (PEG) are frequently used due to their biocompatibility and tunable physical properties. These structures help in effective transport of nutrients, as well as cell penetration and thus help in the viability of cells and tissue integration [31].

Shape memory polymers (SMPs) are another class of smart materials employed in 4D printing. Such polymers have the potential of temporarily changing their shape in response to external stimuli especially thermal stimuli and regain their original shape when the stimulus is removed. Such a reversible change is greatly beneficial in biomedical applications, especially in the field of developing implants, which can be deployed in their compact form and acquire their functional form at the implant site.

9.4.2 Stimuli-Responsive Materials

This aspect of stimuli-responsive materials is a crucial part of the 4D-bioprinted biocomposites development since it can provide the material with dynamic behavior in response to certain conditions, increasing the biomedical relevance of the material. The special characteristic of these materials is that they can undergo a physical or chemical transformation due to the external stimulus, like pH, temperature, light, or magnetic fields. pH-responsive materials are especially useful in drug delivery systems where they can be used to release drugs at a particular site based on the local pH, e.g., in different parts of the GI tract or tumor microenvironments. Thermoresponsive polymers, such as poly(N-isopropylacrylamide)

(PNIPAM), exhibit sol–gel transitions at physiological temperatures, allowing for controlled drug release or scaffold expansion *in situ*. Materials that respond to light, be it ultraviolet or visible light, can be used to initiate a conformational change or a chemical reaction and can be used to create spatial and temporal specificity in areas, such as remote actuation or selective drug release. Moreover, magnetically responsive systems are introduced to use nanoparticles like iron oxide, and its non-invasive manipulation or local heating could be achieved by applying an external magnetic field. The use of such intelligent materials within the bioprinted structures opens a new array of possibilities in creating responsive, patient-specific therapy and devices which can react to changes within the internal environment of the body, or to external control signals in real time [32]. Further comparative study has been depicted in Table 9.2.

Table 9.2 Stimuli-responsive smart materials in 4d bioprinting and their biomedical relevance.

Smart material type	Trigger stimulus	Key functional response	Biomedical applications	References
Shape-memory polymers (SMPs)	Temperature	Shape recovery after deformation	Minimally invasive implants, vascular grafts	[33]
pH-sensitive polymers	Local pH changes	Swelling, solubility change	Targeted drug delivery in GI tract or infection sites	[34]
Photo-responsive polymers	Light (UV/ Visible)	Bond cleavage or crosslinking	On-demand drug release, remote-controlled actuation	[35]
Hydrogel-based moisture-responsive systems	Humidity, hydration	Expansion, contraction	Skin regeneration, wound dressings	[35]
Magneto-responsive nanocomposites	Magnetic fields	Controlled motion or heating	Magnetic navigation, hyperthermia, site-specific drug delivery	[36]

9.4.3 Bioinks and Cell-Laden Materials

First of all, the most advanced type of ink enabling to print functioning tissue constructs directly is an advanced formulation that contains living cells, combined with supportive biomaterials also known as bioinks. The fundamental requirements of bioinks include biocompatibility, suitable rheological properties (notably shear-thinning for extrusion printing), and mechanical stability post-deposition. Such inks can organize the space of cells into a three-dimensional grid, thus leading to tissue maturation and integration. The combinations of alginate and cellulose derivatives, GelMA, and collagen-fibrin mixtures are some of the most commonly employed ingredients of bioinks. These formulations do not only give support to the structure but also reproduce the major features of the ECM so as to promote cell attachment, proliferation, and differentiation [37].

9.4.4 Self-Healing and Self-Assembling Systems

Self-healing materials are a new kind of biocomposites that can automatically recover any microstructural damages and hence improve the durability and functional life of biomedical devices and implants. Examples of such systems include reversible covalent bonds, supramolecular interactions, or micro capsules where healing agents have been absorbed into the system and then released when they need to heal [38].

In contrast, self-assembling materials are made to form predetermined hierarchical structures by means of non-covalent forces, i.e., hydrogen bonding, hydrophobic interactions, and ionic forces. The most spectacular of them is probably the one with peptide amphiphiles, which spontaneously form nanofiberous scaffolds that take the form of the natural ECM at a very high level. They can be used to generate biologically relevant microenvironments, and therefore these materials are targeted to be used in tissue regeneration, wound healing and controlled drug delivery [39].

9.5 Fabrication Techniques and Technologies

9.5.1 Direct Ink Writing

Direct ink writing (DIW) is a widely adopted extrusion-based additive manufacturing method suitable for printing highly viscous bioinks at ambient or physiological temperatures. The technique provides a rather accurate control of filament diameter, pore geometry, and spatial distribution thus being very useful in the fabrication of cell-laden constructs and hydrogel-based scaffolds. Living cells, bioactive molecules and temperature-sensitive materials can also be incorporated using DIW, thus functional tissues, such as skin, cartilage and vasculature can also be made [40].

9.5.2 Fused Deposition Modeling

Fused deposition modeling (FDM) utilizes thermoplastic filaments, such as polylactic acid (PLA) and polycaprolactone (PCL), which are melted and deposited layer-by-layer to form three-dimensional structures. It is an economical technique, which is appropriate in the manufacture of stiff biomedical products such as orthopedic implants and bone scaffolds. FDM has the disadvantage of being unable to work with living cells and thermolabile compounds due to the high temperatures involved in the processing. In spite of this shortcoming, FDM has been one of the major technologies in the manufacturing of structural parts that have to be mechanically robust [41].

9.5.3 Stereolithography

Stereolithography (SLA) is a light-based printing technique that uses ultraviolet (UV) or visible light to selectively polymerize layers of photosensitive resins. It has a good resolution and smooth surface, in comparison to its extrusion-based counterparts and can therefore find uses where fine geometries are needed, e.g. vascular networks, dental prosthetics and microfluidic devices. The reliance of SLA on certain photopolymerizable materials and the possible cytotoxic effects of UV exposure also requires special attention during material selection and post-processing of the material in order to use it in biomedical applications [42].

9.5.4 Multimaterial Printing Strategies

It is also possible to print in multiple materials and cells during the same time and therefore yield structures that are heterogeneous and can replicate the complexity of the native tissues. The technique plays a critical role in the development of constructs that possess grading in mechanical, porosity and biological functions. To illustrate the point, when applying this to osteochondral tissue engineering, multi-material systems can potentially replicate the stiffness of bone and the softness of cartilage interface. Likewise, the tissue constructs with vascularization can be developed through incorporating the sacrificial inks and perfusable channels. Such an approach enhances the functional recapitulation of the printed structures and it allows enabling 4D bioprinting in regenerative medicine [43].

9.6 Dynamic Tissue Engineering Applications

The advancement of 4D bioprinting in tissue engineering has introduced a new dimension to regenerative medicine by enabling the production of constructs that resemble the tissue structures of the native tissue as well as dynamic and time responsive. The bioprinted biocomposites will respond to environmental factors, such as temperature, pH, moisture, and mechanical stress, unlike traditional scaffolds that will not change structurally after implantation. This elasticity allows them to accommodate the evolving bodies better in the case of physiological changes and consequently improves tissue incorporation, quicker recovery, and individualized treatment outcomes. These responsive systems are of particular use when it comes to complex changing tissues, including cardiovascular, musculoskeletal, skin, and soft tissues. The 4D bioprinting can also be applied to make self-adjusting scaffolds, implants that morph, and systems of drug

Table 9.3 Biomedical applications of 4D-bioprinted biocomposites.

Application area	Construct type	Material used	Dynamic feature	References
Cardiovascular grafts	Self-expanding vascular tubes	SMP-based biocomposites	Shape recovery at body temp	[45]
Bone tissue engineering	Osteoconductive scaffolds	PCL + Nano-HA	Fit-to-defect morphology	[45]
Skin and wound healing	Adaptive wound dressings	Hydrogels (GelMA, alginate)	pH/moisture-triggered swelling and healing	[46]
Drug delivery systems	Smart release capsules	PNIPAM, pH-responsive hydrogels	Site-specific or triggered drug release	[47]
Organ-on-chip platforms	Miniaturized tissue mimics	Cell-laden hybrid biocomposites	Simulated tissue motion/ fluidics	[48]

administration that adapt. The practical uses and clinical implications of 4D bioprinted biocomposites in dynamic tissue engineering are also brought out in the next section with the emphasis that these tissues could transform the future of biomedical therapies [44]. Biomedical applications of 4D-bioprinted biocomposites are listed in Table 9.3.

9.6.1 Cardiovascular Tissue Constructs

Cardiovascular diseases constitute one of the leading causes of death worldwide, which is why it is important to have novel tissue engineering approaches that would be able to mimic the dynamic functionality of the native cardiovascular tissues. Here, 4D bioprinted vascular constructs e.g., self-expanding vascular grafts, heart valves, and cardiac patches have proven to have massive potential owing to the ability to respond and adapt to the physiological stimuli. With the use of shape-memory polymers and stimuli-responsive hydrogels, such constructs can be expanded or rather transformed by temperature or pH changes, thus enabling minimal invasive deployment, and post-deployment *in situ* adaptation. This moving behavior is similar to the natural mechanics of cardiovascular tissues, thereby offering a stronger support to the haemodynamics and the integration of tissues [49].

In addition, dynamic properties of the scaffold structure can be used to increase the endothelialization process, enhance mechanical stability, and minimize the possibility of graft failure. Other designs are designed to react to blood pressure or flow changes so that the construct will continue to work in the long term and also respond to changing physiological needs. Such responsiveness is necessary not only in determining success of initial implantation but also in the long-term functioning of the engineered cardiovascular tissues.

9.6.2 Musculoskeletal and Bone Tissues

The mechanical need and regenerative complexity of the musculoskeletal and bone tissues pose a specific challenge to their engineering, and the 4D-bioprinted biocomposites can potentially be beneficial since they enable the fabrication of mechanically stable, osteoconductive constructs that could actively adapt to mechanical and morphological changes. In case of bone regeneration, the biopolymers (nano-hydroxyapatite reinforced) offer mechanical stability, as well as bioactivity. The scaffolds are also able to reshape dynamically according to the irregularities of bone defects and hence enhance the fitting and integration to the host tissue. They can also be used to deliver the osteoinductive factors in a controlled way, which would further boost the regeneration process [50].

Musculoskeletal applications in musculoskeletal applications, 4D bioprinting has the potential to print constructs that can reproduce a contractile behavior and help in muscle repair. The scaffold can be made up of shape-memory materials that can resemble the muscle contraction and relaxation mechanism. In addition, the mechanical stimuli produced by stimuli-responsive scaffolds aid in the alignment and differentiation of progenitor cells especially mesenchymal stem cells and thus aid in the formation of functional musculoskeletal tissue.

9.6.3 Skin and Soft Tissue Regeneration

Dynamic characteristics benefit the 4D bioprinting especially in skin and soft tissue engineering whereby being flexible to the wound conditions is a key to successful healing. Sophisticated forms of wound dressing and grafts made by the means of 4D bioprinting technologies can sense environmental factors like pH, temperature, and moisture. On-demand drug delivery, antimicrobial properties, and real-time adaptability can be provided by these smart materials, and, thus, to enhance a microenvironment that will optimally promote tissue repair [51].

With their large water volume, adjustable mechanical performance, and outstanding biocompatibility, hydrogels are prominently used in the process of shaping dynamic skin scaffolds. These materials have the capability to by themselves adapt to fit the wound geometry, enhance angiogenesis, and epithelialization. Moreover, by including self-healing functions into such constructs, the durability of such products can be increased, and the necessity to apply repeated dressing to the wounded area can be reduced, which will lead to patient comfort and, hence, cost savings to the healthcare system as a whole. These dynamic systems mark a great step forward as compared to the static wound healing platforms [51].

9.6.4 Organ-on-Chip and Drug Testing Platforms

The intersection of 4D bioprinting and microfluidics has created the most advanced organ-on-chip systems, capable of duplicating the important targets of human physiology. These micro-engineered systems are aimed to recreate the structural, mechanical and biochemical context of a particular organ hence provide credible alternatives in the methods of drug testing and modeling diseases. Organ-on-chip devices may be used to recapitulate the effect of tissue strain, fluid percolation, and periodic loading as they use stimuli-responsive materials and components to represent physiological action [52].

They could be manifested in the form of lung-on-chip systems that mimic alveolar contraction and expansion, liver-on-chip systems allowing the virtual evaluation of metabolism, and kidney-on-chip where filtration can be evaluated. Such systems give a real-time surveillance of cellular reactions to medications, and this enhances the possibility of estimating the efficacy and toxicity of medications in human cells. Significantly, utilization of such platforms minimizes the need of animal models, speed up the drug development process, and promotes the provision of personal medicine using individual testing scenario per patient. With the incorporation of 4D bioprinted biocomposites in the platforms, their functionalrealism increases, and they allow the creation of more physiologically relevant models [53].

9.7 Case Studies and Recent Advances

9.7.1 Vascular Grafts with Shape-Memory Biocomposites

Among the most successful novel technologies in 4D bioprinting invented, a way of producing vascular grafts based on using shape-memory biocomposites was proposed. This kind of grafts is shaped as a small roll and expands when subjected to physiological temperature thus creating complete and fully functional, tubular blood conduits. The natural shape-memory ability of shape-memory polymers enables the constructs to fit perfectly to the implantation site making the delivery of the implant highly minimally invasive as well as leading to better surgical results. Besides structural efficacy, the grafts promote endothelialization and neovascularization which improves the patency of the graft and reduces the possibility of thrombosis. New developments have involved the addition of the magnetically responsive nanoparticles to the polymeric structure that can help direct or guide the placement of the scaffold during the implantation process leading to greater precision in the procedure and shortening of the operation time [54].

9.7.2 Self-Morphing Scaffolds for Tissue Reconstruction

The idea of self-folding scaffolds has also obtained major attention in terms of applications that would require geometrical complex three dimensional

(3D) structures of tissues. These scaffolds, have been designed as two-dimensional (2D) sheets, and when stimulated by environmental signals, like hydration, temperature, or pH, they can be programmed to fold into a pre-determined shape. The third-order transformation enables such dynamic reconstructions that can lead to manufacturing of organ mimicry structures which presents a closer resemblance to the native tissue architecture. One such instance is the development of self-folding intestinal models in which the bilayer hydrogels, in which intestinal epithelial cells have been incorporated, swell anisotropically to generate villus-like protrusions. Such bioarchitectures are able to reproduce physiological topographies that are relevant in the constraints of nutrient uptakes and localized drug delivery measurements [55].

9.7.3 Stimuli-Responsive Bioprinted Drug Delivery Systems

The development of 4D bioprinting with smart materials that respond to the surrounding stimuli has transformed the design of drug delivery platforms. Such constructs are designed to deliver treatment agents when they are locally stimulated by some changes in pH, glucose, or temperature levels or due to the external light. As an example, PH-sensing hydrogel matrices provide a selective delivery of the drug in acidic environments of infections, thus, promoting antibacterial action and reducing toxicity in the body. Thermoresponsive vehicles have been used to provide anticancer drugs where hyperthermic tumor microenvironments are present. If a means of specifically and reversibly deactivating the drug can be found, optical activation has also been discussed as a possibility, allowing proxicative release of an ophthalmic drug on exposure to light. A large number of possibilities on a personalized medicine and targeted therapy exist with such spatiotemporal control on drug kinetics [56].

9.7.4 Emerging Frontiers in Bioprinted Organoids

The innovation of 4D printing in metal, composite, and tissue fields breathed the contemporary approach to the organization of organoid development which is a combination of high-tech bioinks, stem cell technologies, and programmable materials. Dynamic biocomposite scaffolds have allowed researchers to realize miniaturized miniaturized functional units (hepatic lobules, renal nephrons, and rhythmic myocardial patches). Disease modeling, pharmacokinetic, and preliminary transplantation are applications being made with these organoids. In these constructions, because of the use of time-evolving materials, the process of maturation and morphogenesis can be very similar to *in vivo* tissue development. As the actual full-sized functional organs are still a long way in the future, 4D printing is the next step toward this goal as it adds the spatial and temporal specificity to the idea of fabrication [57].

9.8 Challenges and Limitations

9.8.1 Material and Structural Constraints

Lack of smart biomaterials to form a combination of biocompatibility, mechanical strength, and biomaterial-based responsiveness to physiological stimulus is a limiting factor to the development of 4D bioprinted constructs. Even some shape-memory polymers and high-potential photo-responsive compounds are cytotoxic or have an inadequate mechanical property to use in high-strain conditions, like musculoskeletal systems. Moreover, to balance good biodegradability and structural stability is a technique that is complex especially when it comes to constructs that are long-term functional integration friendly. Shape transformation has to be precise; during morphing, any drift can cause structural malalignment or interference with other tissues, jeopardizing a treatment process [58].

9.8.2 Cellular Compatibility and Tissue Integration

A constant issue in the 4D printing process is keeping cells viable during the printing procedure, particularly in setups based on thermal or photonic energy. In addition, the special consideration should be given to the integration of the scaffold with host tissue where careful control of such factors as the kinetics of scaffold degradation, vascular infiltration, and immunogenicity is necessary to ensure appropriate integration. The efficacy of some responsive materials is low in longitudinal tissue regeneration due to poor support of the attachment or proliferation of cells. Studies are underway on the methods of surface functionalization and hybrid materials to enhance the interaction between cells and materials [59].

9.8.3 Scale-Up and Economic Barriers

Although the possibilities of 4D bioprinting are clearly proven on laboratory level, bringing the systems to a clinical and industrial-scale manufacturing is fraught with both logistical and financial challenges. Stimuli-responsive bioinks are expensive, and the process requires specialized bioprinters, as well as equipment to post-process the material, a property that prevents scalability. Also, there is reproducibility between dynamic constructs because of dependence on factors like humidity, temperature, and light exposure to environmental conditions [60].

A wider use in fabrication, characterization of the materials, and functional characterization will require establishment of standardized procedures.

9.8.4 Regulatory and Bioethical Complexities

Contact added to 4D bioprinted implants creates new areas of complexity and biological interaction and functions that are autonomous and stand to frustrate regulation. The existing models of biomedical devices lack the complete incorporation of the constructs that develop after implantation or the constructs that incorporate living cells. The ethics also is concerned with the question of what source of human cells are to be obtained especially stem cells, as well, as the fate of the bioengineered implants in the long run. Integration of biomedical scientists, lawyers and ethics committees is needed to establish parameters of safety testing, clinical trials and patient-specific applications [61].

9.9 Future Perspectives

9.9.1 Material and Structural Constraints

Issues are evident in the infrequent biocompatible and mechanically regimented smart biomaterials capable of responding to physiological changes on time, obstructing the exploitation of 4D bioprinted constructs. The materials associated with these materials like shape-memory polymers and photo-responsive compounds are usually cytotoxic or do not give enough mechanical support to act in structures that may be needed to withstand a high load such as musculoskeletal structures. In addition, it is also challenging to strike a balance between biodegradability and structural stability especially in such constructions that are aligned to embarking on long duration functional connectivity. Accuracy of shape change is of essence; any discrepancies that occur during morphing can result in structural discrepancy or alteration of the tissues around the intended structure, thus interfering with therapeutic effect [61].

9.9.2 Cellular Compatibility and Tissue Integration

Maintaining cellular viability during the course of the 4D printing process remains a challenge, particularly in systems that entail the use of thermal energy or photonic energy. Besides, to guarantee the construct to integrate with its host tissue, it is necessary to have tight control over aspects of its degradation kinetics, vascular infiltration and immunogenicity. Certain responsive materials do not support the attachment or proliferation of cells but can facilitate the requirements of using them in long-term tissue regeneration [62].

9.9.3 Scale-Up and Economic Barriers

Although the possibilities of 4D bioprinting have been proven on a laboratory scale, there are still serious logistical and financial challenges regarding the transition of such systems to clinical and industrial production. The scalability is limited by the high price of stimuli-responsive bioinks as well as the necessity of specialized bioprinters and post-processing gear. Also, dynamic construct reproducibility is an issue under this condition because the results are sensitive to the environmental factors like, humidity, temperature, and exposure to light. The development of common methods of fabrication, characterization of materials and functional testing will be a prerequisite to the wider take up [63].

9.9.4 Regulatory and Bioethical Complexities

The implants developed through 4D bioprinting come with new facets of interactions between biological elements and the possibility of functional independence that makes them challenging to regulate. The existing frameworks of biomedical devices do not comprehensively cover frameworks, which change after implantation or have living cells. The issue of ethics also brings about concerns about the sourcing of human cells, especially stem cells and fate of bioengineered implants in the long run. Multidisciplinary approaches in which biomedical scientists collaborate with legal and ethics committees are necessary to determine the guidelines in safety assessment, clinical testing, and patient-specific use [63].

9.10 Conclusion

4D bioprinting brings a dynamic aspect to biomedical printing by enabling the constructs to change their shapes based on the environmental cues. This method has the potential of creating living constructs that will develop as time goes by as in the case in the native tissues and organs unlike traditional 3D systems. When biocomposites are employed in this regard, functional and structural performance are augmented through incorporation of biodegradable matrices and reinforcement, which have biologically and mechanically active strengths. The utilization of these materials finds its application into a wide range of fields such as tissue engineering, regenerative medicine and controlled drug delivery. Shape-changing biocomposites that liberate real-time modulation of cellular interactions or therapeutics may have much greater benefits in terms of improved therapeutic responses as well as patient-specific therapy.

Biofabrication is an emerging scientific field which has been advanced by breakthroughs in material chemistry, nanotechnology, and is entering the realm of solutions that are clinically useful. But in order to bring the above innovations to practice much attention is to be paid to material performance, biological safety, and consistency of the manufacturing process. The existing restrictions concerning materials variety, scalability, and regulatory requirements are the points of critical constraints that should be worked out systematically. They are one of the fascinating areas of innovation in healthcare. As the concept of interdisciplinary collaboration is enhanced, even more, rational yet responsive and clinically feasible constructs are being developed. The given platform has the potential to innovate the sphere of medicine in the future due to high flexibility that could lead to the provision of personalized, adaptive, and combined treatment strategies.

AI disclosure statement: The author acknowledges the use of AI-based tools for minor language editing or grammatical correction. However, all intellectual contributions, analyses, and conclusions presented in this chapter are entirely the author's own.

References

1. Adam, G., Benouhiba, A., Rabenorosoa, K., Clévy, C., Cappelleri, D.J., 4D printing: Enabling technology for microrobotics applications. *Adv. Intell. Syst.*, 3, 5, 2000216, 2021 May.

2. Ahadian, S., Civitarese, R., Bannerman, D., Mohammadi, M.H., Lu, R., Wang, E., Davenport-Huyer, L., Lai, B., Zhang, B., Zhao, Y., Mandla, S., Organ-on-a-chip platforms: a convergence of advanced materials, cells, and microscale technologies. *Adv. Healthcare Mater.*, 7, 2, 1700506, 2018 Jan.
3. Ahmed, A., Arya, S., Gupta, V., Furukawa, H., Khosla, A., 4D printing: Fundamentals, materials, applications and challenges. *Polymer*, 228, 123926, 2021 Jul 16.
4. Amukarimi, S., Ramakrishna, S., Mozafari, M., Smart biomaterials—A proposed definition and overview of the field. *Curr. Opin. Biomed. Eng.*, 19, 100311, 2021 Sep 1.
5. Anderson, D.G., Burdick, J.A., Langer, R., Smart biomaterials. *Science*, 305, 5692, 1923–4, 2004 Sep 24.
6. Bajpai, A., Baigent, A., Raghav, S., Brádaigh, CO., Koutsos, V., Radacsi, N., 4D printing: materials, technologies, and future applications in the biomedical field. *Sustainability*, 12, 24, 10628, 2020 Dec 19.
7. Bhullar, S.K., Lala, N.L., Ramkrishna, S., Smart biomaterials-a review. *Rev. Adv. Mater. Sci.*, 40, 3, 303–14, 2015 Jun 1.
8. Rana, A., Mittal, A., Vashist, C., Rajput, S., Sridhar, S.B., Malviya, R., Exploration of 4D Printing and its Applications in the Biomedical Sciences. *Curr. Pharm. Des.*, 2025.
9. Black, C.R., Goriainov, V., Gibbs, D., Kanczler, J., Tare, R.S., Oreffo, R.O., Bone tissue engineering. *Curr. Mol. Biol. Rep.*, 1, 3, 132–40, 2015 Sep.
10. Bock, N., Riminucci, A., Dionigi, C., Russo, A., Tampieri, A., Landi, E., Goranov, V.A., Marcacci, M., Dediu, V., A novel route in bone tissue engineering: magnetic biomimetic scaffolds. *Acta Biomater.*, 6, 3, 786–96, 2010 Mar 1.
11. Bodaghi, M., Noroozi, R., Zolfagharian, A., Fotouhi, M., Norouzi, S., 4D printing self-morphing structures. *Materials*, 12, 8, 1353, 2019 Apr 25.
12. Bodaghi, M., Wang, L., Zhang, F., Liu, Y., Leng, J., Xing, R., Dickey, M.D., Vanaei, S., Elahinia, M., Hoa, S.V., Zhang, D., 4D printing roadmap. *Smart Mater. Struct.*, 33, 11, 113501, 2024 Nov 1.
13. Bonetti, L. and Scalet, G., 4D fabrication of shape-changing systems for tissue engineering: state of the art and perspectives. *Prog. Addit. Manuf.*, 12, 1–31, 2024 Aug.
14. Borse, K. and Shende, P., 3D-to-4D structures: an exploration in biomedical applications. *Aaps Pharmscitech*, 24, 6, 163, 2023 Aug 3.
15. Castro, N.J., Meinert, C., Levett, P., Hutmacher, D.W., Current developments in multifunctional smart materials for 3D/4D bioprinting. *Curr. Opin. Biomed. Eng.*, 2, 67–75, 2017 Jun 1.
16. Chen, X., Han, S., Wu, W., Wu, Z., Yuan, Y., Wu, J., Liu, C., Harnessing 4D printing bioscaffolds for advanced orthopedics. *Small*, 18, 36, 2106824, 2022 Sep.
17. Choi, J., Kwon, O.C., Jo, W., Lee, H.J., Moon, M.W., 4D printing technology: a review. *3D Print. Addit. Manuf.*, 2, 4, 159–67, 2015 Dec 1.

18. de Marco, C., Pané, S., Nelson, B.J., 4D printing and robotics. *Sci. Rob.*, 3, 18, eaau0449, 2018 May 30.
19. Ding, A., Tang, F., Alsberg, E., 4D printing: A comprehensive review of technologies, materials, stimuli, design, and emerging applications. *Chem. Rev.*, 125, 7, 3663–771, 2025 Mar 19.
20. Fetah, K., Tebon, P., Goudie, M.J., Eichenbaum, J., Ren, L., Barros, N., Nasiri, R., Ahadian, S., Ashammakhi, N., Dokmeci, M.R., Khademhosseini, A., The emergence of 3D bioprinting in organ-on-chip systems. *Prog. Biomed. Eng.*, 1, 1, 012001, 2019 Jul 16.
21. Ghosh, S., Chaudhuri, S., Roy, P., Lahiri, D., 4D Printing in biomedical engineering: A state-of-the-art review of technologies, biomaterials, and application. *Regener. Eng. Transl. Med.*, 9, 3, 339–65, 2023 Sep.
22. Rajput, S., Malviya, R., Sridhar, S.B., Osteoporosis Risk Assessment and Individualized Feature Analysis Using Interpretable XAI and RAI Techniques, in: *Explainable and Responsible Artificial Intelligence in Healthcare*, pp. 89–113, Springer Nature, John Wiley & Sons Ltd., 2025 Mar 13.
23. Joshi, S., Rawat, K., Rajamohan, V., Mathew, A.T., Koziol, K., Thakur, V.K., ASS, B., 4D printing of materials for the future: Opportunities and challenges. *Appl. Mater. Today*, 18, 100490, 2020 Mar 1.
24. Kantaros, A., Ganetsos, T., Piromalis, D., 4D printing: technology overview and smart materials utilized. Kantaros, A., Ganetsos, T., Piromalis, D., 4D Printing: Technology Overview and Smart Materials Utilized. *J. Mechatron. Rob.*, 7, 1, 1–4, 20232023 Feb 21.
25. Karami, D., Richbourg, N., Sikavitsas, V., Dynamic *in vitro* models for tumor tissue engineering. *Cancer Lett.*, 449, 178–85, 2019 May 1.
26. Karami, Z., Maleki, S., Moghaddam, A., Jahandideh, A., Self-healing bio-composites: concepts, developments, and perspective, in: *Sustainable polymer composites and Nanocomposites*, pp. 1323–1343, Springer International Publishing, Cham, 2019 Feb 2.
27. Kaur, K. and Singh, A., 4D Printing: Fundamentals, Materials, Applications and Challenges. *Smart Mater. Appl.*, 359, 2025 Oct 21.
28. Khan, F. and Tanaka, M., Designing smart biomaterials for tissue engineering. *Int. J. Mol. Sci.*, 19, 1, 17, 2017 Dec 21.
29. Khan, H.M., Liao, X., Sheikh, B.A., Wang, Y., Su, Z., Guo, C., Li, Z., Zhou, C., Cen, Y., Kong, Q., Smart biomaterials and their potential applications in tissue engineering. *J. Mater. Chem. B*, 10, 36, 6859–95, 2022.
30. Khoo, Z.X., Teoh, J.E., Liu, Y., Chua, C.K., Yang, S., An, J., KF, L., Yeong, W.Y., 3D printing of smart materials: A review on recent progresses in 4D printing. *Virtual Phys. Prototyping*, 10, 3, 103–22, 2015 Jul 3.
31. Khorsandi, D., Fahimipour, A., Abasian, P., Saber, S.S., Seyedi, M., Ghanavati, S., Ahmad, A., De Stephanis, A.A., Taghavinezhaddilami, F., Leonova, A., Mohammadinejad, R., 3D and 4D printing in dentistry and maxillofacial surgery: Printing techniques, materials, and applications. *Acta Biomater.*, 122, 26–49, 2021 Mar 1.

32. Kim J, D.A.G., Debnath, P., Saha, P., Smart multi-responsive biomaterials and their applications for 4D bioprinting. *Biomimetics*, 9, 8, 484, 2024 Aug 11.
33. Kuang, X., Roach, D.J., Wu, J., Hamel, C.M., Ding, Z., Wang, T., Dunn, M.L., Qi, H.J., Advances in 4D printing: materials and applications. *Adv. Funct. Mater.*, 29, 2, 1805290, 2019 Jan.
34. Lai, J., Liu, Y., Lu, G., Yung, P., Wang, X., Tuan, R.S., Li, Z.A., 4D bioprinting of programmed dynamic tissues. *Bioact. Mater.*, 37, 348–77, 2024 Jul 1.
35. Lau, K.T., Ho, M.P., Au-Yeung, C.T., Cheung, H.Y., Biocomposites: their multifunctionality. *Int. J. Smart Nano Mater.*, 1, 1, 13–27, 2010 Feb 17.
36. Le Duigou, A., Correa, D., Ueda, M., Matsuzaki, R., Castro, M., A review of 3D and 4D printing of natural fibre biocomposites. *Mater. Des.*, 194, 108911, 2020 Sep 1.
37. Leor, J., Aboulafia-Etzion, S., Dar, A., Shapiro, L., Barbash, I.M., Battler, A., Granot, Y., Cohen, S., Bioengineered cardiac grafts. *Circulation*, 102, suppl_3, III56, 2000 Nov 7.
38. Li, C. and Cui, W., 3D bioprinting of cell-laden constructs for regenerative medicine. *Eng. Regener.*, 2, 195–205, 2021 Jan 1.
39. Li, Z., Chen, L., Wu, J., Chen, Y., Zhu, Y., Li, G., Xie, G., Tang, G., Xie, M., A review of 3D bioprinting for organoids. *Med. Rev.*, 2025 Jan 14.
40. Liu, D., Xiang, T., Gong, T., Tian, T., Liu, X., Zhou, S., Bioinspired 3D multilayered shape memory scaffold with a hierarchically changeable micropatterned surface for efficient vascularization. *ACS Appl. Mater. Interfaces*, 9, 23, 19725–35, 2017 Jun 14.
41. Luo, X., Fabrication techniques, in: *Engineering Optics 2.0: A Revolution in Optical Theories, Materials, Devices and Systems*, pp. 179–242, Springer Singapore, Singapore, 2019 Mar 1.
42. Mahmood, A., Perveen, F., Chen, S., Akram, T., Irfan, A., Polymer composites in 3D/4D printing: materials, advances, and prospects. *Molecules*, 29, 2, 319, 2024 Jan 9.
43. Miao, S., Castro, N., Nowicki, M., Xia, L., Cui, H., Zhou, X., Zhu, W., Lee, S.J., Sarkar, K., Vozzi, G., Tabata, Y., 4D printing of polymeric materials for tissue and organ regeneration. *Mater. Today*, 20, 10, 577–91, 2017 Dec 1.
44. Mitchell, A., Lafont, U., Hołyńska, M., Semprimoschnig, C.J., Additive manufacturing—A review of 4D printing and future applications. *Addit. Manuf.*, 24, 606–26, 2018 Dec 1.
45. Rana, A., Mittal, A., Vashist, C., Rajput, S., Sridhar, S.B., Sharma, S., Malviya, R., Advancements in Herbal Drug Loaded Nanoparticulate System for Antimicrobial Effect. *Curr. Top. Med. Chem.*, 25, 25, 2918–36, 2025 Oct.
46. Neumann, M., di Marco, G., Iudin, D., Viola, M., van Nostrum, C.F., van Ravensteijn, B.G., Vermonden, T., Stimuli-responsive hydrogels: the dynamic smart biomaterials of tomorrow. *Macromolecules*, 56, 21, 8377–92, 2023 Oct 18.

47. Pei, E. and Loh, G.H., Technological considerations for 4D printing: an overview. *Prog. Addit. Manuf.*, 3, 1, 95–107, 2018 Jun.
48. Perić Kačarević, Z., Rider, P., Alkildani, S., Retnasingh, S., Pejakić, M., Schnettler, R., Gosau, M., Smeets, R., Jung, O., Barbeck, M., An introduction to bone tissue engineering. *Int. J. Artif. Organs.*, 43, 2, 69–86, 2020 Feb.
49. Qu, G., Huang, J., Gu, G., Li, Z., Wu, X., Ren, J., Smart implants: 4D-printed shape-morphing scaffolds for medical implantation. *Int. J. Bioprint.*, 9, 5, 764, 2023 May 30.
50. Ramaraju, H., Akman, R.E., Safranski, D.L., Hollister, S.J., Designing biodegradable shape memory polymers for tissue repair. *Adv. Funct. Mater.*, 30, 44, 2002014, 2020 Oct.
51. Ramezani, M. and Mohd Ripin, Z., 4D printing in biomedical engineering: advancements, challenges, and future directions. *J. Funct. Biomater.*, 14, 7, 347, 2023 Jun 29.
52. Ronaldson-Bouchard, K. and Vunjak-Novakovic, G., Organs-on-a-chip: a fast track for engineered human tissues in drug development. *Cell Stem Cell*, 22, 3, 310–24, 2018 Mar 1.
53. Sadraei, A. and Naghib, S.M., 4D printing of physical stimuli-responsive hydrogels for localized drug delivery and tissue engineering. *Polym. Rev.*, 65, 1, 104–68, 2025 Jan 2.
54. Shin, D.G., Kim, T.H., Kim, D.E., Review of 4D printing materials and their properties. *Int. J. Precis. Eng. Manuf. Green Technol.*, 4, 3, 349–57, 2017 Jul.
55. Shinde, S., Mane, R., Vardikar, A., Dhumal, A., Rajput, A., 4D printing: From emergence to innovation over 3D printing. *Eur. Polym. J.*, 197, 112356, 2023 Oct 10.
56. Singha, K. and Singha, M., Cardio vascular grafts: existing problems and proposed solutions. *Int. J. Biol. Eng.*, 2, 2, 1–8, 2012 Apr 9.
57. Subeshan, B., Baddam, Y., Asmatulu, E., Current progress of 4D-printing technology. *Prog. Addit. Manuf.*, 6, 3, 495–516, 2021 Aug.
58. Tamay, D.G., Dursun Usal, T., Alagoz, A.S., Yucel, D., Hasirci, N., Hasirci, V., 3D and 4D printing of polymers for tissue engineering applications. *Front. Bioeng. Biotechnol.*, 7, 164, 2019 Jul 9.
59. Thakur, A., Anadebe, V.C., Dagdag, O., Sharma, D., Om, H., Fundamentals and Recent Applications of 3D, 4D, and 5D Printing of Polymer and Its Nanocomposites: Challenges and Its Future Prospects, in: *Polymer Nanocomposites for 3D, 4D and 5D Printing: Fundamental to Applications*, vol. 4, pp. 345–92, Elsevier, 2025 May.
60. Venkatesan, J. (Ed.), *Biocomposite Nanomaterials and Their Applications*, Springer Nature, Switzerland, 2024 Nov 28.
61. Wang, Y., Cui, H., Esworthy, T., Mei, D., Wang, Y., Zhang, L.G., Emerging 4D printing strategies for next-generation tissue regeneration and medical devices. *Adv. Mater.*, 34, 20, 2109198, 2022 May.

62. Rajput, S., Malviya, R., Srivastava, S., Ahmad, I., Rab, S.O., Uniyal, P., Targeted Treatment of Various Cancers using Radionuclide: Advancement and Prospective. *Curr. Cancer Ther. Rev.*, 21, 4, 508–24, 2025 Sep.
63. Yarali, E., Mirzaali, M.J., Ghalayaniesfahani, A., Accardo, A., Diaz-Payno, P.J., Zadpoor, A.A., 4D printing for biomedical applications. *Adv. Mater.*, 36, 31, 2402301, 2024 Aug.

10

Biocomposites in Tissue Engineering

Anjali Rana[1], Shivam Rajput[2], Ajesh Chauhan[1] and Rishabha Malviya[2*]

[1]Department of Pharmacy, Hindu College of Pharmacy, Sonipat, Haryana, India
[2]Department of Pharmacy, School of Medical and Allied Sciences, Galgotias University, Greater Noida, Uttar Pradesh, India

Abstract

Environmental and sustainability concerns have increased this century, leading to a proliferation of innovations in eco-friendly materials, such as biocomposites. Biopolymers combined with reinforcing agents form biocomposites, which have recently attracted interest as possible components of medical devices. The global advancement of high-performance materials derived from renewable resources is flourishing. The potential of biocomposite materials that replaced traditional materials has resulted in a significant rise in their prominence in recent years. Tissue engineering is an intriguing therapeutic approach for healing injured or sick tissues. Biocomposites have numerous uses in tissue engineering and regenerative medicine. One of these uses is scaffold construction, which involves mimicking the extracellular matrix. This allows cells to proliferate and damaged tissues to recover. Biocomposites are extensively used in tissue engineering (TE) techniques for a variety of tissues, including skin, bone, cartilage, neural, and more. This chapter will examine their primary components, manufacturing processes (such as compression, injection, and resin transfer moulding), and their noteworthy applications in these fields.

***Keywords*:** Biocomposite, biopolymers, reinforcing agent, tissue engineering, resin transfer moulding, injection moulding

**Corresponding author*: rishabhamalviya19@gmail.com; rishabha.malviya@galgotiasuniversity.edu.in

Rishabha Malviya and Mukesh Roy (eds.) Biocomposites in 3D and 4D Printing: Materials, Techniques, and Applications in Regenerative Medicine, (271–298) © 2026 Scrivener Publishing LLC

10.1 Introduction

Investigation into tissue engineering (TE) scaffolds is crucial for establishing their efficacy and safety as a minimally invasive regenerative medicine option. To enhance cell adhesion, migration, differentiation, and proliferation, the TE scaffold approach infuses a biocompatible substance into the body. The scaffold is the name given to this component. A lot of work in tissue engineering has gone into developing biocompatible materials that can mimic the structure and function of unhealthy or injured tissues. The regeneration of brain tissue is a prime example of the life-improving potential of biomaterials and tissue engineering applications. Therefore, there would be significant social benefits to creating biomaterials that could repair damaged tissue and encourage the regeneration of healthy tissue, thereby reducing the impacts of neurodegenerative diseases. For the purpose of simulating real tissue, biocomposite scaffolds are essential. The researchers in this study used biocomposite scaffolds made of conductive polymers and self-assembling nanoribbons to promote axonal outgrowth and cortical cell growth and development [1]. Together, regenerative medicine and tissue engineering have given biomedical researchers and practitioners cutting-edge tools for replacing lost or damaged organs and tissues with biomaterials. The unique characteristics of polymers make them an essential component in tissue engineering. Careful design and structuring of polymers into diverse microstructures can provide the requisite performance. They are frequently utilized as a base for tissue engineering and have the potential to transport cells and therapeutic materials to the intended site [2–5]. Films, nanofibers, and hydrogels are some of the polymeric forms used in tissue engineering [6, 7]. The term “biocomposites” denotes a material that integrates two or more substances with differing chemical and physical properties, incorporating at least one bio-based component. On top of that, depending on their specific uses, biocomposites can be grouped into different categories. Consequently, this special issue sought contributions from a wide range of research organizations to explore the many aspects and applications of biocomposite materials. Both solid and soft forms of these materials are utilized in many biological contexts; the former are antibacterial barriers and structural components, while the latter are drug delivery and tissue regeneration scaffolds, respectively. Tissue engineering relies heavily on the development of biocomposite scaffolds with appropriate properties [8]. Environmental sustainability and the development of biocomposites from sustainable resources have propelled polymer research to new heights this century. The importance of lowering pollution, halting environmental damage, and combating climate change has received a lot of attention as of late. Biocomposites do not produce any harmful gases or compounds; they are non-abrasive, biodegradable, renewable, and safe for human consumption [9]. Fiber reinforcement imparts macroscopic stiffness and strength, bearing the bulk of the structural loads; in contrast, the matrix mostly influences the shape, surface appearance, weather resistance, and overall durability of composites. One type of biocomposites uses thermoplastic and thermoset polymers as reinforcements; the other uses renewable resources (such as starch, soy, and cellulosic plastic) and biodegradable polymers (such as polyesteramides and aliphatic co-polyester) derived from petroleum. Variants that are completely biodegradable have biodegradability and/or composability attributes and are made entirely of biobased components [10–12].

10.2 Biocomposites

Biocomposites are materials consisting of natural fibers and a polymer matrix, typically sourced from renewable resources. These materials are attracting interest due to their ecological advantages, including biodegradability and less dependence on fossil fuels. The advancement of biocomposites commenced in the late 1980s. Biocomposites consist of two or more distinct material phases, with one phase serving as the continuous matrix—composed of either biopolymer, metal and alloy, or bioceramic—while the other phase is discontinuous. The integration of fillers inside a biocompatible matrix material. Matrix and reinforcing component are the primary components. The matrix of a biocomposite material is generally composed of bio-based polymers, like starch, cellulose, or vegetable oils. These polymers function as a binder, uniting the reinforcing fibers and imparting the material with its intended shape and characteristics. The matrix functions to unite the reinforcing elements and offer mechanical support [13].

Fibers are the principal structural load-bearing component of composite systems and the reinforcing agents found in composite materials. The reinforcing component serves as the principal load-bearing element, which may consist of fibers, whiskers, particles, or flakes [13]. The reinforcing fibers in biocomposites are derived from renewable materials, including natural fibers (e.g., hemp, flax, bamboo), agricultural byproducts (e.g., straw, husks), or wood particles. These fibers contribute strength, rigidity, and longevity to the material, rendering it appropriate for many applications. The production and deployment of biocomposites represent significant advancements in substituting traditional non-biodegradable petroleum-derived materials. Biopolymers and bio-based reinforcing agents are the main components of biocomposites. Problems with waste management and the depletion of petroleum reserves may have answers in bio-based products. Biodegradable polymers such as Biopolm, Mater Bi, Bioceta, polycaprolactone, and Scona cell have been studied for their properties, especially as matrix materials for biocomposites [14, 15]. Matrix polymers must have low processing temperature viscosity, reinforcement fibers and matrix must have favorable mechanical properties, and there must be effective fiber-matrix adhesion for biocomposites to be of good quality.

It is possible to classify biocomposites in a number of ways based on their biological and environmental performance and the types of natural fibers and matrices that make them up. Sustainable materials with high mechanical performance and low environmental impact can be created through the interaction of living tissues with bioceramics and biocomposites, respectively, using different types of natural fibers (such as flax, hemp, or sisal) and biodegradable matrices (such as PLA or PHA). Biocomposites are made more robust, flexible, and stable by adding natural fibers [16]. These natural fibers are a great substitute for synthetic ones because they are more accessible, cheaper, and better for the environment. They may have a mineral, plant, or animal origin. Biodegradable matrices are crucial for reinforcing the natural fibers and maintaining the integrity of the biocomposites. Matrixes serve to both insulate the fibers from environmental hazards and facilitate the transfer of stress between them. Using a biodegradable matrix that can naturally decompose in nature, biocomposites are engineered to reduce their impact on the environment after usage. Biodegradable polymers, either naturally occurring or synthesized, can make up these matrices.

10.2.1 Types of Biocomposites

10.2.1.1 Biocomposites Derived from Cellulose Fibers

The incorporation of cellulosic materials into polymer composites has been acknowledged as a noteworthy field of study for more than ten years. Fillers, panels, composites, alloys, mixes, and cellulose derivatives are just a few of the many uses for cellulosic materials in the polymer industry [17]. Cellulose is a common component of tissue engineering composite scaffolds because to its mechanical durability, affordability, and biodegradability [18–20]. A great deal of research from different parts of the globe has focused on composite materials that use cellulose fibers for reinforcement. Natural fibers mostly comprise cellulose, composed of D-glucopyranose units linked by 1,4-b-D-glycosidic linkages (C1 and C4). Cellulose is hydrophilic and has a degree of polymerisation of roughly 10,000, which varies by fiber type. Amorphous and crystalline regions alternate in microfibrils made of biosynthesized chains of cellulose molecules [21]. Microfibrils of cellulose, which give the fiber its mechanical strength, have a diameter of 10 to 30 nm and include 30 to 100 cellulose molecules in an extended chain shape [22]. Cellulose polymer composites have had their mechanical characteristics improved by using matrix graft copolymers and adding a polar group [23, 24]. Common thermoset polymers that include cellulosic fibers as reinforcement include epoxy, polyester, amino resins, and phenolic resins [25–27]. The rubber industry has shown practical and economic interest in short cellulosic fiber-reinforced elastomer composites [28]. However, because they contain synthetic matrix components that do not biodegrade, none of the cellulose-based composites discussed above are completely biodegradable. Cellulose has several uses and is a versatile substance with many different derivatives and composites. Green biocomposites made from renewable resources are becoming more popular, and cellulose is a key ingredient because of its many desirable properties, including its biodegradability, actuation potential, light weight, high chemical modification capacity, and environmental friendliness [29]. The remarkable efficacy of these composites in wound dressing, tissue creation, and drug administration suggests great promise for their use in biomedical engineering.

10.2.1.2 Flax, Hemp, and Ramie-Based Biocomposites

Polymers, whether biodegradable or not, are mixed with natural fibers, including sugarcane, palm, kenaf, bamboo, hemp, sisal, Carlota, banana, pineapple, jute, and coir [30, 31]. Researchers are showing a lot of interest in natural fiber-reinforced composites because of all the great things they can do for the environment, how affordable they are, and how lightweight they are. Flax is among the most robust natural fibers utilized in biocomposites. In addition to being very light, it has great tensile strength and can absorb a lot of water. Flax's structure provides considerable stiffness, rendering it suitable for applications necessitating great mechanical strength, such as automobile parts and construction. Linen fabric is derived from the flax plant and, owing to its considerable strength, has extensive applications. Due to its susceptibility to moisture, improperly processed flax can deteriorate in humid environments, limiting its use in harsh locations.

Nonwoven flax fiber reinforced polypropylene composites were studied for their static and dynamic mechanical properties in relation to the effect of the zein coupling agent [32]. One of the several soluble proteins found in maize is zein. A zein coupling agent improved the mechanical characteristics of composites when used in their construction. The storage modulus of the composites was raised by applying a layer of zein coupling agent, which improved interfacial adhesion. To evaluate their tensile mechanical qualities, the diameter and placement of flax fibers within the stems are taken into consideration [33]. The thermal deterioration and fire resistance of flax fiber composites reinforced with polyester resin, as well as the influence of chemical treatments on surface properties, adhesion, water absorption, and mechanical qualities, have all been studied [34]. Using three soybean oil-based resins—methacrylic anhydride modified, methacrylated, and acetic anhydride modified—as matrices, biocomposites reinforced with flax fiber were created [35].

Hemp belongs to the family Cannabis and is a major crop for bast fiber. This plant is best enjoyed in mild climates, where it blooms only once a year. In addition to having great tensile and impact resistance, hemp is a very flexible and strong fiber. Due to its low pesticide and fertiliser needs, it is easy to cultivate and has little environmental impact. Due to its ecological and mechanical properties, hemp finds widespread use in construction, furnishings, and automobiles. Additional treatments may be necessary to standardize the fiber's qualities if hemp exhibits variable quality. The phase behavior and thermal stability of composites were significantly influenced by modifications to the matrix and fiber. The recyclability of polypropylene composites reinforced with hemp fibers is very significant [36]. According to studies, mechanical qualities of composites made of hemp fiber and polypropylene are preserved even after subjected to several reprocessings.

The ramie plant is part of the Urticaceae family, also referred to as Boehmeria, which comprises over one hundred distinct species. The chemical composition and manufacturing sites of Ramie have mostly constrained its utilisation as a textile fiber, necessitating more extensive pre-treatment than other commercially important bast fibers. Ramie fiber-reinforced polypropylene composites were made using a hybrid process that combined melt blending and injection moulding [37]. The length, composition, and pre-treatment of the ramie fibers were varied to create a wide range of PP composites. Tensile, flexural, and compressive strengths were all found to be improved as the fiber content and length were increased, according to the results. As a matrix, Ramie fibers were reinforced with epoxy-bioresin, polyester, soy protein, and epoxy.

10.2.1.3 Jute-Based Biocomposites

Jute comes from more than a hundred different types of plants that are members of the genus Corchorus [38–40]. At present, it is the most produced bast fiber, and it is also one of the most cost-effective natural fibers. Growing jute is most suited to the climates of Bangladesh, India, and China. Jute is a widely utilized agro-fiber due to its elevated tensile modulus and minimal elongation at break [23]. Jute is a cost-effective and biodegradable natural fiber, extensively utilized in diverse applications such as packaging, furniture components, and technical textiles. Although it is considered one of the lightest natural fibers, this one possesses exceptional tensile strength. Jute may be more susceptible to moisture than other natural fibers like hemp or flax, and it possesses lower mechanical strength relative to these fibers [41, 42]. The particular stiffness and strength of this fiber are analogous to those of glass fiber [43–46], despite its low density of 1.45 grammes per cubic centimetre. Jute possesses a modulus that exceeds that of glass fiber, and when evaluated by the ratio of its modulus to its cost, jute emerges as a significantly more beneficial material. Composites reinforced with jute fibers were tested in a polyester resin matrix. Elasticity, notched strength, fracture criteria, impact damage characterisation, thermal behavior, weathering, and the effects of silane treatment were also investigated [47].

10.2.2 Processing Methods of Biocomposites

In both the commercial and residential sectors, there has been a considerable increase in the demand for items that are created from biocomposite materials over the course of the previous decade. Researchers have devised various processing strategies for the effective manufacture of biocomposite components [48, 49]. Compression moulding, Injection moulding, vacuum infusion, hand layup, resin transfer moulding (RTM), and direct extrusion are some of the production methods that are utilized in the manufacturing of biocomposites [50–52]. Multiple researchers have made substantial use of these methodologies and have developed them over time. The manufacturing process can be influenced by various elements, including humidity, temperature, and pressure, among others [53].

The majority of the preforms are made up of fibers, textiles, or nonwoven materials. During the manufacturing process of composites, prepregs are employed to a significant degree [54]. In the manufacture of composites, Equation (10.1) is utilized quite frequently.

$$Vf = \frac{Wf/\rho f}{(Wf/\rho f)+(Wm/\rho m)} \tag{10.1}$$

where the fiber-volume percentage is denoted by Vf, the fiber weight is Wf, and the matrix weight is Wm. The fiber density is represented by ρf, while the matrix density is represented by ρm.

The composite manufacture is optimized concerning temperature, pressure, and moulding duration. Preheating natural fibers to diminish moisture content prior to composite processing is typically essential. High temperatures cause cellulose to degrade, which negatively affects the mechanical characteristics of the composites. Fiber aggregation, caused by inefficient dispersion of fibers within the matrix, lowers tensile strength [54]. Typically, biocomposites are formulated or produced with filler ingredients utilizing a corotating twin-screw extruder. The corotating twin screw extruder is commonly employed for its superior mixing capabilities. The material is subjected to significant shearing forces, resulting in the abrasion of the filler components, occasionally causing damage. This approach involves the continual feeding of particles into the extruder using gravimetric dosing. Generally, mild processing techniques should be employed for producing biocomposites. This processing entails a brief residence time, a mild screw configuration, reduced processing temperatures, and, if necessary, a predrying of the materials. Failure to address this issue may result in damage that adversely affects the succeeding injection moulding process due to the material's recurrent exposure to heat stress.

10.2.2.1 Injection Moulding

Injection moulding is the preferred technique for large-scale production of composites (Figure 10.1) [55]. The inaugural machine is documented to have been patented in the United States in 1872, explicitly for application with celluloid. This procedure involves injecting ingredients into a mould to create a biocomposite, enabling the production of various thermosetting or thermoplastic polymers. The injection moulding technique is employed for specialized machinery and comprises three primary components: the clamping unit, the injection unit, and the moulding unit. Two principal varieties of injection moulding are ram injection and screw injection. Screw injection facilitates accelerated melting and enhanced mixing, frequently employed for intricate structures. The process of injection moulding involves inserting molten plastic or rubber into a mold's chamber using a ram or a screw-type plunger. The material assumes the form of the mould as it cools. Mould cavities are filled with molten plastic to create unique shapes. The principle of injection moulding involves heating and inserting materials into a mould [56, 57]. As an advantage over direct extrusion, injection moulding can create materials with three-dimensional shapes for a variety of industrial uses [58]. Conventional injection moulding machines are capable of processing natural fiber-reinforced polymers. Fundamentally, injection moulding is an uncomplicated procedure. In a standard injection moulding procedure, fiber composites in pellet form are introduced *via* a hopper and subsequently transported by a screw within a heated barrel. During this transportation, the thermal energy from the regions and the pressure exerted by the screw liquefies the plastic and compress it to approximately 2.0104 psi (1.4105 kPa). Once the necessary amount of material has been melted in a barrel, it is forced into the mould through a nozzle by the screw. Once cooled, the material takes on the desired shape. When it comes to thermoplastic encapsulation of electrical equipment used in healthcare, injection moulding is dominant.

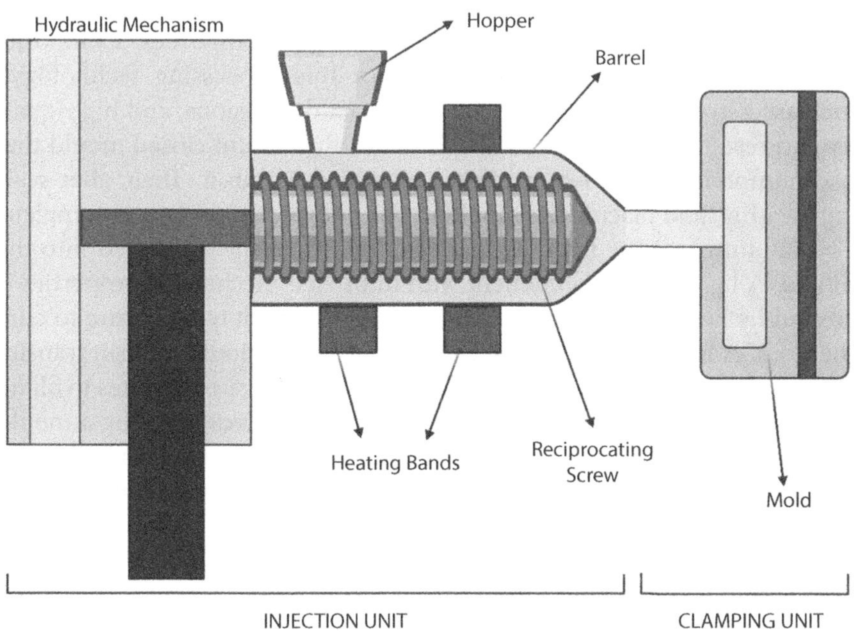

Figure 10.1 Diagrammatic representation of injection moulding process.

Surface treatments of biocomposites improve compatibility between fibers and matrix materials and ensure that fibers are evenly distributed throughout the matrix [59]. The primary benefits of the process encompass its adaptability in shaping a diverse array of goods, the simplicity of integrating automation, the potential for elevated production rates, and the capability to produce items with precise tolerances. The fundamental concept of injection moulding can also be modified for application with thermosetting materials. Injection moulding is employed to produce composite items that necessitate intricate, large-volume forms. The benefits of employing the technique include superior dimensional tolerance, reduced cycle times, and minimal post-processing requirements [60]. Equipment for injection moulding is expensive, and the method has limitations due to a lack of top members or short fiber reinforcement [61]. The use of injection moulding has multiple positive effects on the environment, including less waste, improved recyclability, and long product life.

10.2.2.2 Resin Transfer Moulding

Thermosetting composites made by resin transfer moulding (RTM) have great performance and are used extensively (Figure 10.2) [62]. The RTM process is one method for fabricating technical composites. Composite parts used in aircraft may benefit from this processing technology's increased strength, complex structures, exact dimensions, and high-quality end result. This method made use of an aluminium closed mould that had reinforcing fibers like graphite embedded within it. Then, after sealing, heating, and placing the mould under hoover, the process is complete. To fully immerse the fiber layup, heated plastic resin is injected into the mould. A lightweight component with improved mechanical properties is created by heating the enclosed mould to a sufficient temperature to cure the resin. A fiber preform is placed within a sealed mould in resin transfer moulding (RTM), and then resin is injected through specific gates to fill up all the spaces between the fibers. One method involves injecting a mould with natural fibres into a biodegradable resin mixture, and then letting the mixture harden. Whether the fiber reinforcement is long or man-made, it is cut using a template and a knife or scissors in this operation. Using instruments like vacuum or pressure techniques, the reinforcing parts are put within the mould cavity after being bonded with a thermoplastic binder. Resins typically employed in this procedure comprise polyester, epoxy, vinyl ester, and phenolic compounds [63]. The procedure is rapid and facilitates the fabrication of intricate components. When injecting

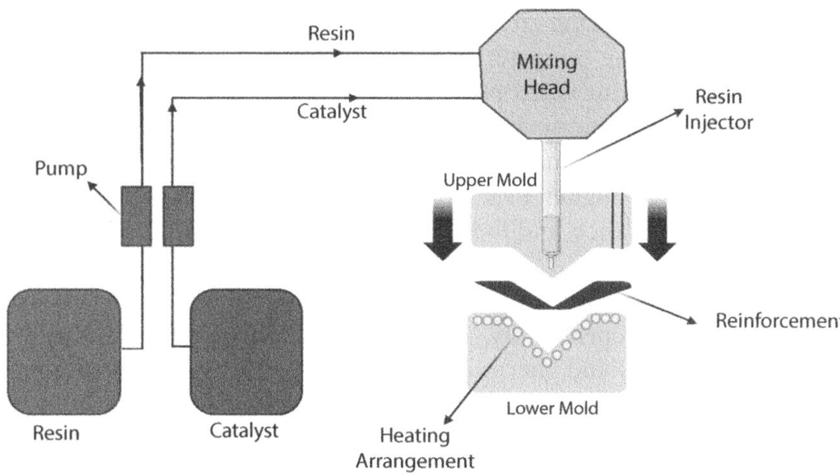

Figure 10.2 Resin transfer moulding.

biodegradable polymers (PHA, PLA, etc.) into biocomposites, it is crucial to tightly control the temperature and pressure to guarantee equitable distribution and avoid thermal degradation of the material.

10.2.2.3 Compression Moulding

The conventional method of compression moulding involves placing the hot moulding material into an open, preheated mould cavity. Its primary use is the fabrication of larger flat or somewhat curved sections, but it was originally developed for use in metal substitution applications as a composite component. Compression moulding is a process that uses heat and pressure inside a mould to compact a layer of natural fibers and a matrix. Biocomposites, employed for substantial components like automotive panels or structural building parts, are generally manufactured in this manner. Compression moulding is a manufacturing technique that produces sheets of fiber-reinforced composites, plastics, and rubber by forcing heated rubber and plastic materials into an open mould chamber. Then, the open mould is sealed using pressure, ensuring that the heated plastic materials reach every portion of the mould. Curing, sometimes called vulcanisation, occurs when heated materials are subjected to controlled heat and pressure. This processing technique utilises plastic components in the shapes of pellets, powder, preforms, or putty-like substances. In order to modify this technique for biocomposites, it was necessary to make precise adjustments to the pressure and temperature parameters. This was done in order to accommodate the qualities of biodegradable matrices, which may exhibit lower melting temperatures than traditional synthetic materials. Compression moulding is an optimal high-volume, high-pressure technology for fabricating intricate, high-strength reinforcements. In comparison to thermoplastics, these materials typically have superior thermophysical properties, such as higher heat flux levels, better thermal conductivity, higher impact resistance, and mechanical capability retention at high temperatures [64].

Compression moulding provides a high level of automation, brief cycle durations, strong repeatability, and superior dimensional stability for both thermoplastics and thermosets, which accounts for its widespread uses across different industrial sectors, including the automotive industry [63]. The compression moulding technique is among the most economical methods for mass production, owing to its cost-effectiveness and efficiency. This approach is highly efficient, resulting in minimal material waste. The technology generates high-volume components of diverse lengths, thicknesses, and complexities at a superior cost per unit compared to alternative production methods. They possess dimensional accuracy, thermal resistance, and superior surface quality.

10.2.2.4 Extrusion

Extrusion is a continuous method employed for the compounding of biocomposite materials utilizing thermoplastic resin. A thermoplastic resin, such as polyethylene (PE) or polypropylene (PP) combined with biofiber, is processed utilizing the extrusion technique. A feed screw is used to continuously feed solid plastic (beads or pellets) into a heated chamber during the extrusion process [65]. Single-screw or twin-screw extruders are employed, with the screws rotating either clockwise or anticlockwise, contingent upon the final products. The processing screws' arrangement determines the fiber-reinforced material's shape, final dimensions, and properties, much like the injection moulding technique. Extrusion is a thermal melting method employed for the continuous fabrication of biocomposites. In the production of biocomposites, thermoplastic resin beads or pellets are combined with 30 to 40% of short or long natural fibers, then compressed, melted, and extruded through a die. The solid resin is crushed, melted, and extruded from the chamber through a die during transportation. The extruded product is cooled and trimmed to the specified specifications. Upon completion of the cooling process, the molten component yields a continuous profile. The ultimate characteristics of biocomposites produced through extrusion are dependent upon two criteria [66]:

1) Type of components used:
   All materials used in thermo-mechanical processing must be in an ideal physical condition. When components are subjected to high shear forces and temperatures during extrusion, they may break down and release volatile chemicals.
2. Cylinder temperature, shear pressures, residence time distribution, screw speed, and other factors impact the biocomposites' homogeneity during extrusion processing.

The extrusion method is utilized to manufacture products with a uniform cross-sectional shape by expelling molten material *via* a specific aperture (die). This technique offers advantages over other processing methods, including as compression and thermoforming, because to its capacity to create highly intricate cross-sectional shapes.

10.3 Applications of Biocomposites in Tissue Engineering

The processing technologies currently in use present a wide range of innovative opportunities for the research and industry sectors, particularly for tissue engineering applications. Biocomposites reduce the likelihood of implant rejection and enhance compatibility in orthopaedic implants. Regenerative medicine utilises biocompatibility to facilitate the production of novel tissues and organs. Biocomposites are essential in tissue engineering as they offer scaffolding for cellular development. Biocompatible and photoprotective tissue engineering applications utilize silk fibroin and melanin. When making scaffolds and incorporating them into living organisms, polycaprolactone and collagen are fine-tuned for maximum flexibility. The versatile matrix of gelatin, alginate, and fibrinogen can be tailored to meet various tissue engineering requirements. Collagen type I and gelatin methacryloyl, which have photopolymerization characteristics, are used for accurate scaffold modeling. Biocompatible scaffolds that promote cell proliferation are fabricated using fibrin, alginate, and genipin. Conductive PEDOT, chitosan, and gelatin composites are utilized in applications requiring electrical stimulation. The scaffolds must exhibit biocompatibility, possess a regulated degradation rate that aligns with cellular proliferation in the damaged region, and facilitate both scaffold renewal and the restoration of the injured tissue or area. In addition, the scaffold's mechanical characteristics, such as its rigidity, should be very similar to those of the damaged tissue [16, 67–69]. Depending on the rate of regeneration in the damaged area, biocomposites are utilized in tissue engineering to fabricate artificial scaffolds that encourage cell proliferation and tissue regeneration. Biomedical applications include tissue engineering, wound healing, and drug delivery systems benefit from biocomposites due to their biocompatibility and biodegradability (Figure 10.3).

Tissue engineering includes various biomaterials and applications, such as neural tissue regeneration, which can enhance individuals' quality of life. The development of biomaterials that can repair damaged tissue and foster the formation of healthy tissue yields significant socioeconomic advantages, therefore mitigating the effects of neurodegenerative disorders. Natural fiber composites are applicable in biomedical fields for the healing and rebuilding of bone and tissue [70]. Biocomposites are made stronger and more rigid by adding a dispersed phase of natural fiber reinforcing components to a biopolymer matrix [71]. Historically, biocomposite scaffolds have depended on natural fibers including ramie, coir, flax, silk, and jute for structural support. Hemp, kenaf, bamboo, wheat, sugarcane, cotton, coconut, oil palm, and sisal are increasingly recognized as natural fibers with potential applications in the fabrication of scaffolds for bone tissue engineering. Highly reinforced, environmentally acceptable scaffolds can be made from agricultural wastes including walnut waste,

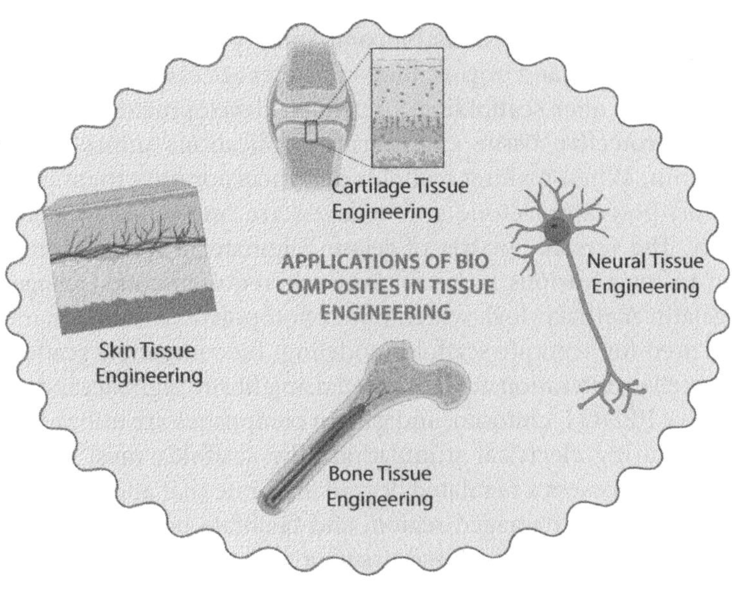

Figure 10.3 Various applications of biocomposite in tissue engineering.

sugarcane residues, and almond shells by incorporating these materials into a variety of biopolymer matrices [72]. The tensile strength and other mechanical properties of biocomposites composed of natural fibers are very fiber-specific. The biological mechanism of the TE scaffold approach entails the introduction of a biocompatible element, referred to as the scaffold, into the organism, facilitating cell attachment, migration, differentiation, and proliferation. Biocompatible materials that can mimic the look and function of injured or diseased tissues have been the subject of substantial investigation by tissue engineering researchers. When biocompatible materials are incorporated into neighbouring tissues, it triggers the release of the body's own extracellular matrix (ECM), which aids in the repair of injured tissue. In tissue engineering, the creation of biocomposite scaffolds with appropriate properties is of utmost importance.

10.3.1 In the Field of Bone Tissue Engineering

The hierarchical architecture of bone, spanning from the nanoscale to the macroscale, is defined by the inclusion of both organic (collagen) and inorganic (nano-hydroxyapatite) constituents [73]. The dynamic tissue referred to as bone safeguards essential organs, generates blood cells, and facilitates movement, among its numerous other activities. Furthermore, it serves as a repository for minerals throughout the body. In addition to cells, bones also include a collagenous extracellular matrix and substances that do not contain collagen. Bone cells can be classified into three types: osteocytes, osteoblasts, and osteoclasts. Osteoblasts are cells that help build bone, while osteoclasts are cells that break down bone [74]. Mature osteoblasts, or osteoocytes, make up the bulk of bone cells and are embedded in the matrix of the bone. As mechanosensors, these structures detect changes in the mechanical tension on bone [75, 76]. Creating implantable bone substitutes for large, non-healing skeletal lesions is the goal of bone tissue engineering (BTE). One polysaccharide that has recently come to the fore in BTE is chitosan (CS), which is used to make bone scaffolds. Since its discovery twenty years ago, the natural polymer chitosan—derived from chitin, the main component of crab exoskeletons—has made substantial contributions to bone tissue engineering. The chitosan-hydroxyapatite composite is well regarded for its biodegradability and biocompatibility. Biocomposites comprise two components: reinforcement and matrix.

BTE experts have created new materials to address issues with conventional grafting methods [77]. The "scaffold" is shorthand for the building block of BTE. The three-dimensional matrix known as the "scaffold" provides an environment favorable for cell adhesion, proliferation, and differentiation. A bone scaffold is another name for it [78]. Bone tissue engineering is an intriguing field because it seeks to develop scaffold materials that can mimic the structure and mechanical qualities of natural bone. Metals, ceramics, and polymers are common components of scaffolds. Scaffolds that are optimal for cell proliferation and osteogenic differentiation should possess the necessary mechanical qualities, including biocompatibility, non-toxicity, pore size, surface topology, and load-bearing capability. Orthopaedic and dental applications in particular have shown promise for biocomposites made of biodegradable polymer matrices and bioceramics. One of the many possible applications of tissue engineering—which makes use of materials such as cells, growth hormones, ceramics, metals, and polymers—is the creation of artificial bones. When it comes to mimicking the normal characteristics and functions of bone, polymer ceramic composites work well. Alginate, an anionic polymer with numerous biomedical applications, is gaining popularity as a biocompatible and gel-forming component in bone tissue manufacturing. Scaffolds for bone and cartilage tissue are made of hydroxyapatite (HA) materials, which are biocompatible and osteoconductive. This is due to the fact that HA materials mimic the natural extracellular matrix of bone and improve cell adhesion [79]. The biomedical applications of alginate and its hydroxyapatite composite scaffold are extensive and varied, spanning from bone tissue engineering to cell and growth factor delivery to wound healing. A primary hurdle in bone tissue engineering is the development of an appropriate scaffold material. The biodegradability and biocompatibility of poly (lactic acid) (PLA) render it a highly valued matrix material for bone tissue engineering applications. PLA/HA composites have garnered significant interest because to their biodegradability, superior mechanical strength, osteoconductivity, and osteoinductivity [80–82].

10.3.2 In Neural Tissue Engineering

The nervous system is an essential component of the body, and any disruption to this system, whether it be the result of an injury or a disease, can have severe and sometimes fatal implications. By virtue of its intricate physiology and restricted capacity for regeneration, the damaged nervous system presents a significant obstacle in the way of its restoration. Neural tissue engineering constitutes a distinct sub-discipline of tissue engineering. Neural tissue engineering primarily seeks techniques to mitigate inflammation and fibrosis following the implantation of foreign materials. When the CNS and PNS suffer damage, it is common practice to implant foreign material like grafts and scaffolds to aid in nerve regeneration and recovery. Recently, there has been a rise in the amount of research conducted on the engineering of brain tissue by the utilisation of novel electroactive biomaterials. The restricted potential of brain tissues to renew themselves presents a significant challenge for tissue engineering, which is why the process of regenerating neural systems is so difficult. Damage to the nervous system can be caused by a variety of factors, including neurodegenerative disorders like Alzheimer's, multiple sclerosis, or Parkinson's disease, as well as accidents [83–85].

The CNS and PNS have recently seen significant advancements in treatment and regeneration. The unique properties of natural polymers have made them extensively used in nerve tissue regeneration. Reinforcement of mechanical strength for neurite growth, regulation of biological processes that guide axonal development, protection of newborn tissue formation, and improvement of integration with healthy tissue are all functions performed by polymeric structures. Neural tissue engineering has numerous significant hurdles, including the intricate nature of the tissue and the inadequacy of traditional therapies. The advancement of tissue engineering is essential for both fostering neuroprotection against infection and inflammation and for augmenting neurite growth and the functionality of newly produced tissues [86]. Biocompatibility, adaptability, degradation rate, bioactivity, and longitudinal channels to improve axon regeneration are some of the numerous qualities that should be present in the ideal neuronal biomaterial.

Recent evidence indicates that neural tissue engineering can facilitate the recovery of neural cells through constructs like scaffolds, sheets, and nanoparticles developed *via* bioengineering. These structures provide biological, physiochemical, and topological signals that promote neuronal regeneration. The integration of growth factors and other therapeutic compounds into biocompatible polymers accomplishes this objective [87, 88]. A novel method for addressing neuronal injuries and illnesses utilises graphene's distinctive electrical and surface characteristics to replicate *in vivo* signals and facilitate stem cell neuronal differentiation. Studies have shown that graphene-based nanomaterials (GN) can speed up the process of neural cell differentiation and increase the proliferation rate of neural cells [89–91]. However, the specific mechanism *via* which graphene affects the maturation of neural stem cells remains a mystery. On top of that, GNs can induce neuronal differentiation for tissue repair and directly enhance neural regeneration when used as electrodes. Electrical stimulation is essential for promoting cell growth and mimicking *in vivo* brain networks [92]. This is because brain cells in real tissues are connected through neurites. Electrical stimulation is also essential for enhancing cell proliferation.

In addition, GNs were employed in the process of neural regeneration by loading medications and distributing them across the blood-brain barrier, which is generally a barrier that prevents drug molecules from passing through. The importance of graphene in the growth and multiplication of brain cells has been demonstrated in numerous studies [93, 94]. Materials based on graphene have come a long way in the last few decades [95]. Thus far, this initial preclinical research is promising and inspiring; yet, significant hurdles persist that impede the achievement of future clinical applications.

10.3.3 In Skin Tissue Engineering

The skin, which covers the entire body and accounts for around 8% of its total mass, is the biggest organ. As a protective barrier, the skin prevents potentially dangerous chemicals and germs from penetrating the body. In order to keep the elements out, the epidermis is the skin's outermost layer, while the dermis, composed of connective tissue, is the skin's innermost layer [96, 97]. There are various types of skin wounds, including superficial (affecting only the outer layer), partial thickness (affecting only part of the dermis), full-thickness (affecting all of the dermis, including sweat glands and hair follicles), and complex (affecting all layers). In two to three months, a chemical injury, burn, or surgical incision will heal. In contrast to common wounds, chronic wounds including pressure sores, diabetic ulcers, and infectious wounds take much longer to heal. Investigations into innovative regenerative medicine concepts, particularly those focused on skin tissue regeneration for the treatment of chronic or acute wounds, can significantly benefit from utilizing skin as a model organ. A growing number of patients are facing high morbidity and amputation risks due to chronic wounds, which are becoming an increasingly pressing worldwide health and economic concern due to the inadequacy of present treatment options and the resulting enormous societal and economic burden. Extensive dermal defects have recently been repaired using skin replacements, such as xenografts, autografts and allografts [98].

The three main elements of tissue engineering—growth factors, scaffolds, and cells—make up a unique scientific approach to replacing or repairing damaged organs with biological alternatives [99]. The main building blocks that interact with cells and growth factors to form new tissues are scaffolds. Consequently, tissue-engineering approaches to improve organ function restoration and tissue regeneration have been the subject of substantial investigation. As a crucial component of skin tissue engineering, three-dimensional scaffolds offer structural support and an extracellular matrix (ECM) analogue. This allows host cells to infiltrate and, *via* cell proliferation and differentiation, produce new tissues or organs [100]. Skin tissue engineering relies heavily on biocomposites, which facilitate wound healing, cell adhesion, and tissue regeneration by acting as scaffolds. To improve mechanical strength and biological activity, these materials include reinforcing agents such as nanoclay, silver nanoparticles, or bioactive glass with biocompatible polymers including gelatin, chitosan, alginate, or collagen. To facilitate skin regeneration and wound healing, it is essential to comprehend the impact of skin tissue engineering scaffolds on healing mechanisms [101]. Wound healing is aided by chitosan, which increases the migration of PMN cells. We postulate two possible mechanisms: (1) activation of complement and (2) stimulation of neutrophil IL-8 excretion. The amount of acetylation is strongly related to the overexpression of IL-8, a neutrophil chemotactic agent, which chitosan promotes.

10.3.4 In Cartilage Tissue Engineering

Articular cartilage deterioration in synovial joints is the hallmark of osteoarthritis (OA), a common joint condition. It is the predominant form of arthritis that induces discomfort and rigidity, particularly in the hip, knee, and thumb joints. Cartilage tissue engineering has arisen as a viable strategy to overcome the shortcomings of existing osteoarthritis treatments, with the objective of regenerating cartilage and recovering the normal function of compromised joints. Similar to other domains of tissue engineering, cartilage tissue engineering employs various manufacturing processes and biomaterials to create structures. To address existing limits in osteoarthritis treatment, the research industry has concentrated on tissue engineering-based regenerative medicine, which has emerged as a potential method for cartilage regeneration and functional restoration. The integration of growth factors, chondrocytes, and mesenchymal stem cells (MSCs) into scaffolds that promote cell formation and proliferation is a key component of cartilage tissue engineering [102]. The effectiveness of a scaffold for cartilage tissue engineering is mostly affected by the biomaterials used to make it. Progress in biomaterials includes a variety of formulations designed to promote natural tissue (cartilage) regeneration. When it comes to healing damaged or sick tissues, tissue engineering is still a promising therapeutic option. Biomaterial scaffolds that can regulate the size and shape of the developing tissue while also encouraging the attachment, proliferation, and specialisation of the cells that are integrated with them and bioactive compounds like DNA snippets, cytokines, and growth factors have been the primary focus of research up to this point [103].

In addition to influencing the designed construct's form and metabolic function, scaffolds force the mechanical characteristics of the natural tissue to be replaced to endure [104, 105].

10.4 Conclusion

Due to the fact that biocomposites are made from a finely balanced blend of ceramics, polymers, and naturally occurring fibers, they have the potential to drastically transform patient care as well as the outcomes of healthcare in general. Their specific set of qualities makes them ideal for many different kinds of medical uses. Biocomposites are an important part of sustainable development because they provide a viable substitute for conventional materials created from fossil fuels. Tissue engineering seeks to repair injured tissues by combining cells with biomimetic substances that promote the growth of new tissues. They substantially mitigate environmental impact owing to their biodegradability and utilisation of renewable materials. Numerous opportunities for novel advancements in the field of biocomposites are presented by the techniques described in this chapter. This section concentrates on potential biocomposite materials and their applications in tissue engineering. Biocomposites and their use in tissue engineering are explored extensively in this chapter.

AI disclosure statement: The author acknowledges the use of AI-based tools for minor language editing or grammatical correction. However, all intellectual contributions, analyses, and conclusions presented in this chapter are entirely the author's own.

References

1. Smith, A.M., Pajovich, H.T., Banerjee, I.A., Development of self-assembled nanoribbon bound peptide-polyaniline composite scaffolds and their interactions with neural cortical cells. *Bioengineering*, 5, 1, 6, 2018 Jan 13.
2. Wells, C.M., Harris, M., Choi, L., Murali, V.P., Guerra, F.D., Jennings, J.A., Stimuli-responsive drug release from smart polymers. *J. Funct. Biomater.*, 10, 3, 34, 2019 Jul 31.
3. Mohapatra, A., Wells, C., Jennings, A., Ghimire, M., Mishra, S.R., Morshed, B.I., Electric stimulus-responsive chitosan/MNP composite microbeads for a drug delivery system. *IEEE Trans. Biomed. Eng.*, 67, 1, 226–33, 2019 Apr 16.

4. Wang, W., Wei, J., Lei, D., Wang, S., Zhang, B., Shang, S., Bai, B., Zhao, C., Zhang, W., Zhou, C., Zhou, H., 3D printing of lithium osteogenic bioactive composite scaffold for enhanced bone regeneration. *Composites, Part B*, 256, 110641, 2023 May 1.
5. Markandan, K. and Lai, C.Q., Fabrication, properties and applications of polymer composites additively manufactured with filler alignment control: A review. *Composites, Part B*, 256, 110661, 2023 May 1.
6. Comotto, M., Saghazadeh, S., Bagherifard, S., Aliakbarian, B., Kazemzadeh-Narbat, M., Sharifi, F., Mousavi Shaegh, S.A., Arab-Tehrany, E., Annabi, N., Perego, P., Khademhosseini, A., Breathable hydrogel dressings containing natural antioxidants for management of skin disorders. *J. Biomater. Appl.*, 33, 9, 1265–76, 2019 Apr.
7. Tamayol, A., Akbari, M., Zilberman, Y., Comotto, M., Lesha, E., Serex, L., Bagherifard, S., Chen, Y., Fu, G., Ameri, S.K., Ruan, W., Flexible pH-sensing hydrogel fibers for epidermal applications. *Adv. Healthcare Mater.*, 5, 6, 711–9, 2016 Mar.
8. Kumar, S., Malviya, R., Uniyal, P., Srivastava, S., Sridhar, S.B., Rab, S.O., Singh, B., Plant Fiber Composites for Biomedical Applications: Advances and Prospective. *Curr. Pharm. Biotechnol.*, 26, 13, 2111–26, 2025 Oct.
9. Zwawi, M., A review on natural fiber bio-composites, surface modifications and applications. *Molecules*, 26, 2, 404, 2021 Jan 14.
10. Faruk, O., Bledzki, A.K., Fink, H.P., Sain, M., Biocomposites reinforced with natural fibers: 2000–2010. *Prog. Polym. Sci.*, 37, 11, 1552–96, 2012 Nov 1.
11. John, M.J. and Thomas, S., Biofibres and biocomposites. *Carbohydr. Polym.*, 71, 3, 343–64, 2008 Feb 8.
12. Fowler, P.A., Hughes, J.M., Elias, R.M., Biocomposites: technology, environmental credentials and market forces. *J. Sci. Food Agric.*, 86, 12, 1781–9, 2006 Sep.
13. Rudin, A. and Choi, P., Biopolymers, in: *The elements of polymer science & engineering*, vol. 3, pp. 521–35, Elsevier, Amsterdam, Netherlands, 2013.
14. Herrmann, A.S., Hanselka, H., Niederstadt, G., Faserverbundwerkstoffe und Verfahren zur Herstellung von Strukturbauteilen. DE 44 20 817, 1994 Jun.
15. Riedel, U. and Chowanietz, M., Faserverbundwerkstoff und Verfahren zu seiner Herstellung. *DE Akz.*, 196, 47, 671–2, 1996 Nov.
16. Bălțatu, M.S., Vizureanu, P., Sandu, A.V., Achitei, D.C., Nergis, D.D., Perju, M.C., Nabiałek, M., Biocomposites: Materials, Properties, and Applications.
17. Maldas, D. and Kokta, B.V., Current trends in the utilization of cellulosic materials in the polymer industry. *Trends Polym. Sci.*, 1, 6, 174–8, 1993.
18. Trache, D., Hussin, M.H., Haafiz, M.M., Thakur, V.K., Recent progress in cellulose nanocrystals: sources and production. *Nanoscale*, 9, 5, 1763–86, 2017.
19. Fu, L.H., Qi, C., Ma, M.G., Wan, P., Multifunctional cellulose-based hydrogels for biomedical applications. *J. Mater. Chem. B*, 7, 10, 1541–62, 2019.

20. Dutta, S.D., Patel, D.K., Lim, K.T., Functional cellulose-based hydrogels as extracellular matrices for tissue engineering. *J. Biol. Eng.*, 13, 1, 55, 2019 Jun 20.
21. Haafiz, M.M., Eichhorn, S.J., Hassan, A., Jawaid, M., Isolation and characterization of microcrystalline cellulose from oil palm biomass residue. *Carbohydr. Polym.*, 93, 2, 628–34, 2013 Apr 2.
22. Kalia, S., Dufresne, A., Cherian, B.M., Kaith, B.S., Avérous, L., Njuguna, J., Nassiopoulos, E., Cellulose-based bio-and nanocomposites: a review. *Int. J. Polym. Sci.*, 2011, 1, 837875, 2011.
23. Mohanty, A.K., Misra, M.A., Hinrichsen, G.I., Biofibres, biodegradable polymers and biocomposites: An overview. *Macromol. Mater. Eng.*, 276, 1, 1–24, 2000 Mar 1.
24. Kishi, H., Yoshioka, M., Yamanoi, A., Shiraishi, N., Composites of wood and polypropylenes I.
25. Chauhan, A., Rajput, S., Malviya, R., Sridhar, S.B., Production and Multidisciplinary Applications of Bioplastic: A Biodegradable and Eco-Friendly Alternative to Plastic. *Curr. Green Chem.*, 2025.
26. Zadorecki, P., Karnerfors, H., Lindenfors, S., Cellulose fibers as reinforcement in composites: determination of the stiffness of cellulose fibers. *Compos. Sci. Technol.*, 27, 4, 291–303, 1986 Jan 1.
27. Hua, L., Flodin, P., Rönnhult, T., Cellulose fiber-polyester composites with reduced water sensitivity (2)—surface analysis. *Polym. Compos.*, 8, 3, 203–7, 1987 Jun.
28. Gejo, G., Kuruvilla, J., Boudenne, A., Sabu, T., Recent advances in green composites. *Key Eng. Mater.*, 425, 107–66, 2010 Mar 4.
29. Ali, S.N. and Arafat, M.T., Cellulose-based biocomposites, in: *Green Biocomposites for Biomedical Engineering*, pp. 135–195, Woodhead Publishing, Cambridge, United Kingdom, 2021 Jan 1.
30. Etaati, A., Pather, S., Rahman, M., Wang, H., Ground hemp fibers as filler/reinforcement for thermoplastic biocomposites. *Adv. Mater. Sci. Eng.*, 2015, 1, 513590, 2015.
31. Etaati, A., Pather, S., Fang, Z., Wang, H., The study of fibre/matrix bond strength in short hemp polypropylene composites from dynamic mechanical analysis. *Composites, Part B*, 62, 19–28, 2014 Jun 1.
32. John, M.J. and Anandjiwala, R.D., Chemical modification of flax reinforced polypropylene composites. *Composites, Part A*, 40, 4, 442–8, 2009 Apr 1.
33. Charlet, K., Baley, C., Morvan, C., Jernot, J.P., Gomina, M., Bréard, J., Characteristics of Hermès flax fibres as a function of their location in the stem and properties of the derived unidirectional composites. *Composites, Part A*, 38, 8, 1912–21, 2007 Aug 1.
34. Alix, S., Philippe, E., Bessadok, A., Lebrun, L., Morvan, C., Marais, S., Effect of chemical treatments on water sorption and mechanical properties of flax fibres. *Bioresour. Technol.*, 100, 20, 4742–9, 2009 Oct 1.

35. Adekunle, K., Åkesson, D., Skrifvars, M., Biobased composites prepared by compression molding with a novel thermoset resin from soybean oil and a natural-fiber reinforcement. *J. Appl. Polym. Sci.*, 116, 3, 1759–65, 2010 May 5.
36. Akter, M., Uddin, M.H., Tania, I.S., Biocomposites based on natural fibers and polymers: A review on properties and potential applications. *J. Reinf. Plast. Compos.*, 41, 17-18, 705–42, 2022 Sep.
37. He, L.P., Tian, Y., Wang, L.L., Study on ramie fiber reinforced polypropylene composites (RF-PP) and its mechanical properties. *Adv. Mater. Res.*, 41, 313–6, 2008 May 17.
38. Lodha, P. and Netravali, A.N., Characterization of stearic acid modified soy protein isolate resin and ramie fiber reinforced 'green'composites. *Compos. Sci. Technol.*, 65, 7-8, 1211–25, 2005 Jun 1.
39. Kumar, R. and Zhang, L., Aligned ramie fiber reinforced arylated soy protein composites with improved properties. *Compos. Sci. Technol.*, 69, 5, 555–60, 2009 Apr 1.
40. Kishi, H. and Fujita, A., Wood-based epoxy resins and the ramie fiber reinforced composites. *Environ. Eng. Manage. J.*, 7, 5, 517–23, 2008 Sep 1.
41. Vaiani, L., Boccaccio, A., Uva, A.E., Palumbo, G., Piccininni, A., Guglielmi, P., Cantore, S., Santacroce, L., Charitos, I.A., Ballini, A., Ceramic materials for biomedical applications: an overview on properties and fabrication processes. *J. Funct. Biomater.*, 14, 3, 146, 2023 Mar 4.
42. LeGeros, R.Z., Properties of osteoconductive biomaterials: calcium phosphates. *Clin. Orthop. Relat. Res. (1976-2007)*, 395, 81–98, 2002 Feb 1.
43. Zafeiropoulos, N.E. and Baillie, C.A., A study of the effect of surface treatments on the tensile strength of flax fibres: Part II. Application of Weibull statistics. *Composites, Part A*, 38, 2, 629–38, 2007 Feb 1.
44. Mohanty, S., Nayak, S.K., Verma, S.K., Tripathy, S.S., Effect of MAPP as a coupling agent on the performance of jute–PP composites. *J. Reinf. Plast. Compos.*, 23, 6, 625–37, 2004 Apr.
45. Mishra, S., Naik, J.B., Patil, Y.P., The compatibilising effect of maleic anhydride on swelling and mechanical properties of plant-fiber-reinforced novolac composites. *Compos. Sci. Technol.*, 60, 9, 1729–35, 2000 Jul 1.
46. Yang, H.S., Kim, H.J., Park, H.J., Lee, B.J., Hwang, T.S., Effect of compatibilizing agents on rice-husk flour reinforced polypropylene composites. *Compos. Struct.*, 77, 1, 45–55, 2007 Jan 1.
47. Fraga, A.N., Frullloni, E., De la Osa, O., Kenny, J.M., Vázquez, A., Relationship between water absorption and dielectric behaviour of natural fibre composite materials. *Polym. Test.*, 25, 2, 181–7, 2006 Apr 1.
48. Dash, B.N., Rana, A.K., Mishra, H.K., Nayak, S.K., Tripathy, S.S., Novel low-cost jute–polyester composites. III. Weathering and thermal behavior. *J. Appl. Polym. Sci.*, 78, 9, 1671–9, 2000 Nov 28.
49. Sever, K., Sarikanat, M., Seki, Y., Erkan, G., Erdoğan, UH., The mechanical properties of γ-methacryloxypropyltrimethoxy silane-treated jute/polyester composites. *J. Compos. Mater.*, 44, 15, 1913–24, 2010 Jul.

50. Shahinur, S., Sayeed, M.A., Hasan, M., Sayem, A.S., Haider, J., Ura, S., Current development and future perspective on natural jute fibers and their biocomposites. *Polymers*, 14, 7, 1445, 2022 Apr 1.
51. Arivendan, A., Chen, X., Zhang, Y.F., Keerthiveettil Ramakrishnan, S., Gao, W., Winowlin Jappes, J.T., Syamani, F.A., Alagumalai, V., Continuous natural fiber-reinforced polymer composites (CNFRPCs) manufacturing: additive manufacturing pathways to a sustainable future—A critical review. *Polym. Compos.*, 47, 3, 2050–71, 2026 Feb 10.
52. Bourmaud, A., Shah, D.U., Beaugrand, J., Dhakal, H.N., Property changes in plant fibres during the processing of bio-based composites. *Ind. Crops Prod.*, 154, 112705, 2020 Oct 15.
53. Khalid, M.Y., Imran, R., Arif, Z.U., Akram, N., Arshad, H., Al Rashid, A., García Márquez, F.P., Developments in chemical treatments, manufacturing techniques and potential applications of natural-fibers-based biodegradable composites. *Coatings*, 11, 3, 293, 2021 Mar 4.
54. Njuguna, J., Wambua, P., Pielichowski, K., Kayvantash, K., Natural fibre-reinforced polymer composites and nanocomposites for automotive applications, in: *Cellulose fibers: bio-and nano-polymer composites: Green Chemistry and Technology*, pp. 661–700, Springer, Berlin, Germany, 2011.
55. Zamora-Mendoza, L., Gushque, F., Yanez, S., Jara, N., Álvarez-Barreto, J.F., Zamora-Ledezma, C., Dahoumane, S.A., Alexis, F., Plant fibers as composite reinforcements for biomedical applications. *Bioengineering*, 10, 7, 804, 2023 Jul 5.
56. Ho, M.P., Wang, H., Lee, J.H., Ho, C.K., Lau, K.T., Leng, J., Hui, D., Critical factors on manufacturing processes of natural fibre composites. *Composites, Part B*, 43, 8, 3549–62, 2012 Dec 1.
57. Mahmud, S., Hasan, K.F., Jahid, M.A., Mohiuddin, K., Zhang, R., Zhu, J., Comprehensive review on plant fiber-reinforced polymeric biocomposites. *J. Mater. Sci.*, 56, 7231–64, 2021 Apr.
58. Wang, J., Wang, D., Mao, Q., Chen, J., Fabric insert injection molding for the preparation of ultra-high molecular weight polyethylene/high-density polyethylene two-component self-reinforced composites. *Polymers*, 14, 20, 4384, 2022 Oct 17.
59. Nautiyal, S., Dimri, S., Riyal, I., Sharma, H., Dwivedi, C., Natural fibres and their composites: a review of chemical composition, properties, retting methods, and industrial applications. *Cellulose*, 32, 6, 3497–527, 2025 Apr.
60. Wirawan, W.A., Setyabudi, S.A., Widodo, T.D., Choiron, M.A., Surface modification with silane coupling agent on tensile properties of natural fiber composite. *J. Energy Mech. Mater. Manuf. Eng.*, 2, 2, 97–104, 2017 Dec 28.
61. Tavares, T.D., Antunes, J.C., Ferreira, F., Felgueiras, H.P., Biofunctionalization of natural fiber-reinforced biocomposites for biomedical applications. *Biomolecules*, 10, 1, 148, 2020 Jan 16.
62. Das, P.P., Muduli, M., Borah, S., Chaudhary, V., Principle of Green Chemistry: A modern perspective for development of sustainable textile fiber-based

green nanocomposites, in: *Green Chemistry for Sustainable Textiles*, pp. 121–136, Woodhead Publishing, Cambridge, United Kingdom, 2021 Jan 1.
63. Surid, S.M., Maraz, K.M., Shahida, S., Ahmed, A., Khan, R.A., A review on the properties of natural fibres and manufacturing techniques of fibre reinforced biocomposites. *Mod. Concepts Mater. Sci.*, 4, 4, 1–5, 2021.
64. Bloom, P.D., Baikerikar, K., Otaigbe., G., Sheares, J.U.V.V., Development of novel polymer/quasicrystal composite materials. *Mater. Sci. Eng., A*, 294–6, 2000.
65. Suryanto, H., Jenuardy, S., Kustono, D., Dharmabintara, R.A., Maulana, J., Yanuhar, U., The influence of ultrasonic wave treatment on the mechanical properties of nanoclay-reinforced starch biocomposite. *IOP Conf. Ser.: Mater. Sci. Eng.*, 515, 1, 012021, 2019 Apr 1, IOP Publishing.
66. Formela, K., Hejna, A., Haponiuk, J., Tercjak, A., *In situ* processing of biocomposites *via* reactive extrusion, in: *Biocomposites for high-performance applications*, pp. 195–246, Woodhead Publishing, Cambridge, United Kingdom, 2017 Jan 1.
67. Rajput, S., Malviya, R., Uniyal, P., Advances in the treatment of kidney disorders using mesenchymal stem cells. *Curr. Pharm. Des.*, 30, 11, 825–40, 2024 Mar 1.
68. Swetha, M., Sahithi, K., Moorthi, A., Srinivasan, N., Ramasamy, K., Selvamurugan, N., Biocomposites containing natural polymers and hydroxyapatite for bone tissue engineering. *Int. J. Biol. Macromol.*, 47, 1, 1–4, 2010 Jul 1.
69. Vallet-Regí, M. and Navarrete, D.A., *Nanoceramics in clinical use: from materials to applications*, Royal Society of Chemistry, Cambridge Science Park, England, 2015 Aug 24.
70. Dhandayuthapani, B., Yoshida, Y., Maekawa, T., Kumar, D.S., Polymeric scaffolds in tissue engineering application: a review. *Int. J. Polym. Sci.*, 2011, 1, 290602, 2011.
71. Awad, S., Siakeng, R., Khalaf, E.M., Mahmoud, M.H., Fouad, H., Jawaid, M., Sain, M., Evaluation of characterisation efficiency of natural fibre-reinforced polylactic acid biocomposites for 3D printing applications. *Sustainable Mater. Technol.*, 36, e00620, 2023 Jul 1.
72. Ortega, F., Versino, F., López, O.V., García, M.A., Biobased composites from agro-industrial wastes and by-products. *Emergent Mater.*, 5, 3, 873–921, 2022 Jun.
73. Guo, L., Liang, Z., Yang, L., Du, W., Yu, T., Tang, H., Li, C., Qiu, H., The role of natural polymers in bone tissue engineering. *J. Controlled Release*, 338, 571–82, 2021 Oct 10.
74. Zhou, H., Lu, S.S., Dempster, D.W., Bone remodeling: cellular activities in bone, in: *Osteoporosis in men*, Second Edition, vol. 30, pp. 15–24, Elsevier Amsterdam, Netherlands, 2010, 2009 Nov.

75. El-Rashidy, A.A., Roether, J.A., Harhaus, L., Kneser, U., Boccaccini, A.R., Regenerating bone with bioactive glass scaffolds: A review of *in vivo* studies in bone defect models. *Acta Biomater.*, 62, 1–28, 2017 Oct 15.
76. Saravanan, S., Leena, R.S., Selvamurugan, N., Chitosan based biocomposite scaffolds for bone tissue engineering. *Int. J. Biol. Macromol.*, 93, 1354–65, 2016 Dec 1.
77. Meng, Z.X., Wang, Y.S., Ma, C., Zheng, W., Li, L., Zheng, Y.F., Electrospinning of PLGA/gelatin randomly-oriented and aligned nanofibers as potential scaffold in tissue engineering. *Mater. Sci. Eng.: C*, 30, 8, 1204–10, 2010 Oct 12.
78. Salehi, S., Tavakoli, M., Mirhaj, M., Varshosaz, J., Labbaf, S., Karbasi, S., Jafarpour, F., Kazemi, N., Salehi, S., Mehrjoo, M., Emami, E., A 3D printed polylactic acid-Baghdadite nanocomposite scaffold coated with microporous chitosan-VEGF for bone regeneration applications. *Carbohydr. Polym.*, 312, 120787, 2023 Jul 15.
79. Venkatesan, J., Bhatnagar, I., Manivasagan, P., Kang, K.H., Kim, S.K., Alginate composites for bone tissue engineering: A review. *Int. J. Biol. Macromol.*, 72, 269–81, 2015 Jan 1.
80. Hong, Z., Zhang, P., He, C., Qiu, X., Liu, A., Chen, L., Chen, X., Jing, X., Nano-composite of poly (L-lactide) and surface grafted hydroxyapatite: mechanical properties and biocompatibility. *Biomaterials*, 26, 32, 6296–304, 2005 Nov 1.
81. Qiu, X., Chen, L., Hu, J., Sun, J., Hong, Z., Liu, A., Chen, X., Jing, X., Surface-modified hydroxyapatite linked by L-lactic acid oligomer in the absence of catalyst. *J. Polym. Sci., Part A: Polym. Chem.*, 43, 21, 5177–85, 2005 Nov 1.
82. Verheyen, C.C., Klein, C.P., de Blieck-Hogervorst, J.M., Wolke, J.G., Van Blitterswijn, C.A., De Groot, K., Evaluation of hydroxylapatite/poly (l-lactide) composites: physico-chemical properties. *J. Mater. Sci.: Mater. Med.*, 4, 58–65, 1993 Jan.
83. Esmaeili, S., Motamed, M., Mirzaasgari, Z., Zamani, B., Joghatayi, M.T., Mojtahed, M., Real-world experiment of rituximab in multiple sclerosis. *J. Neurol. Sci.*, 405, 53, 2019 Oct 15.
84. Esmaeili, S., Shahidi, G., Hajiakhondi, F., Apomorphine induced immune hemolytic anemia. *J. Neurol. Sci.*, 405, 249, 2019 Oct 15.
85. Zarrintaj, P., Saeb, M.R., Ramakrishna, S., Mozafari, M., Biomaterials selection for neuroprosthetics. *Curr. Opin. Biomed. Eng.*, 6, 99–109, 2018 Jun 1.
86. Jabbari, F., Babaeipour, V., Bakhtiari, S., Bacterial cellulose-based composites for nerve tissue engineering. *Int. J. Biol. Macromol.*, 217, 120–30, 2022 Sep 30.
87. Bedian, L., Villalba-Rodríguez, A.M., Hernández-Vargas, G., Parra-Saldivar, R., Iqbal, H.M., Bio-based materials with novel characteristics for tissue engineering applications–A review. *Int. J. Biol. Macromol.*, 98, 837–46, 2017 May 1.

88. Rajput, S., Malviya, R., Bahadur, S., Puri, D., Recent Updates on the Development of Therapeutics for the Targeted Treatment of Alzheimer's Disease. *Curr. Pharm. Des.*, 29, 35, 2802–13, 2023 Oct 1.
89. Li, N., Zhang, Q., Gao, S., Song, Q., Huang, R., Wang, L., Liu, L., Dai, J., Tang, M., Cheng, G., Three-dimensional graphene foam as a biocompatible and conductive scaffold for neural stem cells. *Sci. Rep.*, 3, 1, 1604, 2013 Apr 3.
90. Park, S.Y., Park, J., Sim, S.H., Sung, M.G., Kim, K.S., Hong, B.H., Hong, S., Enhanced differentiation of human neural stem cells into neurons on graphene. *Adv. Mater.*, 23, 36, H263, 2011 Sep 22.
91. Akhavan, O. and Ghaderi, E., Differentiation of human neural stem cells into neural networks on graphene nanogrids. *J. Mater. Chem. B*, 1, 45, 6291–301, 2013.
92. Rojas, R., *Neural networks: a systematic introduction*, Springer Science & Business Media, Springer Berlin, Heidelberg, 2013 Jun 29.
93. Servant, A., Leon, V., Jasim, D., Methven, L., Limousin, P., Fernandez-Pacheco, E.V., Prato, M., Kostarelos, K., Graphene-based electroresponsive scaffolds as polymeric implants for on-demand drug delivery. *Adv. Healthcare Mater.*, 3, 8, 1334–43, 2014 Aug.
94. Pandey, P.K., Sharma, A.K., Gupta, U., Blood brain barrier: An overview on strategies in drug delivery, realistic *in vitro* modeling and *in vivo* live tracking. *Tissue Barriers*, 4, 1, e1129476, 2016 Jan 2.
95. Akhavan, O., Graphene scaffolds in progressive nanotechnology/stem cell-based tissue engineering of the nervous system. *J. Mater. Chem. B*, 4, 19, 3169–90, 2016.
96. Petreaca, M., Martins-Green, M., Atala, A., Lanza, R., Thomson, J.A., Nerem, R., Cell-ECM interactions in repair and regeneration, in: *Principles of regenerative medicine*, vol. 28, pp. 19–65, Academic Press, London, UK, 2011 Apr.
97. Nilforoushzadeh, M.A., Khodadadi Yazdi, M., Baradaran Ghavami, S., Farokhimanesh, S., Mohammadi Amirabad, L., Zarrintaj, P., Saeb, M.R., Hamblin, M.R., Zare, M., Mozafari, M., Mesenchymal stem cell spheroids embedded in an injectable thermosensitive hydrogel: an *in situ* drug formation platform for accelerated wound healing. *ACS Biomater. Sci. Eng.*, 6, 9, 5096–109, 2020 Aug 27.
98. Gandhimathi, C., Venugopal, J.R., Bhaarathy, V., Ramakrishna, S., Kumar, S.D., Biocomposite nanofibrous strategies for the controlled release of biomolecules for skin tissue regeneration. *Int. J. Nanomed.*, 8, 4709–22, 2014 Oct.
99. MacNeil, S., Progress and opportunities for tissue-engineered skin. *Nature*, 445, 7130, 874–80, 2007 Feb 22.
100. Ma, L., Gao, C., Mao, Z., Zhou, J., Shen, J., Hu, X., Han, C., Collagen/chitosan porous scaffolds with improved biostability for skin tissue engineering. *Biomaterials*, 24, 26, 4833–41, 2003 Nov 1.

101. Yazdi, M.K., Zare, M., Khodadadi, A., Seidi, F., Sajadi, S.M., Zarrintaj, P., Arefi, A., Saeb, M.R., Mozafari, M., Polydopamine biomaterials for skin regeneration. *ACS Biomater. Sci. Eng.*, 8, 6, 2196–219, 2022 Jun 1.
102. Jelodari, S., Ebrahimi Sadrabadi, A., Zarei, F., Jahangir, S., Azami, M., Sheykhhasan, M., Hosseini, S., New insights into cartilage tissue engineering: improvement of tissue-scaffold integration to enhance cartilage regeneration. *Biomed Res. Int.*, 2022, 1, 7638245, 2022.
103. Griffin, T.M. and Guilak, F., The role of mechanical loading in the onset and progression of osteoarthritis. *Exerc. Sport Sci. Rev.*, 33, 4, 195–200, 2005 Oct 1.
104. Hollister, S.J., Levy, R.A., Chu, T.M., Halloran, J.W., Feinberg, S.E., An image-based approach for designing and manufacturing craniofacial scaffolds. *Int. J. Oral Maxillofac. Surg.*, 29, 1, 67–71, 2000 Feb.
105. HQ, D.W. and Lee, E.H., *In vivo* mesenchymal cell recruitment by a scaffold loaded with transforming growth factor beta1 and the potential for *in situ* chondrogenesis. *Tissue Eng.*, 8, 469482, 2002.

11

Designing Biocomposites for Tissue Engineering

Deepak Kumar, Sonali Sundram, Shairy Priya, Md. Azhar and Rishabha Malviya*

Department of Pharmacy, School of Medical and Allied Sciences, Galgotias University, Greater Noida, Uttar Pradesh, India

Abstract

Biocomposites are gaining ground in tissue engineering due to the fact that they mimic the extracellular matrix (ECM) that provides structure and biological support. Recent innovations include the combination to overcome the limitations of the traditional scaffolds of natural and synthetic polymers, ceramics, and nanomaterials. This chapter emphasizes design strategies, fabrication approaches, and new technologies for biocomposite scaffolds for tissue engineering. Smart biomaterials, along with other methods like electrospinning, 3D printing, freeze-drying, and stereolithography, make it possible to create customized scaffolds with customized architecture and bioactivity. Combining the use of natural and synthetic polymers provides them with improvements in biocompatibility, mechanical strength, and degradation profiles, while conductive polymers and bioactive molecules can further induce cellular response. Nonetheless, there are still challenges that are hindering clinical translations, including scalability, immune response, reproducibility, and regulatory approval. The solution to these problems involves rational material design and the creation of multifunctional scaffolds. Biocomposites have disruptive potential in regenerative medicine as they provide new opportunities for engineering structures with adequate functions that facilitate successful tissue repair and regeneration.

***Keywords*:** Biocomposites, tissue engineering, extracellular matrix, biocompatibility, biodegradability, scaffold design, natural polymers, synthetic polymers

**Corresponding author*: rishabhamalviya19@gmail.com

Rishabha Malviya and Mukesh Roy (eds.) Biocomposites in 3D and 4D Printing: Materials, Techniques, and Applications in Regenerative Medicine, (299–322) © 2026 Scrivener Publishing LLC

11.1 Introduction

The research and development of tissue engineering (TE) scaffolds is an important arena that has been explored by scientists to determine their potential as a less intrusive and effective mechanism of therapeutic approach in regenerative medicine (RM). The biological mechanism of the TE scaffold approach involves the introduction of a biocompatible element called a scaffold into an organism such that it assists in the attachment, migration, and differentiation of cells, as well as their proliferation [1, 2]. Biocompatible materials that mimic the cellular structure and function of diseased/trauma tissues have been extensively studied in the field of TE research. The secretion of the natural extracellular matrix (ECM) is done to restore the injured tissue as the surrounding tissue assimilates the biocompatible resources [3]. The majority of cell types can stick to the extracellular matrix, and this process is used to build TE scaffolds using biodegradable as well as biocompatible polymers. Presently, there are hindrances to the use of biological alternatives for the development of tissue frameworks. As a result, creating and developing the ideal tissue scaffold that is non-poisonous and does not disrupt the immunity and cellular transportation mechanisms is crucial. When adequately manufactured, biodegradable parts may temporarily replace the extracellular matrix and have characteristics that promote the release of biomolecules and/or drugs for anti-inflammation, as well as antibacterial effects in the regeneration of tissue [4].

Organs and tissues are crucial for organisms, and any injury to these tissues disrupts normal bodily functions and diminishes the quality of life. While tissues like bones and skeletal muscles possess the ability to self-repair after damage, severe injuries need therapeutic intervention [5]. Given the conditions, traditional therapeutic modalities, such as autografts or donor grafts, may be ineffective due to prolonged recovery times, heightened infection risks, or potential immunological rejection [6, 7]. Consequently, TE is the most efficacious option to ensure complete recovery of a patient in instances of significant tissue injury. TE rehabilitates, enhances, and sustains tissues that have been compromised owing to injury or impairment resulting from congenital disorders. Effective tissue regeneration (TR) is accomplished by the utilization of natural and synthetic polymers [8].

Biocomposites in TE are engineered materials containing two or more distinct parts, usually a blend of natural polymers, synthetic polymers, ceramics, or nanomaterials, intended to synergistically emulate the mechanical, structural, and biological characteristics of native tissues.

TE must develop biocomposite scaffolds with the right qualities. The multidisciplinary discipline of RM uses engineering concepts and biological sciences to encourage the regeneration of injured tissues and organs [9, 10].

RM can repair or even substitute tissues and organs that have been injured by illness, trauma, or ageing [11]. TE is essential to RM. In the discipline of TE, new tissues and organs are developed using biological and technical concepts. It addresses the significant disparity between the high demand for tissues and organs and their restricted availability. Tissue engineering is essential since it delivers an alternative supply of biocompatible and biodegradable tissues and organs. For tissue engineering (TE) to function well, a combination of cells, scaffolds, and growth factors is frequently employed in the approach of tissue and organ regeneration (Figure 11.1). The objective is to replace the original tissue or encourage its regeneration. In addition to offering a structure similar to that of an ECM, a scaffold can function as a framework for the addition, infusion, or utilization of biological signals to encourage regeneration and replicate a setting present in healthy tissues as well as organs [12, 13].

The use of scaffolds in tissue engineering (TE) has evolved as time went on. Scaffolds are constructed out of various materials—catgut in Europe, linen in Egypt, ant heads in India, and assorted sources in South Africa. These materials shape a structure to facilitate tissue healing or regeneration [11]. A major challenge for many researchers is developing scaffolds that not only provide a structural framework but also generate cell

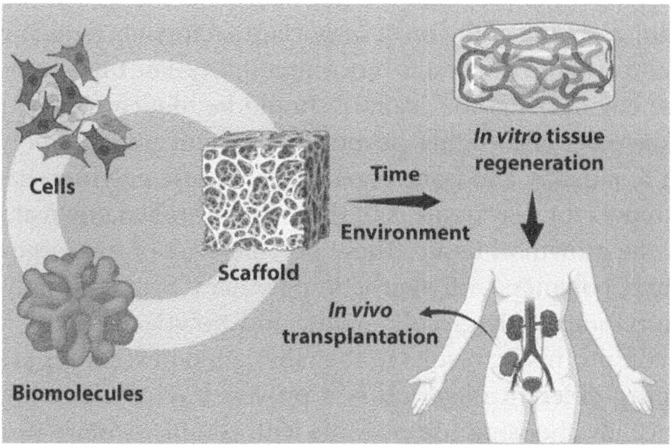

Figure 11.1 Diagram representation of the TE construct for regeneration and transplantation.

differentiation, adhesion, and proliferation. To succeed, the scaffold has to enable key *in vivo* interactions between cells and cells with the matrix [15].

The 3D, porous scaffolds produce a suitable environment for tissue repair (TR) and healing [14]. An ideal scaffold should meet several requirements: high porosity, suitable mechanical properties, biodegradability, biocompatibility, and economic feasibility [16]. Cell adhesion and proliferation depend on the biocompatibility of the scaffold, and immunological response is one more indicator of the compatibility. An unimportant immune response is desirable, because a chronic inflammation interferes with healing or causes rejection [17].

This study describes design methodologies for the design of biocomposites in TE, accounting for the appropriate material choice, manufacturing methods, and innovations, including nanostructured composites, smart, and 3D printing bio-ink systems. It also explores critical challenges associated with issues of scalability, safety, and regulation, providing future insights for advancements in these areas. In order to carry out an extensive review, a systematic search was carried out in the most reliable databases: ScienceDirect, PubMed, Scopus, and Google Scholar, in which a well-defined search strategy enabled to identification of studies with a close relationship with the chapter's objectives.

11.2 Biomaterials for Tissue Engineering Scaffolds

Chemical, physical, and structural tests for the scaffold materials are needed to ensure that the scaffolds can withstand changes of shape under living organisms. One of the most important factors for reducing immune reactions and risk of rejection is biocompatibility. For the best outcome of bone tissue engineering, the scaffold should be biodegradable, physically strong, non-toxic, and highly porous [18]. Biodegradable polymers are often used to aid skin and bone regeneration; these polymers can provide physical support for new tissue, and they can be broken down at a rate that is equal to the growth of tissue. These polymers are of two large classifications: natural polymers and synthetic polymers. Synthetic polymers like poly(caprolactone), poly(glycolic acid), and poly(lactic acid) have been found to have promising results in the fabrication of scaffolds [19].

Combining different materials is a proven way of improving scaffold performance. By choosing components with complementary properties, it is possible to modify wettability, rate of degradation, mechanical strength, and cell adhesion. Hydroxyapatite (HA), considered as a major component of bone, can be combined with synthetic biopolymers to form composite scaffolds that more closely mimic the structure and function of natural bone, e.g., when nano hydroxyapatite (nHA) is used with a nanofibrous polymer matrix [20]. Synthetic polymers tend to have less attachment by cells than natural ones. Polycaprolactone, for one, is hydrophobic, breaks down slowly, and has no cell recognition sites, making it limited in use even though it is a recyclable material that is safe. Gelatin, on the other hand, is hydrophilic, breaks down faster, and is a collagen-based natural polymer that is the primary structural protein in the extracellular matrix of the skin. By blending gelatin with PCL, flexible, wettable, and slowly degradable porous supports resembling the extracellular matrix have been produced for improving cell adhesion and growth [21].

The natural polymers chitosan, collagen, and chitin are frequently utilized in TE scaffolds. Despite being porous, biocompatible, and biodegradable, collagen, a vital component of the extracellular matrix, lacks enough mechanical strength. To get around this, it is frequently mixed with bioceramics like beta-tricalcium phosphate (β-TCP) and HA to develop composite scaffolds that have better structural integrity. Moreover, β-TCP accelerates scaffold disintegration and improves osteoconductivity, making it easier for freshly regenerated tissue to replace the scaffold [22, 23].

Gelatin, a protein made up of from the hydrolysis of collagen, is frequently utilized in TE scaffolds. It has the arginyl–glycyl–aspartic acid sequence, which promotes cell adhesion, spreading, and proliferation while also improving biocompatibility. However, gelatin by itself is not enough to heal bone abnormalities because of its poor mechanical qualities. Gelatin-based scaffolds are frequently reinforced with silica nanoparticles, as well as polymer microparticles to increase their strength [24].

Chitosan is a polysaccharide whose structure is linear, which is commonly obtained from the shells of crustaceans, molluscs, and insect cuticles, and is another common natural polymer in TE scaffolds. Chitosan is renewable, biocompatible, biodegradable, and sticky. It has protonation-capable free amino groups that allow for simple chemical modification and electrostatic interactions with negatively charged molecules like synthetic polymers, proteins, lipids, and DNA. To compensate for its limited mechanical strength, chitosan is sometimes coupled with mechanical reinforcing agents. A recent study has investigated the uses of pectin and chitosan, two naturally occurring polymers, as a scaffold material for skin and bone TR. An increased biocompatibility, swelling action, porous microstructure, and mechanical durability scaffold may be produced by combining chitosan with pectin to form a polyelectrolyte complex [25].

Biodegradable matrix polymer nanocomposites exhibit significant potential in bone TE owing to their capability to facilitate bone and TR.

Natural biopolymers, including elastin, collagen, alginate, gelatin, chitosan, hyaluronic acid, and glycosaminoglycans, are extensively utilized and thoroughly studied for their biocompatibility, low toxicity, limited immunogenicity, accessibility, and cost-effectiveness [26].

Synthetic scaffolds, on the other hand, circumvent the moral dilemmas surrounding animal-derived sources while addressing the heterogeneity found in natural materials. They provide benefits including reduced production costs, uniformity, and adaptability. PGA and PLA are examples of synthetic polymers that are often utilized. PGA promotes osteoblast activity and exhibits regulated degradation despite its high crystallinity. In applications involving bone regeneration, PLA is prized for its mechanical and thermal qualities, as well as its biocompatibility and osteogenic potential. PCL is a semicrystalline, biocompatible synthetic material that finds extensive application in TE and drug administration because of its advantageous manufacturing properties and flexibility [18]. It may be made into a variety of shapes, counting fibers, films, and nanoparticles. Its weak cell affinity, hydrophobicity, and low stiffness, however, restrict its use in bone TR. PCL is frequently mixed with bioactive substances as nHA, biphasic calcium phosphate (BCP), or bioactive glass to solve these issues. Furthermore, surface modification methods are used to increase its hydrophilicity and boost cell contact, such as chemical or plasma treatments, ECM protein coatings, or mixing with bioactive compounds [27].

Plasma treatment improves the surface characteristics of PCL fibers by incorporating oxygen-containing species such as air, oxygen (O_2), ammonia (NH_3), sulphur dioxide (SO_2), and carbon dioxide (CO_2) or polar functional groups like hydroxyl, carboxyl, amino, and sulphate. Research indicates that plasma treatment enhances biological interactions; for instance, Prabhakaran *et al.* observed heightened Schwann cell proliferation on plasma-treated PCL fibers in comparison to untreated PCL or PCL collagen fibers [28].

Sodium hydroxide (NaOH) is often used in modifying the surface of PCL to induce carboxylate and hydroxyl groups to enhance wettability. Another way is to add adhesion-promoting proteins, such as laminin, gelatin, collagen, vitronectin, or fibronectin to the PCL fibers, to provide biochemical cues for enhanced cell attachment. Koh *et al.* found that neurite outgrowth was enhanced by laminin-coated fibers. The addition of biologically active elements to PCL is an additional method to overcome its inherent weakness in mechanical strength and poor bioactivity [29].

Composite scaffolds are a combination of numerous biomaterials such as biopolymers, bioceramics, biodegradable metals, and natural ingredients. Since bone is itself a composite, these scaffolds better mimic the tissue and increase biological performance. Turnbull *et al.* mention that for a normal scaffold, various types of components are a polymer hydrogel, metal, ceramics, and bio-glasses. These combinations enhance the mechanical properties and ensure beneficial cell–material interaction [30].

Biopolymers are already known to have poor mechanical strength, too high hydrophilicity, and poor conductivity. The use of conductive polymers (CPs) could compensate for these problems when embedded. CPs contain alternating double bonds, which provide good biocompatibility as well as the ability to provide electrical stimulation needed for activating cells (as neurons require). Tissues, such as the brain, heart, and muscle, depend on electrical signals, so CPs can serve as a link between bioelectronic devices and biology in order to create therapeutic interfaces. Mechanical strain can also cause piezoelectric effects, where endogenous electric fields are produced, which actually improve mechanisms, such as osteoblast adhesion and proliferation [31]. Research has shown that the addition of CPs has enhanced scaffold biodegradability and strength whilst retaining the biocompatibility. However, CPs are not only slow to degrade in many instances but can also result in lingering inflammation. This problem can be overcome by mixing them with biodegradable polymers. Common conductive polymers utilized in thermoelectrics encompass poly(3,4-ethylenedioxythiophene): poly(styrene sulfonate), polypyrrole, and polyaniline. Scaffolds developed by integrating PEDOT: PSS with nHA and chitosan by lyophilization exhibited improved cell attachment. Electrospun PLA/Poly(3-hydroxybutyrate-co-3-hydroxyvalerate) membranes covered with PEDOT: PSS exhibited enhanced roughness of surface and wettability, resulting in superior cell survival, proliferation, and adhesion, relative to uncoated scaffolds [32].

Polypyrrole (PPy) has been widely investigated in the area of tissue engineering (TE) and drug delivery applications. One study developed a scaffold made of polycaprolactone fumarate and PPy, which enhanced the growth of PC12 cells and proved to be a suitable conductive nerve conduit [33]. On the other hand, the investigations relating to PANI have proceeded at a more languid pace [34, 35]. Studies on fiber shape, chemical composition, and electroactivity showed promising potential for TE. Nonetheless, there remain obstacles, particularly in optimizing the PHB: PANI solvent ratio to ensure that the spin is adequate. Even slight amounts of PANI can have a significant effect on the structural organization of PHB/PANI nanofibers [36].

Every scaffold material used in the art of TE has its own advantages and disadvantages. Natural substances are preferred, as they have low toxicity, bio-compatibility and regenerating capabilities, but they often have a batch-to-batch variability. Synthetic materials offer improved uniformity and modification suitability with the drawback that they are often lacking in the bioactivity naturally present in the scaffolds. Composite scaffolds aim at combining the advantages of both types; however, their manufacture is often complex and costly [37].

A major problem of scaffold design is achieving ideal properties such as higher porosity, appropriate degradation rate, mechanical strength, and biocompatibility. The choice of material has a strong influence on these characteristics, especially degradation behaviour, and must be associated with the timescale for TR. In nanofiber scaffolds, the combination of materials can promote the hydrophilicity and degradation rates; however, careful control of the production conditions is important for consistent quality production. The immune response is very important for the success of scaffolds. Initial reactions to implanted materials may develop into persistent immunological responses, potentially resulting in scaffold rejection. Strategies for managing immune responses encompass material design, immune cell regulation, anti-inflammatory administration, and surface alterations, including the adjustment of electrospun scaffold morphology to more accurately replicate native bone tissue [38].

11.3 Designing Smart Biomaterials

Designing bio-functional materials is very important in biological applications. There are many ways to fabricate hybrid smart biomaterials using synthetic and natural polymers, so that their applications are widened. These approaches enable the addition of bio-recognition moieties to polymers, which influences the self-assembly in exact 3D structures [39]. Smart polymer-based biomaterials with the required network structure and characteristics are designed on the basis of functional monomers, feed ratio, kinetics, polymerization process, molecular architecture at synthesis, and crosslink networks. During the design, synthesis, processing, and production steps, physical and chemical parameters must be controlled to produce complex functions of biomaterials. The 3D network topology of smart material is related to synthesis techniques, polymer, monomer, and oligomer properties [40]. Chemical, physical, and radiation-induced crosslink processing can be used to fabricate network biomaterials, or hydrogels, of 3D shapes and properties. The 3D polymer network renders the hydrogel insoluble in water and has an equilibrium swelling. Hydrogels are 3D scaffolds that are created using covalent and hydrogen-ionic bonds between polymer chains. Crosslinked network gels are not soluble in water because cohesive and dispersion forces of polymer chains are balanced in their structure. These compounds can be used as biomaterials, delivery systems, functional TR scaffolds, and therapeutic devices [41].

Factors which influence the expansion of the smart materials 3D network comprise the following: (i) Swelling is enhanced when the solvent and polymer undergo a positive shift in thermal energy; and (ii) The conformations of the polymer chains are related to this shift in thermal energy, thus a negative movement in energy reduces swelling [42].

These networks of polymers may be categorized based on how they develop, including physical mixing and chemical processes. Examples of hydrogel chemical reactions include block copolymer, graft-copolymer, and polymer crosslinking reactions [43–45]. The processes for the hydrogel formation can be started using either chemical initiation techniques or techniques triggered by ionizing radiation. Impurities are introduced into the finished product *via* the chemical initiation procedures, which call for the addition of a bifunctional reagent and chemical catalyst. It can occasionally be quite challenging to exclude contaminants from the hydrogel networks. As an alternative to chemical start methods, ionizing radiation-induced technologies have benefits that enable effective crosslinking and provide a much cleaner result. Polymer hydrogels can be developed using high-energy (such as electron beams and γ-radiation), plasma, and UV-induced techniques [46].

The physical amalgamation of several polymers, achieved by either solution processing or thermal methods, is a practical and cost-effective strategy for the development of novel structural smart materials, while preserving the distinct features of each polymer within the mix. This technology offers numerous advantages, particularly in the facilitation of managing the processing and production of smart devices [47]. This technology permits the change of process parameters, beginning polymer ratios, and the selection of appropriate solvents to achieve certain properties, like porosity, pore diameters, biodegradability, swelling ratio, as well as mechanical qualities. Therapeutic substances can be easily dispersed in micro-phase reservoirs or included in the blended solution [48].

Copolymerizing different chemical functional building blocks allows scientists to create polymeric biomaterials with complex structures that may be fine-tuned in terms of biocompatibility, hydrophilicity, mechanical characteristics, biodegradation, and reactions to environmental stimuli [42]. This biomaterial system is designed in a manner that permits copolymer chains to self-organize in the aqueous phase by means of intramolecular and intermolecular interactions and that produce structural diversity. Designing model copolymer systems to be experimentally challenging, which results in slow responses to small stimuli. Despite these challenges, smart materials are biologically and industrially relevant given their capability to react to many inputs. Material scientists have to keep researching these systems to discover their fields.

By embedding peptides and proteins in the polymer matrix, smart biomaterials can be used to provide a 3D scaffold for tissue engineering applications [49, 50]. Ito *et al.* utilized a genetically engineered collagen-type II protein to synthesize 3D scaffolds that improved chondrocyte proliferation and migration and promoted the development of artificial cartilage [50]. Separate studies showed that employing Michael-type addition to conjugate cysteine-tagged fibronectin using poly(ethylene glycol) derivatives formed a 3D hydrogel matrix. Then, the matrix was crosslinked with thiol-modified hyaluronic acid. They proved that these hydrogels were compatible with human cells by allowing fibroblasts to adhere, multiply, and migrate vigorously; furthermore, they showed improved dermal wound healing after being implanted in pig skin [46]. Similarly, TE applications have seen the development of several elastin-based block copolymers showing a variety of mechanical features [51, 52]. To increase cell adherence, the signaling molecules in scaffold materials can undergo chemical conjugation. It has been extensively researched that functionalizing synthetic polymers with PNIPAm-(Arg-Gly-Asp (RGD)) peptide improves cell adherence, functioning, and migration [53].

Self-assembly methods for brain TE and human dermal fibroblast proliferation have been developed recently, employing chemically and biologically synthesized peptides, exhibiting remarkable biocompatibility [54–56]. The advance of novel biomaterials utilizing self-assembling peptides and proteins is an ongoing and future-focused area of study, as this system reacts to external environmental factors, specifically pH, as well as temperature. While the synthesis of 3D-structure scaffolds utilizing, self-assembling peptides is highly repeatable, it encounters several hurdles, including process scale-up, manufacture, and purification in a cost-effective way. Undoubtedly, significant effort is necessary, concentrating on discovery, rational design, manufacturing, and process development [57].

11.4 Advances in Biocomposites Scaffold Fabrication Techniques

Tissue consists of 3D units that are recursive structures, of size 100 to 1000 μm. This repeating 3D arrangement enables an array of mechanical features to emerge, the ability of multicellular processes, and the capacity for organ systems to communicate with one another and the microcirculation. The scaffolds are thought to impart artificial support to allow the 3D cells to grow as the tissue heals [58]. Another important part of the tissue, which is inside the body, is the local cellular environment. Cellular destiny processes, such as proliferation, migration, differentiation, and death, are mediated by cellular, physical, and biochemical signaling pathways found in the microbiological environment (approximately 10 mm). To make fully functional scaffolds, you should look at it from two points of view: (a) at the microscale level it should have condition that helps cells to survive and work; and (b) at the macroscale level the tissue construction should be able to (i) coordinate the multicellular processes, (ii) be able to transport the nutrients well as well as, and (iii) have mechanical properties. A diagram of the techniques of biocomposite scaffold fabrication is mentioned in Figure 11.2.

The methods for constructing 3D scaffolds are categorized into traditional as well as rapid prototyping (RP) procedures, each yielding a scaffold with distinct properties [59]. Traditional scaffolding fabrication approaches comprise the development of porous polymer constructs that serve as substrates for cell adhesion; however, achieving complex structures with adjustable microscale along with macroscale characteristics using these traditional methods is challenging [58]. The RP scaffold construction process offers several prospects for TE. The autonomous regulation

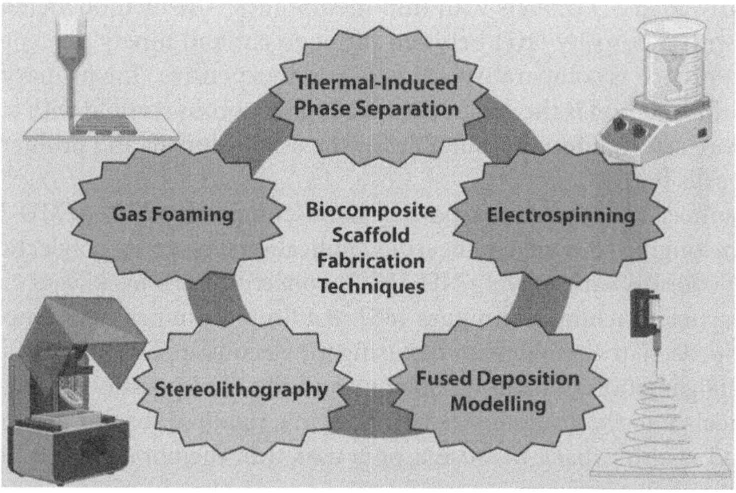

Figure 11.2 Overview of biocomposite scaffold fabrication techniques.

of macroscale as well as microscale characteristics facilitates the construction of multicellular structures essential for intricate tissue functions. Second, the development of 3D vascular beds would facilitate the support of substantial tissue growth that would not have been feasible otherwise. Interestingly, the integration of clinical imaging data with 3D manufacturing methods enables the development of both customized scaffolds and the mass creation of scaffold designs [60, 61].

11.4.1 Traditional Scaffold Fabrication Methods

Scaffolds were initially developed and designed to aid in bounty delivery of medicines, but are now also applied for 3D cell growth in tissue engineering. The traditional methods of scaffold fabrication, such as solvent casting and particle leaching, are mostly limited by the arrangement and connection of the void space. The idea is to regulate the geometry and the size of the pores of the scaffolds [62].

11.4.2 Solvent Casting Along with Particle Leaching

This technique, which dissolves polymer solution, uses a solvent that carries uniformly distributed salt particles of a specific size. As the solvent disappears, the matrix is salt-laden. The next step in the process is to try to immerse that matrix in water because the water dissolves the salt and forms a very porous structure. To recover the soluble particles from the polymer matrix, solvent casting having particle leaching is only applicable to thin-walled 3D models with thin membranes. The method yields scaffolds with a porosity level between fifty percent and ninety percent [63]. This approach is comparatively simple and inexpensive. The primary benefit of this method is the generation of a highly porous scaffold with adjustable pore size, rendering it suitable for the establishment and proliferation of 3D cells [64].

Thanh *et al.* proposed a porous nano-composite of PLA/MD-HAP/PEO belonging to bone engineering applications, utilizing polylactic acid (PLA) coupled with varying NH_4HCO_3 concentrations by solvent casting and particle leaching techniques [65]. Additional studies have employed this approach in scaffold manufacturing for various applications, including the amalgamation of natural polymers and the incorporation of bioactive chemicals into the scaffold [66–69]. The time required for this production method is a drawback because it only uses thin membranes. The limited pore size of porous sheet layers may limit their usefulness since they only offer a certain number of pore networks between them [67]. This approach employs several hazardous solvents that require extensive time to evaporate, ranging from days to weeks.

11.4.3 Freeze-Drying

A manufactured polymer is dissolved in a suitable solvent in order to begin the lyophilization (freeze-drying) process. The solution is then cooled below its freezing point, and ice crystals form. As the ice sublimates, it forms a porous solid with interconnecting pores [70]. The surrounding polymer is like a "fence" around the ice, and allows the solutes to be immobilized in the ice lattice. The scaffolds are formed by a procedure of staged drying, and the macroporosity produced relates to the position of the former ice crystals. This approach prevents exposing the material to excessive temperatures, which may damage living components. By changing the protocol used to freeze, the size of the pores can be adjusted [71]. For example, a BG-collagen-phosphatidylserine (BGP-SP) scaffold with interconnected pores with about 300 μm in diameter was fabricated using this technique. It has been demonstrated that it may form compounds with calcium and phosphate and initiate hydroxyapatite production. Numerous studies have shown that the technology is flourishing in scaffold development for diverse applications [72]. Min and Lee utilized this technology for the development of a 3D scaffold employing chitosan nanoparticles [73]. In addition, Jayachandran and colleagues used lyophilized fucoidan to obtain chitosan-alginate biocomposites for bone tissue engineering. Their scaffold was easily water-absorbing and had a porous structure [74]. A similar approach has been described by Aranaz *et al.* Freeze-drying often occurs for scaffold fabrication but has disadvantages like production and process time, cytotoxicity of solvent residues, and creation of small and not very regular pores (15–35 μm) [71]. Geraldine *et al.* proposed to reduce this freezing temperature from −10°C to −70°C and increase annealing time to increase the rate of ice crystal growth. Because of the fact that freeze-drying uses water and ice instead of organic solvents, it is well suited to biomedical scaffolds but less effective in creating other types of complex material architectures such as vascular systems [58, 75]. This approach employs cytotoxic solvents for polymer mixing; hence, the resulting scaffold must be thoroughly washed to eliminate the solvent and reduce cell mortality.

11.4.4 Thermal-Induced Phase Separation

Thermal-induced phase separation (TIPS) works in low temperatures and relies on a difference in temperatures to separate phases. This is done by putting a homogeneous polymer solution that is at a high temperature in a lower temperature setting. This provides for the possibility of the development of a poor phase and a polymer-rich phase [63]. With this method, a flexible crystalline polymer scaffold can be made. It forms a kind of porous and nano-scale fiber network after the removal of the solvent by freeze-drying. The incorporation of bioactive chemicals in fibrous scaffolds can be done using low temperatures. Blaker *et al.* proposed a TIPS-based fabrication process for fabricating microspheres for tissue engineering and drug delivery. This method allows researchers to change the size of the pores to include fillers and medicines. Studies showed that the addition of proteins in the pores resulted in an even distribution when exposed to the solvent–water interface. In addition, increasing the exposure times improved this distribution [76]. Smith *et al.* combined some of the different components to obtain pores of size 50 to 500 nm to make nanofibrous scaffolds to mimic the tissue structure. However, the method has serious disadvantages - the materials used for production are hard to obtain, and the quality is poor.

Phase separation is a much more promising route towards the generation of 3D nanofibrous scaffolds featuring a homogeneous pore structure than electrospinning. This is because it is possible to use two or more phase separation techniques [58]. In addition, the phase separation process may be employed with other production methods in TE, such as solid free-form.

11.4.5 Gas Foaming

The use of organic cytotoxic solvents at high temperatures prompted the development of the gas foaming technique. Using relatively inert gas foaming agents like carbon dioxide or nitrogen, this technique pressurises model biologically degradable polymers with water or fluoroform until they are saturated or full of gas bubbles. This method often yields sponge-like structures with hole sizes ranging from 30 to 700 μm and porosities of up to 85%. One disadvantage of this method is that the final product may occasionally have a solid polymeric skin or a closed pore structure [77, 78]. Harris *et al.* enhanced the procedure to produce a highly porous product with enhanced porous connectivity in order to address this issue. According to material investigations conducted *in vitro*, seeded cells will stick to the matrix and keep growing into 3D tissues [79].

11.4.6 Electrospinning

The development of nanofibrous scaffolds in TE requires the use of the electrospinning technology that allows the creation of fibers from a solution by means of electricity. A liquid is loaded with a high voltage during the very complicated process of electrospinning, which makes the droplets explode and stretch on the spinneret by surface tension and electrostatic repulsion. There are four major components in a typical electrospinning arrangement: a metal needle, a high voltage power source, a grounded collector, and a spinner that consists of a syringe pump. A liquid jet forms when the electric field is strong enough and the droplet's surface tension is low enough. As the jet gets closer to the grounded collector, it gets longer and breaks apart. As it gets closer, electrostatic repulsion makes it smaller until it hits the collector. During this process, the solvent disappears, leaving behind a layer that is hard, not woven, and fibrous. To create a homogeneous mixture of fibers with high tensile strength, researchers used electrospinning to make composite fibers made of collagen as well as spider dragline silk protein [80]. Other studies have shown that such a resultant matrix can be used to support the growth of human choroid decidua parietalis placental stem cells (hdpPSCs) *in vitro*. Sarhan *et al.* used the same method to produce a wound dressing material nanofiber by the addition of an antibacterial component. Against *Staphylococcus aureus*, methicillin-resistant *S. aureus*, *Escherichia coli*, and multidrug-resistant *Pseudomonas aeruginosa*, the synthesized scaffolds demonstrated active antimicrobial agents [81]. The benefits of both natural and synthetic polymers have been combined in chitosan-based mix nanofibers, which are created by electrospinning chitosan alongside synthetic biodegradable polymers such as polylactic acid (PLA) [82]. Although electrospinning is a rapid and easy way to develop nanofibrous scaffolds, it still presents a problem when developing scaffolds with intricate architecture, including uniform pore distribution, which limits its use in biomedicine [58].

11.4.7 Stereolithography

The stereolithography process is employed to fabricate solid, 3D structures by sequentially depositing tiny layers of ultraviolet (UV) curable material. A moving build platform, a UV laser for irradiating the resin, a dynamic mirror system, and a tank filled with photosensitive liquid resin are the four main parts of a stereolithography system. A UV laser applies a layer of photosensitive liquid resin on the platform to begin the process. After the first layer solidifies, the platform drops vertically. The process is repeated until a three-dimensional scaffold is created. After that, a second coating is applied to the previous layer. The scaffold is post-cured by being exposed to UV light after the uncured resin has been removed. The problems with material waste in subtractive manufacturing methods are addressed by this technology [83]. Melchels *et al.* utilized the stereolithographic technique to fabricate scaffold structures with intricate designs at high resolution. In accordance with the research, the mechanical growth of the scaffolds is in agreement with the finite element's expectations [84]. Similarly, Robert *et al.* have used a comparable technique to construct 3D scaffolds using gelatin methacrylate. This made it possible to design a special scaffold that had the requisite mechanical properties. Persistent pore interconnectivity was observed from *in vitro* tests of the scaffold. This suggests that it can support the growth and dispersion of human umbilical vein endothelial cells to making it useful for tissue engineering applications [85]. Stereolithography only has limitations in the photopolymerization process. The high amounts of monomers needed and the need for post-polymerization treatment to convert monomers were found to be a vital disadvantage. Nevertheless, the technique was found to be effective in the design of scaffolds in several studies [86].

According to research, the stereolithography technique is a powerful biofabrication technology that can be used to develop a variety of cellular machines suitable for a variety of applications, such as drug discovery, environmental remediation, energy harvesting, and biosensing. The beam width of the laser dictates the feature size of the possible scaffold [87].

11.4.8 Fused Deposition Modeling

FDM technology involves melting and extruding a solid polymer through a hot extrusion nozzle above a 3D shape. The process is processed by a computer. There are several layers of microfilaments which is next to each other that make up the scaffold. Using biodegradable polyesters, Hutmacher *et al.* made a microfiber support to help the growth of cells in TE. Thermoplastic biopolymers are produced by FDM [88]. Furthermore, Zein *et al.* showed that the flexibility of this FDM is for modeling novel scaffold architectures with regulated mechanical characteristics. The study also produced a priming material called polymer bioresorbable poly (e-caprolactone), which is utilized to develop porous scaffolds [89]. According to several studies, FDM is generally helpful for scaffold design under various scaffold manufacturing features [90]. The requirement for prepared, uniformly sized fibers to pass through rollers and nozzles is the primary disadvantage of FDM. Its applicability to biodegradable polymers is therefore restricted.

Several modified FDM procedures have been put forth to get around these issues. Xiong *et al.* verified that low-temperature deposition manufacturing is used to develop composite scaffolds for bone TE. The study also said that the support was biocompatible, biodegradable, and had good bone connectivity for bone healing [91].

11.4.9 Selective Laser Sintering

The University of Texas at Austin developed SLS in the year 1986. A laser gives this technology its energy, which sinters the powdered material in thin layers following the model of a 3D object. With this technique, the use of a laser is used to work with metals, plastics, and ceramics, among other things [92]. This technique works well for quality creation of bio-nano composite microspheres from poly(L-lactide) (PLLA) as well as for creating scaffolds from ultrahigh molecular weight polyethylene used in tissue engineering. The method was able to produce carbonated hydroxyapatite (CHAP) nanospheres in a PLLA matrix. Handlers can achieve the desired qualities of the scaffolds and have more control over the microstructure of the scaffold by adjusting a few factors in the selective laser sintering (SLS) process, such as the percentages of physically mixed polymer/composite powder. One of the biggest problems with SLS is the removal of the powder that is required to be pumped out after the phase spin process [93].

11.5 Conclusion

Biocomposites have become attractive candidates in TE due to the ability to mimic properties of native extracellular matrix while providing structural and functional support for TR. By combining natural polymers, synthetic materials, ceramics, and nano-structures, scaffolds with optimum biocompatibility, porosity, degradation rates, and mechanical strength can be created. Advances in smart biomaterials and fabrication technology, such as electrospinning, solvent casting, freeze-drying, 3D printing, and stereolithography, have increased the options for scaffold design with the desired architectures and biological functions. Furthermore, the introduction of conductive polymers, bioactive substances, and self-assembling systems improves the interaction with the cells and their capacity to regenerate. Despite these advancements, there are still significant challenges related to the scalability, reproducibility, immune response, and regulatory approval needed to advance these types of therapies into widespread clinical adoption. Future directions should put more emphasis on rational material design and on the integration of multifunctional properties and cost-effective fabrication strategies. Overall, biocomposites are a transformational path to TE, one that spans from material science to RM to address critical needs in healthcare.

AI disclosure statement: The author acknowledges the use of AI-based tools for minor language editing or grammatical correction. However, all intellectual contributions, analyses, and conclusions presented in this chapter are entirely the author's own.

References

1. Bin, A.Z. and Das, G.A., Advancements within Molecular Engineering for Regenerative Medicine and Biomedical Applications an Investigation Analysis towards A Computing Retrospective. *J. Electron. Electrochem. Med. Inform.*, 6, 2024.
2. Kumar, S., Malviya, R., Sridhar, S.B., Shareef, J., Wadhwa, T., Biomaterials and tissue engineering strategies for management of neurodegeneration: A critical perspective. *Med. Nov. Technol. Devices*, 26, 100369, 2025.
3. Adhikari, B., Stager, M.A., Krebs, M.D., Cell-instructive biomaterials in tissue engineering and regenerative medicine. *J. Biomed. Mater. Res. Part A*, 111, 2023.
4. Kuperkar, K., Atanase, L.I., Bahadur, A., Crivei, I.C., Bahadur, P., Degradable *Polymer*ic Bio(nano)materials and Their Biomedical Applications: A Comprehensive Overview and Recent Updates. *Polymer*, 16, 2024.
5. Garot, C., Bettega, G., Picart, C., Additive Manufacturing of Material Scaffolds for Bone Regeneration: Toward Application in the Clinics. *Adv. Funct. Mater.*, 31, 2021.
6. Gregor, A., Filová, E., Novák, M., Kronek, J., Chlup, H., Buzgo, M., *et al.*, Designing of PLA scaffolds for bone tissue replacement fabricated by ordinary commercial 3D printer. *J. Biol. Eng.*, 11, 2017.
7. Aldana, A.A. and Abraham, G.A., Current advances in electrospun gelatin-based scaffolds for tissue engineering applications. *Int. J. Pharm.*, 523, 2017.
8. Socci, M.C., Rodríguez, G., Oliva, E., Fushimi, S., Takabatake, K., Nagatsuka, H., *et al.*, Polymeric Materials, Advances and Applications in Tissue Engineering: A Review. *Bioengineering*, 10, 2023.
9. Alkaron, W., Almansoori, A., Balázsi, C., Balázsi, K., A Critical Review of Natural and Synthetic Polymer-Based Biological Apatite Composites for Bone Tissue Engineering. *J. Compos. Sci.*, 8, 523, 2024.
10. Radha, G., Manjubaashini, N., Balakumar, S., Nano-hydroxyapatite/natural polymer composite scaffolds for bone tissue engineering: a brief review of recent trend. *In Vitro Models*, 2, 2023.

11. Mao, A.S. and Mooney, D.J., Regenerative medicine: Current therapies and future directions. *Proc. Natl. Acad. Sci.*, 112, 2015.
12. Sindhi, K., Pingili, R.B., Beldar, V., Bhattacharya, S., Rahaman, J., Mukherjee, D., The role of biomaterials-based scaffolds in advancing skin tissue construct. *J. Tissue Viability*, 34, 100858, 2025.
13. Porwal, S., Sridhar, S.B., Talath, S., Wali, A.F., Warsi, M.H., Malviya, R., 3D printable sustainable hydrogel formulations for tissue engineering applications. *J. Drug Deliv. Sci. Technol.*, 101, 106308, 2024.
14. Al-Zyoud, W., Haddadin, D., Hasan, S.A., Jaradat, H., Kanoun, O., Biocompatibility Testing for Implants: A Novel Tool for Selection and Characterization. *Materials*, 16, 2023.
15. Kaul, H. and Ventikos, Y., On the genealogy of tissue engineering and regenerative medicine. *Tissue Eng. Part B: Rev.*, 21, 2015.
16. Flores-Rojas, G.G., Gómez-Lazaro, B., López-Saucedo, F., Vera-Graziano, R., Bucio, E., Mendizábal, E., Electrospun Scaffolds for Tissue Engineering: A Review. *Macromol*, 3, 2023.
17. Krishani, M., Shin, W.Y., Suhaimi, H., Sambudi, N.S., Development of Scaffolds from Bio-Based Natural Materials for Tissue Regeneration Applications: A Review. *Gels*, 9, 2023.
18. Wong, S.K., Yee, M.M.F., Chin, K.Y., Ima-Nirwana, S., A Review of the Application of Natural and Synthetic Scaffolds in Bone Regeneration. *J. Funct. Biomater.*, 14, 2023.
19. Tavares, T.D., Antunes, J.C., Ferreira, F., Felgueiras, H.P., Biofunctionalization of natural fiber-reinforced biocomposites for biomedical applications. *Biomolecules*, 10, 2020.
20. Damiri, F., Fatimi, A., Magdalena Musuc, A., Paiva Santos, A.C., Paszkiewicz, S., Igwe Idumah, C., *et al.*, Nano-hydroxyapatite (nHAp) scaffolds for bone regeneration: Preparation, characterization and biological applications. *J. Drug Deliv. Sci. Technol.*, 95, 105601, 2024.
21. Backes, E.H., Harb, S.V., Beatrice, C.A.G., Shimomura, K.M.B., Passador, F.R., Costa, L.C., *et al.*, Polycaprolactone usage in additive manufacturing strategies for tissue engineering applications: A review. *J. Biomed. Mater. Res. Part B: Appl. Biomater.*, 110, 2022.
22. Marques, C.F., Diogo, G.S., Pina, S., Oliveira, J.M., Silva, T.H., Reis, R.L., Collagen-based bioinks for hard tissue engineering applications: a comprehensive review. *J. Mater. Sci. Mater. Med.*, 30, 2019.
23. Fan, J., Abedi-Dorcheh, K., Vaziri, A.S., Kazemi-Aghdam, F., Rafieyan, S., Sohrabinejad, M., *et al.*, A Review of Recent Advances in Natural PolymerBased Scaffolds for Musculoskeletal Tissue Engineering. *Polymers*, 14, 2022.
24. Kuttappan, S., Mathew, D., Nair, M.B., Biomimetic composite scaffolds containing bioceramics and collagen/gelatin for bone tissue engineering - A mini review. *Int. J. Biol. Macromol.*, 93, 2016.

25. Islam, S., Bhuiyan, M.A.R., Islam, M.N., Chitin and Chitosan: Structure, Properties and Applications in Biomedical Engineering. *J. Polym. Environ.*, 25, 2017.
26. Dong, C. and Lv, Y., Application of collagen scaffold in tissue engineering: Recent advances and new perspectives. *Polymers*, 8, 2016.
27. Sowmya, S., Bumgardener, J.D., Chennazhi, K.P., Nair, S.V., Jayakumar, R., Role of nanostructured biopolymers and bioceramics in enamel, dentin and periodontal tissue regeneration. *Prog. Polym. Sci.*, 38, 2013.
28. Prabhakaran, M.P., Venugopal, J., Chan, C.K., Ramakrishna, S., Surface modified electrospun nanofibrous scaffolds for nerve tissue engineering. *Nanotechnology*, 19, 2008.
29. Koh, H.S., Yong, T., Chan, C.K., Ramakrishna, S., Enhancement of neurite outgrowth using nano-structured scaffolds coupled with laminin. *Biomaterials*, 29, 2008.
30. Turnbull, G., Clarke, J., Picard, F., Riches, P., Jia, L., Han, F., *et al.*, 3D bioactive composite scaffolds for bone tissue engineering. *Bioact. Mater.*, 3, 2018.
31. Guo, B. and Ma, P.X., Conducting Polymers for Tissue Engineering. 19, 2018.
32. Lari, A., Sun, T., Sultana, N., PEDOT:PSS-Containing Nanohydroxyapatite/ Chitosan Conductive Bionanocomposite Scaffold: Fabrication and Evaluation. *J. Nanomater.*, 2016, 2016.
33. Habibzadeh, F., Sadraei, S.M., Mansoori, R., Singh Chauhan, N.P., Sargazi, G., Nanomaterials supported by polymers for tissue engineering applications: A review. *Heliyon*, 8, 2022.
34. El-Hadi, A.M., Al-Jabri, F.Y., Altaf, W.J., Higher dielectric properties of semiconducting biopolymer composites of poly(3-hydroxy butyrate) (PHB) with polyaniline (PANI), carbon black, and plasticizer. *Polym. Bull.*, 75, 2018.
35. Daraeinejad, Z. and Shabani, I., Tailor-made conductive PANI-coated nanofibers for tissue engineering applications. *Synth. Met.*, 286, 2022.
36. Abd Razak, S.I., Wahab, I.F., Fadil, F., Dahli, F.N., Md Khudzari, A.Z., Adeli, H., A review of electrospun conductive polyaniline based nanofiber composites and blends: Processing features, applications, and future directions. *Adv. Mater. Sci. Eng.*, 2015, 2015.
37. Rana, A., Mittal, A., Vashist, C., Rajput, S., Sridhar, S.B., Malviya, R., Exploration of 4D Printing and its Applications in the Biomedical Sciences. *Curr. Pharm. Des.*, 2025.
38. Sultana, N., Cole, A., Strachan, F., Biocomposite Scaffolds for Tissue Engineering: Materials, Fabrication Techniques and Future Directions. *Materials*, 17, 5577, 2024.
39. Das, S., Datta, S., Barman, A., Barua, R., Smart biodegradable and biobased polymeric biomaterials for biomedical applications, in: *Modeling, Characterization, and Processing of Smart Materials*, IGI Global Scientific Publishing, Hershey, Pennsylvania, USA, 2023.

40. Nadgorny, M. and Ameli, A., Functional Polymers and Nanocomposites for 3D Printing of Smart Structures and Devices. *ACS Appl. Mater. Interfaces*, 10, 2018.
41. Farag, M.M., Recent trends on biomaterials for tissue regeneration applications: review. *J. Mater. Sci.*, 58, 2023.
42. Khan, F. and Tanaka, M., Designing smart biomaterials for tissue engineering. *Int. J. Mol. Sci.*, 19, 2018.
43. Ehrick, J.D., Stokes, S., Bachas-Daunert, S., Moschou, E.A., Deo, S.K., Bachas, L.G., *et al.*, Chemically tunable lensing of stimuli-responsive hydrogel microdomes. *Adv. Mater.*, 19, 2007.
44. Lee, S.H., Oh, J.M., Son, J.S., Lee, J.W., Kim, B.S., Khang, G., *et al.*, Controlled preparation of poly(ethylene glycol) and poly(L-lactide) block copolymers in the presence of a monomer activator. *J. Polym. Sci.: Part A: Polym. Chem.*, 47, 5917–22, 2009.
45. Yoshida, R., Uchida, K., Kaneko, Y., Sakai, K., Kikuchi, A., Sakurai, Y., *et al.*, Comb-type grafted hydrogels with rapid deswelling response to temperature changes. *Nature*, 374, 1995.
46. Satarkar, N.S. and Zach Hilt, J., Hydrogel nanocomposites as remote-controlled biomaterials. *Acta Biomater.*, 4, 2008.
47. El-Husseiny, H.M., Mady, E.A., Hamabe, L., Abugomaa, A., Shimada, K., Yoshida, T., *et al.*, Smart/stimuli-responsive hydrogels: Cutting-edge platforms for tissue engineering and other biomedical applications. *Mater. Today Bio*, 13, 2022.
48. Alexandre-Franco, M.F., Kouider, R., Kassir Al-Karany, R., Cuerda-Correa, E.M., Al-Kassir, A., Recent Advances in Polymer Science and Fabrication Processes for Enhanced Microfluidic Applications: An Overview. *Micromachines*, 15, 1137, 2024.
49. Anderson, D.G., Burdick, J.A., Langer, R., Materials science. Smart biomaterials. *Smart Biomater.*, 305, 2004.
50. Ito, H., Steplewski, A., Alabyeva, T., Fertala, A., Testing the utility of rationally engineered recombinant collagen-like proteins for applications in tissue engineering. *J. Biomed. Mater. Res. Part A*, 76, 2006.
51. Nagapudi, K., Brinkman, W.T., Thomas, B.S., Park, J.O., Srinivasarao, M., Wright, E., *et al.*, Viscoelastic and mechanical behavior of recombinant protein elastomers. *Biomaterials*, 26, 2005.
52. Ghosh, K., Ren, X.D., Shu, X.Z., Prestwich, G.D., Clark, R.A.F., Fibronectin functional domains coupled to hyaluronan stimulate adult human dermal fibroblast responses critical for wound healing. *Tissue Eng.*, 12, 2006.
53. Hersel, U., Dahmen, C., Kessler, H., RGD modified polymers: Biomaterials for stimulated cell adhesion and beyond. *Biomaterials*, 24, 2003.
54. Koss, K.M. and Unsworth, L.D., Neural tissue engineering: Bioresponsive nanoscaffolds using engineered self-assembling peptides. *Acta Biomater.*, 44, 2016.

55. Kyle, S., Felton, S.H., McPherson, M.J., Aggeli, A., Ingham, E., Rational molecular design of complementary self-assembling peptide hydrogels. *Adv. Healthc. Mater.*, 1, 2012.
56. Kyle, S., Aggeli, A., Ingham, E., McPherson, M.J., Recombinant self-assembling peptides as biomaterials for tissue engineering. *Biomaterials*, 31, 2010.
57. Kyle, S., Aggeli, A., Ingham, E., McPherson, M.J., Production of self-assembling biomaterials for tissue engineering. *Trends Biotechnol.*, 27, 2009.
58. Lu, T., Li, Y., Chen, T., Techniques for fabrication and construction of three-dimensional scaffolds for tissue engineering. *Int. J. Nanomedicine*, 8, 2013.
59. Roseti, L., Parisi, V., Petretta, M., Cavallo, C., Desando, G., Bartolotti, I., *et al.*, Scaffolds for Bone Tissue Engineering: State of the art and new perspectives. *Mater. Sci. Eng.: C*, 78, 2017.
60. Peltola, S.M., Melchels, F.P.W., Grijpma, D.W., Kellomäki, M., A review of rapid prototyping techniques for tissue engineering purposes. *Ann. Med.*, 40, 2008.
61. Chia, H.N. and Wu, B.M., Recent advances in 3D printing of biomaterials. *J. Biol. Eng.*, 9, 2015.
62. Dutta, R.C., Dey, M., Dutta, A.K., Basu, B., Competent processing techniques for scaffolds in tissue engineering. *Biotechnol. Adv.*, 35, 2017.
63. Li, Z., Xie, M.B., Li, Y., Ma, Y., JS, L., Dai, F.Y., Recent progress in tissue engineering and regenerative medicine. *J. Biomater. Tissue Eng.*, 6, 2016.
64. Sanz-Herrera, J.A., García-Aznar, J.M., Doblaré, M., On scaffold designing for bone regeneration: A computational multiscale approach. *Acta Biomater.*, 5, 2009.
65. Thanh, D.T.M., Trang, P.T.T., Thom, N.T., Phuong, N.T., Nam, P.T., Trang, N.T.T., *et al.*, Effects of porogen on structure and properties of poly lactic acid/hydroxyapatite nanocomposites (PLA/HAp). *J. Nanosci. Nanotechnol.*, 16, 2016.
66. Preethi Soundarya, S., Sanjay, V., Haritha Menon, A., Dhivya, S., Selvamurugan, N., Effects of flavonoids incorporated biological macromolecules based scaffolds in bone tissue engineering. *Int. J. Biol. Macromol.*, 110, 2018.
67. Jahed, E., Khaledabad, M.A., Almasi, H., Hasanzadeh, R., Physicochemical properties of Carum copticum essential oil loaded chitosan films containing organic nanoreinforcements. *Carbohydr. Polym.*, 164, 2017.
68. Fonseca-Santos, B. and Chorilli, M., An overview of carboxymethyl derivatives of chitosan: Their use as biomaterials and drug delivery systems. *Mater. Sci. Eng. C*, 77, 2017.
69. Agrawal, P., Sonali, Singh, R.P., Sharma, G., Mehata, A.K., Singh, S., *et al.*, Bioadhesive micelles of D-α-tocopherol polyethylene glycol succinate 1000: Synergism of chitosan and transferrin in targeted drug delivery. *Colloids Surf. B: Biointerfaces*, 152, 2017.

70. Qian, L. and Zhang, H., Controlled freezing and freeze drying: A versatile route for porous and micro-/nano-structured materials. *J. Chem. Technol. Biotechnol.*, 86, 2011.
71. Aranaz, I., Gutiérrez, M.C., Ferrer, M.L., Del Monte, F., Preparation of chitosan nanocompositeswith a macroporous structure by unidirectional freezing and subsequent freeze-drying. *Mar. Drugs*, 12, 2014.
72. Nasiri, B. and Mashayekhan, S., Fabrication of porous scaffolds with decellularized cartilage matrix for tissue engineering application. *Biologicals*, 48, 2017.
73. Kim, M.Y. and Lee, J., Chitosan fibrous 3D networks prepared by freeze drying. *Carbohydr. Polym.*, 84, 2011.
74. Venkatesan, J., Bhatnagar, I., Kim, S.K., Chitosan-alginate biocomposite containing fucoidan for bone tissue engineering. *Mar. Drugs*, 12, 2014.
75. Guex, A.G., Puetzer, J.L., Armgarth, A., Littmann, E., Stavrinidou, E., Giannelis, E.P., *et al.*, Highly porous scaffolds of PEDOT:PSS for bone tissue engineering. *Acta Biomater.*, 62, 2017.
76. Blaker, J.J., Knowles, J.C., Day, R.M., Novel fabrication techniques to produce microspheres by thermally induced phase separation for tissue engineering and drug delivery. *Acta Biomater.*, 4, 2008.
77. Eltom, A., Zhong, G., Muhammad, A., Scaffold Techniques and Designs in Tissue Engineering Functions and Purposes: A Review. *Adv. Mater. Sci. Eng.*, 2019, 2019.
78. Smith, I.O., Liu, X.H., Smith, L.A., Ma, P.X., Nanostructured polymer scaffolds for tissue engineering and regenerative medicine. *Wiley Interdiscip. Rev. Nanomed. Nanobiotechnol.*, 1, 2009.
79. Harris, L.D., Kim, B.S., Mooney, D.J., Open pore biodegradable matrices formed with gas foaming. *J. Biomed. Mater. Res.*, 42, 1998.
80. Zhu, B., Li, W., Lewis, R.V., Segre, C.U., Wang, R., E-spun composite fibers of collagen and dragline silk protein: Fiber mechanics, biocompatibility, and application in stem cell differentiation. *Biomacromolecules*, 16, 2015.
81. Sarhan, W.A., Azzazy, H.M.E., El-Sherbiny, I.M., Honey/Chitosan Nanofiber Wound Dressing Enriched with Allium sativum and Cleome droserifolia: Enhanced Antimicrobial and Wound Healing Activity. *ACS Appl. Mater. Interfaces*, 8, 2016.
82. Abdul Khalil, H.P.S., Saurabh, C.K., Adnan, A.S., Nurul Fazita, M.R., Syakir, M.I., Davoudpour, Y., *et al.*, A review on chitosan-cellulose blends and nanocellulose reinforced chitosan biocomposites: Properties and their applications. *Carbohydr. Polym.*, 150, 2016.
83. Zhou, G., Liu, S., Ma, Y., Xu, W., Meng, W., Lin, X., *et al.*, Innovative biodegradable poly(L-lactide)/collagen/hydroxyapatite composite fibrous scaffolds promote osteoblastic proliferation and differentiation. *Int. J. Nanomedicine*, 12, 2017.

84. Melchels, F.P.W., Bertoldi, K., Gabbrielli, R., Velders, A.H., Feijen, J., Grijpma, D.W., Mathematically defined tissue engineering scaffold architectures prepared by stereolithography. *Biomaterials*, 31, 2010.
85. Gauvin, R., Chen, Y.C., Lee, J.W., Soman, P., Zorlutuna, P., Nichol, J.W., *et al.*, Microfabrication of complex porous tissue engineering scaffolds using 3D projection stereolithography. *Biomaterials*, 33, 2012.
86. Landers, R. and Mülhaupt, R., Desktop manufacturing of complex objects, prototypes and biomedical scaffolds by means of computer-assisted design combined with computer-guided 3D plotting of polymers and reactive oligomers. *Macromol. Mater. Eng.*, 282, 2000.
87. Bajaj, P., Schweller, R.M., Khademhosseini, A., West, J.L., Bashir, R., 3D biofabrication strategies for tissue engineering and regenerative medicine. *Annu. Rev. Biomed. Eng.*, 16, 2014.
88. Rajput, S., Malviya, R., Sridhar, S.B., Osteoporosis Risk Assessment and Individualized Feature Analysis Using Interpretable XAI and RAI Techniques, in: *Explainable and Responsible Artificial Intelligence in Healthcare*, pp. 89–113, Wiley-Scrivener (John Wiley & Sons, Inc. and Scrivener Publishing LLC), Hoboken, New Jersey, USA, and Beverly, Massachusetts, USA, 2025 Mar 13.
89. Zein, I., Hutmacher, D.W., Tan, K.C., Teoh, S.H., Fused deposition modeling of novel scaffold architectures for tissue engineering applications. *Biomaterials*, 23, 2002.
90. Wang, L., Gramlich, W.M., Gardner, D.J., Improving the impact strength of Poly(lactic acid) (PLA) in fused layer modeling (FLM). *Polymer*, 114, 2017.
91. Xiong, Z., Yan, Y., Wang, S., Zhang, R., Zhang, C., Fabrication of porous scaffolds for bone tissue engineering *via* low-temperature deposition. *Scr. Mater.*, 46, 771–76, 2002.
92. Rimell, J.T. and Marquis, P.M., Selective laser sintering of ultra high molecular weight polyethylene for clinical applications. *J. Biomed. Mater. Res.*, 53, 2000.
93. Tan, K.H., Chua, C.K., Leong, K.F., Cheah, C.M., Cheang, P., Abu Bakar, M.S., *et al.*, Scaffold development using selective laser sintering of polyetheretherketone-hydroxyapatite biocomposite blends. *Biomaterials*, 24, 2003.

12

Biocompatibility and Immunogenicity of Biocomposites

Ajesh Chauhan[1,2], Shivam Rajput[2] and Rishabha Malviya[2*]

[1]*Department of Pharmacy, Hindu College of Pharmacy, Sonipat, Haryana, India*
[2]*Department of Pharmacy, School of Medical and Allied Sciences, Galgotias University, Greater Noida, Uttar Pradesh, India*

Abstract

Biocomposites are constructed of natural polymers that are made stronger with bio-based or synthetic fillers. They are becoming more popular in biomedical applications because they are safe for living things, good for the environment, and can be altered to match varied needs. Two major things that determine how well they work in the clinic are bioavailability and immunogenicity. Bioavailability is the amount and speed at which the active biological agents in a biocomposite are taken up and rendered available at the site of action. The material's degradation profile, porosity, surface chemistry, and how it interacts with biological fluids all affect this. It is highly crucial to make sure that bioavailability is as high as feasible, especially for drug delivery systems, scaffolds, and wound healing matrices, where controlled release and targeted action are needed. Immunogenicity, on the other hand, is about how likely the biocomposite is to make the immune system react, which could be a positive or negative thing. A little bit of immune activation can help tissues heal, but too much or too long an immunological response might cause inflammation, fibrosis, or implant rejection. The immunogenic characteristics of biocomposites are closely linked to their composition, the origin of biomaterials (e.g., plant versus animal sources), the presence of residual antigens, and the surface topography. Finding the correct balance between bioavailability and immunogenicity is a big difficulty when creating biocomposites for medicinal use. Recent advancements in surface modification, nanotechnology, and biofunctionalization are enhancing the efficacy of biocomposites by facilitating targeted drug delivery and mitigating adverse immune reactions. This abstract makes it

**Corresponding author*: rishabhamalviya19@gmail.com, rishabha.malviya@galgotiasuniversity.edu.in

Rishabha Malviya and Mukesh Roy (eds.) Biocomposites in 3D and 4D Printing: Materials, Techniques, and Applications in Regenerative Medicine, (323–348) © 2026 Scrivener Publishing LLC

clear that both factors need to be looked at during the development phase to make sure that biocomposite based medical devices and therapies are safe and effective.

Keywords: Biocomposite, bioavailability, biofunctionalization, immunogenicity, scaffolds.

12.1 Introduction

Innovations in materials science have brought about a new age of revolutionary medicine by bridging the gap between biology, engineering, and technology. One example of a new material is a biocomposites, which combine biodegradable polymers with reinforcing elements and show great promise for use in medicine [1]. Biocomposites have come a long way in terms of mechanical excellence, environmental friendliness, and human-friendly integration. Replicating natural tissue properties reduces risks of immunological rejection, while using materials like natural fibers, ceramics, and nanoparticles improves mechanical and thermal properties [2]. Several medical uses, including as implants and tissue scaffolding, are made possible by this convergence. Because of their biocompatibility and strength, these materials find use in cardiovascular stents and grafts. In orthopedics, they treat bone because of its load-bearing capacities.

In the realm of biomedicine, biocomposites, which include both biodegradable polymers and reinforcing components, are indispensable. In cardiovascular surgery, they are used to make stents that gradually dissolve into the body; thus, there is no longer any need to insert permanent foreign objects into blood vessels. When it comes to orthopedic implants, biocomposites improve compatibility and decrease the possibility of implant rejection. Biocompatibility is a key component of regenerative medicine, which aims to promote the growth of new organs and tissues. Tissue engineering relies heavily on biocomposites because of the scaffolding they provide for cellular development. They also improve therapeutic efficacy, decrease adverse effects, and provide controlled release mechanisms, all of which are necessary components of pharmaceutical delivery systems. The biocompatibility and adaptability of these materials are causing a dramatic shift in medical practice [3].

By incorporating biocompatible components, a new class of biocomposites has been created. These biocomposites have the possibility to be used in living tissues since they are very biocompatible, biodegradable, and have low toxicity. With an emphasis on four key areas: tissue engineering, orthopedic and dental implants, wound healing, and bone regeneration, this comprehensive overview details the most recent advancements in the use of biocomposite materials in biomedical engineering. Biocomposite porosity is enhancing cell proliferation and organ regeneration in the quickly growing field of regenerative medicine, which is enhancing tissue engineering. Biocomposites are becoming more important in personalized healthcare as they allow for the development of customized medical instruments and treatments. Current problems, such as better production processes, complicated biological-material interactions, and complete biocompatibility investigations, necessitate interdisciplinary collaboration [4].

Tissue engineering relies on biomaterials, which provide a three-dimensional scaffolding mimicking the original extracellular matrix's (ECM) structure and biological functions. Tissue regeneration and repair rely on this framework, which can be either natural or manufactured, to support and guide important cellular behaviors. In particular, it facilitates lineage-specific differentiation, encourages cell adhesion, and stimulates cell proliferation as it directs the formation of new, functional tissue [5].

In order to be suitable for these functions, a scaffold must fulfil several demanding biological and physicochemical requirements. To allow nutrients, gases, and metabolic waste to be exchanged, as well as for cells to infiltrate and migrate, it should have the right amount of porosity and pore interconnectivity. Tissue type-specific pore sizing is necessary because various cells and tissues have varied spatial demands. The scaffold also needs to be structurally sound, but it should be able to degrade slowly enough to keep up with the rate at which new tissue is being formed. So long as no foreign material is left behind, the scaffold can give initial mechanical support without causing fibrosis or chronic inflammation [6].

The immunological profile and biocompatibility of the substance are of equal importance. To prevent problems with tissue integration or implant rejection caused by unfavorable immune reactions, the scaffold should not be poisonous or immunogenic. It should also facilitate the ECM's natural molecular signaling capabilities, such as binding to surface receptors like integrins that regulate cellular processes. Natural polymers such as alginate, chitosan, hyaluronic acid and collagen have garnered considerable interest among the various materials studied for the production of hydrogel-based scaffolds. Because of their structural similarity to the original ECM, intrinsic bioactivity, and biodegradability, these biopolymers are highly esteemed. Because of their hydrophilic properties, they are able to hold a large amount of water, resulting in a hydrated and pliable matrix that is very similar to the biochemical and mechanical properties of real tissue. They are also very amenable to functionalization and cross-linking due to their chemical versatility, which allows for the customization of mechanical strength, degradation rate, and biological signaling qualities. The ability to treat these polymers under moderate circumstances is crucial because it allows for easy construction into different scaffold structures while keeping the activity of integrated cells or bioactive chemicals [7].

If tissue regeneration is to be achieved, biomaterials and hydrogel-forming natural polymers in particular are required. These materials can mimic the extracellular matrix (ECM), sustain important cellular functions, and blend in with biological systems without a hitch [8]. Biomedical researchers, especially those working in tissue engineering, have taken an interest in two naturally occurring biomaterials, chitosan and hydroxyapatite, because of the complementary structural and biological features they exhibit. Regenerative medicine makes use of two materials—hydroxyapatite (HA) sourced from minerals and chitosan from crustaceans' exoskeletons, which provide distinct but complementary advantages [9].

Calcium and phosphate form the mineral hydroxyapatite (HAp), which resembles the inorganic component of actual bone in appearance. Its chemical similarity to human hard tissues significantly improves its biocompatibility and osteoconductivity. This makes it an ideal substrate for osteoblast growth and adhesion, and it also blends in naturally with the body. One of the most important aspects of HAp is its ability to form direct chemical interactions with real bone tissue. This improves implant durability and allows for better long-term integration. Surface apatite layers are essential for initiating bone regeneration, and HAp can promote their formation in physiological contexts. This quality is what makes HAp bioactive. Its exceptional blend of properties makes it a popular choice for bone graft replacements, dental restorations, load-bearing scaffolds, and orthopedic implants [10]. But deacetylating chitin yields chitosan, a linear polysaccharide and the second most prevalent natural polymer after cellulose. It has great physicochemical and biological qualities, which are very useful for applications involving soft tissues. Through electrostatic interactions with negatively charged cell membranes and biomolecules, cationic polymer chitosan enhances cell adhesion, migration, and proliferation, essential processes in tissue repair and regeneration. Beyond that, being biodegradable means that chitosan can be easily eliminated from the body as harmless, bioresorbable by-products. It is immune-system safe for implantation because of its low toxicity, biocompatibility, and absence of antigenicity. The versatility of chitosan in tissue engineering is enhanced by its wide array of biofunctional characteristics, which go beyond its fundamental characteristics [11]. It aids hemostasis, the process of quickly stopping bleeding, by encouraging the aggregation of platelets and the production of clots. This helps with angiogenesis, which is essential for delivering oxygen and nutrients to the healing tissue because it creates new blood vessels. Its activation of macrophages and influence on their polarization toward a regenerative (M2) phenotype are additional functions beyond wound healing and immunological control. Additionally, its inherent antibacterial properties lessen the likelihood of infection as tissues heal by preventing the colonization of affected areas by bacteria [12].

12.1.1 *In Vitro* Mucotoxicity Tests

There is no consideration for the living organism during biocompatibility testing, which is conducted using *in vitro* protocols. The point of performing lab studies in a controlled setting is to mimic the effects that substances would have on live creatures [13]. One can quantify the item's influence by measuring the amount, growth rate, metabolic activity, or other biological activities of the cells exposed to it [14]. When assessing new products, these tests are a good substitute for the time-consuming and expensive animal testing. They do not raise any ethical questions, are quick, easy to replicate, and manageable in the lab. Limitations of *in vitro* testing include questionable clinical relevance and the lack of *in vivo* modeling [15]. Biomaterials can be tested for *in vitro* biocompatibility using a variety of methods. A biological system, an interaction between cells and materials, and a biological outcome are the three mainstays of any cytotoxicity testing method.

12.2 Biological System

Oral mucosal cell monolayer cultures are one potential biological system for *in vitro* assessment of mucosal toxicity caused by resin-based dental materials. Utilizing three-dimensional tissue-engineered models of human oral mucosa is an additional alternative to consider [16].

12.2.1 Cytoplasmic Monolayer System

The goal of monolayer cell culture is to cultivate and maintain certain cell types in a controlled environment. Dental materials close to the gingival tissue can be evaluated for cytotoxicity using gingival fibroblasts and keratinocytes [17]. When testing dental materials for biocompatibility, human gingival fibroblasts are often used. They can be quickly cultured in normal media and are easy to isolate from patients, which are two of their main advantages. When tested for cytotoxicity, they also show a high degree of sensitivity. L-929 murine fibroblasts are among the most popular cell lines; however, 3T3 mouse fibroblasts are also employed. It is the biological endpoint that dictates the cell type that is used for the cytotoxicity experiment. Some biological tests need certain kinds of cells. The release of cytokines by these cells can be studied using THP-1 monocytes and fibroblasts [18].

The sensitivity of numerous cell lines to different dental materials has been investigated. According to research, mouse macrophages are more sensitive than normal human fibroblasts, while hamster fibroblasts BHK-21 (C-13), are more sensitive than normal human fibroblasts. Researchers discovered that human periodontal ligament (PDL) and pulp fibroblasts were more responsive than 3T3 cells. from mice and gingival fibroblasts from humans. The sensitivity order varies among testing methods, even though numerous cell lines display unique biomaterial sensitivities [19]. To determine the sensitivity of *in vitro* biocompatibility systems, researchers tested 12 standardized cell lines with 20 different materials. They found that different testing procedures provide widely varying degrees of sensitivity in established cell lines derived from different tissues [20]. Testing with fresh cells was not as repeatable as testing with established cell lines, according to another study. The lack of therapeutic value is the major limitation of these methods, even though there is some batch-to-batch variability in monolayer cell cultures. Because normal human oral mucosa contains cells with specialized functions and barrier characteristics, the findings may not be generalizable to normal human oral mucosa or patients in general when epithelial cells are cultured in a monolayer. A precise risk assessment and near-perfect representation of the real environment are achieved with three-dimensional, tissue-built oral mucosa models [21].

12.2.2 3-D Tissue-Engineered Models

New tissue-generated, three-dimensional models of the human oral mucosa have revealed that resin-based dental materials are more biocompatible *in vitro*. This study aimed to determine whether dental metals irritate the mucosa by using a three-dimensional co-culture system that included nylon mesh, human fibroblasts, and keratinocytes. There was also the option of using a pulp chamber in an *in vitro* setting, which involved growing three-dimensional fibroblasts on nylon mesh [22]. To determine the cytotoxicity of dental materials, investigators utilized a 3-D culture of transfected bovine pulp-derived cells on polyamide meshes. These *in vitro* models are better than single-layer cell culture test models for evaluating the biocompatibility of dental biomaterials because they can perform numerous endpoint studies of the oral mucosa's response to restorative materials and because they are more similar to the clinical context. This allows them to be more accurate without using animals for testing [23, 24].

12.2.3 Cell/Material Contact

For materials to be biologically evaluated, there must be enough cell-material interaction during testing. Contact through extracts, indirect contact, and direct contact are the three modes by which cells and materials can interact [25].

Direct contact: Substances are physically handled by the cells or culture media in a direct contact test. Soluble chemicals are dissolved in the culture medium directly, which allows for the best possible cell-material contact and increased sensitivity.

Here are some ways to get non-water-soluble chemicals into direct contact with cells:

- The material to be tested is positioned near the tissue explant.
- On top of an already-established cell monolayer, the material is placed for testing.
- To create a cell monolayer around the specimen, a cell suspension is added to a culture vessel with the specimen placed at its base.
- Cells are grown on the specimens themselves.

One way to achieve the optimal cell-material interaction is to grow the cells directly on the specimen. In this case, the material's surface properties are critical; a lack of surface energy will lead to poor development since it inhibits cellular attachment [26].

Contact through intermediate: The specimen is separated from the cells *via* a permeable intermediate in a testing approach that uses indirect cell/material contact. This method works with materials in any state of matter, including solids, semi-solids, and liquids. The lack of a culture medium around the specimen allows for testing to take place in an unset condition, which is useful for materials evaluation [51]. The agar overlay methodology, the first indirect cell/material contact method, to assess the detrimental effects of leachable components disseminated *via* an agar layer over a monolayer cell culture. This method has long been used for the purpose of determining whether dental materials are biocompatible [27].

To avoid direct cell–material interaction, researchers instead used a dentine slice or synthetic filter to separate the specimen from the culture liquid. As a roundabout way of testing, scientists used cells grown on Millipore filters. The little filter makes it easier for the test substance to come into intimate contact with the cells, which means that even components that are not soluble in water can still permeate and reach the cells [28].

Contact through extracts and elutes: Helping insoluble compounds interact with cells can be achieved by using an emulsifying agent or by removing leachable components using a solvent. If you want to study cytotoxicity, you can dissolve dental resin monomers in ethanol or dimethyl sulfoxide (DMSO) and then dilute them with culture medium. These experiments did not employ doses of DMSO or ethanol high enough to induce a cytotoxic response. In order to make sure that the monomer cytotoxicity was evaluated accurately, a control group was made using the culture media that included the greatest concentration of ethanol or DMSO. The cytotoxicity of different dental materials was traditionally assessed using the extraction method. Dentine adhesives, dental cements, amalgams, and denture base resins are all part of this category. Culture media, distilled water, saline, balanced salt solution, ethanol/acetone in saline, and balanced salt solution are among the media utilized for extraction.

Researchers seldom compare various extraction processes. Investigators contrasted culturing media with saline when it came time for extraction. There was cytotoxicity in the saltwater extract, as opposed to the medium extract, according to their testing. The choice of extraction media and the duration of analysis have a significant impact on the detection of monomers released into different water-based media from experimental composite resins. Consequently, false-negative results could be produced by cytotoxicity evaluations of dental materials [29].

12.3 Biological Endpoint

The morphological and numerical characterization of cytotoxicity experiments can be done with a focus on cell survival, proliferation, adhesion, migration, secretion of specific chemicals, and activities such as apoptosis [30].

12.3.1 Morphological Evaluation

This approach is based on pathological cellular alterations such as vacuolation, nuclear abnormalities, binucleation, or expansion of the nucleus. Nuclear disintegration and pyknosis are observed in cells that have died. Researchers were able to observe the direct harm that resin monomers caused to multiple layers of the oral mucosa using a three-dimensional model of the human mouth. However, there is a lack of information regarding how much harm resin monomers cause to a 3D *in vitro* oral mucosal model compared to normal human oral mucosa. For this reason, to conduct a biological evaluation of dental restorations made of resin, it is essential to validate these tissue-engineered models of the oral mucosa in a clinical setting [31].

12.3.2 Cell Viability and Proliferation Assays

In clinical situations, the affected area of the oral mucosa has a reduced metabolic rate and proliferative power compared to healthy oral mucosa, which is a result of injured tissue having fewer viable cells. Various assays can be employed to detect the proliferation and vitality of cells exposed to test substances *in vitro*, allowing one to ascertain the relative toxicity of materials [32]. Figure 12.1 illustrates the tabular depiction of cell viability and proliferation assays

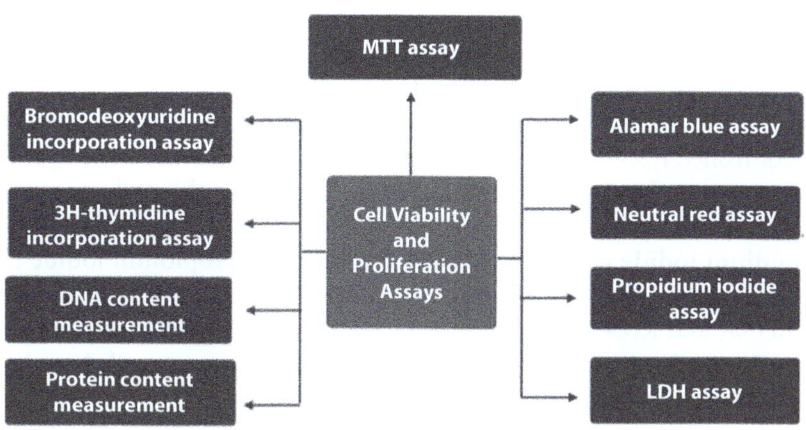

Figure 12.1 Cell viability and proliferation assays.

MTT assay: Mossman conducted a colorimetric MTT [(3-(4,5-dimethylthiazol-2-yl)-2,5-diphenyltetrazolium bromide] experiment to measure cell viability by observing fluctuations in mitochondrial dehydrogenase activity. Methylthiazole tetrazolium is a water-soluble compound that becomes an insoluble purple formazan by this process. The formazan can be dissolved in water and its concentration can be measured using spectrophotometry [33].

Alamar blue assay: The Alamar blue assay may detect metabolic activity using a non-toxic aqueous dye called Alamar Blue. This dye is stable over time, changes color, and fluoresces in response to the chemical decrease of the growth media induced by cell multiplication. Alamar Blue allows for continuous cell monitoring in a culture because it is soluble in culture fluid and is non-toxic. Specifically, there is no discernible impact on the viability of cells cultivated as time progresses [34].

Two benefits are offered by Alamar Blue in comparison to the MTT assay: Spectrophotometric methods can also detect the color shift, however fluorometric methods are more exact. Second, as it does not harm cells in any way, it can be used for multiple cell viability tests. The MTT method is more popular among researchers, although the Alamar Blue assay is more expensive. Dental ceramic biocompatibility testing with the Alamar Blue assay. Several studies have used the Alamar blue assay and either monolayer or three-dimensional cultured human oral epithelial cells to assess the biological characteristics of resin-based dental materials [35].

Neutral red assay: An important pigment called neutral red is released into the surrounding media when live cells' membranes are disrupted. It serves as a marker for cell viability. The neutral red assay is a reliable measure of cell viability % because it uses photometric measurement following controlled lysis to determine whether cells have absorbed the dye. The cytotoxicity evaluation of dental materials has frequently made use of this assay [36].

Propidium iodide assay: In this test, a solution of propidium iodide (PI) dye is left out. Upon entry into cells, propidium iodide (PI) intercalates with DNA and RNA in response to a harmful agent's weakened cell membrane. Using flow cytometry, one can determine how many cells have been labelled with the fluorescent dye. Since PI is an exclusion dye, the proportion of glowing cells indicates the number of dead cells. Whether dental materials are toxic to cell cultures has been the subject of some research using this method [37].

LDH assay: When membranes are severely damaged or undergo cytolysis, cells release the stable cytosolic enzyme lactate dehydrogenase (LDH). This colorimetric cytotoxicity test determines its concentrations. The number of damaged or lysed cells directly correlates with the amount of lactate dehydrogenase released. The maximal amount of LDH released by controlled cell lysis can be compared to the amount of LDH released by exposed cells to assess the degree of cytotoxicity [38].

Bromodeoxyuridine incorporation assay: Since 5-bromo-2′-deoxyuridine (BrdU) is incorporated into newly synthesized DNA during cell division, it is possible to detect rapidly proliferating cells using fluorescently labeled anti-BrdU antibodies or particular nucleic acid stains [39].

3H-thymidine incorporation assay: An alternative technique for assessing cell proliferation is to introduce 3H-thymidine into cells during cell division and then use radioactivity detection to identify newly synthesized DNA. Researchers examined the effects of dental composite resin components on the growth of human and rat mononuclear cells using a 3H-thymidine incorporation test [40].

Measurement of DNA content: Through the measurement of fluorescence intensity, DNA-intercalating dyes enable the evaluation of cellular DNA concentration. One of three nucleic acid stains, Hoechst 33,258, Hoechst 33,342, or DAPI is used in the majority of cell-cycle investigations. At AT-rich sites, these stains attach to the minor groove of DNA. The biocompatibility of different glass ionomer cements was assessed by researchers using this methodology [41].

Protein content measurement: Protein content quantification in cytotoxicity assays now makes use of modifications to existing techniques. Dental composite components and mercury compounds were evaluated for their cytotoxicity using protein quantification [42].

12.4 *In Vivo* Tests

Evaluating biocompatibility in a living creature is known as *in vivo* biocompatibility. The most common kind of *in vivo* experiment is the use of animals in tests.

12.4.1 Animal Tests

Research on animals often involves inserting a substance into an animal's body and then seeing how it reacts locally. Compared to *in vitro* investigations, this one is more relevant because it allows researchers to look at how the chemical interacts with the biological system. Animal experimentation, however, is fraught with ethical problems, takes a lot of time and money, and is difficult to control for factors. In addition, there are still unanswered problems about which animal species best exemplifies the human response [43].

12.4.2 Implantation Testing

An inert substance, such as silicone rubber, serves as a negative control in implantation experiments, which involve inserting material specimens into an animal's connective tissue, muscle or bone [44]. After a predetermined amount of time has passed, the test and control materials' effects on the tissues are evaluated. A number of biological factors come into play, including the presence or absence of necrosis, inflammation, infiltration, fibrogenic cell activity, and the encapsulation's structural state [45].

12.5 PLA Biocomposites

A biodegradable composite, made by combining CFF with biopolymers like PLA, is an excellent material for use in plastic goods and implants. Feathers from chickens were soaked in alcohol for a day, then rinsed with an organic solvent that was water-soluble, and finally dried at 60°C for another day to make the composites. The quill was used to extract CFFs that were around 5 μm in diameter and 10 to 30 mm long with a CFF level of 5 wt %. The CFF/PLA composite's modulus of elasticity increases by 35.6% to 4.38 GPa, reaching its maximum modulus with an increasing CFF concentration. According to the results, adding CFF to the matrix has a significant positive effect on reinforcement. By establishing 5 wt % CFF as the critical threshold, we can see that the CFF and matrix are not adhering optimally, which results in a decrease in modulus for the composite when the CFF content exceeds 5 wt %. There is still a problem with adhesion between CFF and PLA, even though CFF and PLA are both hydrophobic and have a rough surface. Matrix stress could not be transferred to stronger fibers. The phenomenon could have occurred because the carbon fiber filaments (CFFs) were not uniformly aligned throughout the composite. If the matrix failed first, the composite could have failed as a whole, and the fibers would have broken off at that point. These findings may be supported by the microstructures' fractured morphology, which was observed using SEM [46].

12.6 Nanoparticle-Reinforced Biocomposites

A polymer matrix combined with a trace quantity (less than 10%) of nano-sized fillers produces a new material with significantly enhanced characteristics; this material is known as a nanocomposite. The utilization of bio-nanocomposites in biomedical and tissue engineering applications allows for the creation of materials with enhanced strength, stiffness, and controllable biodegradability and bioresorbability rates. To create these materials, biopolymers are combined with appropriate and/or bioresorbable nano-fillers. In order to improve the thermal and mechanical characteristics of biodegradable polymers, numerous studies have looked at the usage of nanoparticles, such as nano-apatite, carbon nanotubes, and nanoclay. The utilization of these particles to control their biodegradability is clearly a priority. Epoxies, which are thermoset polymers commonly used in aeronautical engineering, can have their Young's modulus and glass transition temperature improved by adding these particles, according to numerous studies. In order for them to do their job, they either block the bond between polymer chains or act as nanoscale fibers, allowing the particles to transfer stress from the polymer matrix [47].

When compared to more traditional materials, carbon nanotubes are generally considered to be among the strongest on the planet. Not only that, but their shape makes them perfect for transporting pharmaceuticals and hydrogen, respectively, for use in fuel cells and biomedical engineering. There are three distinct kinds of nanotubes: zigzag, armchair, and chiral. Different mechanical, electrical, and fractural characteristics are exhibited by each of these. There have been claims that nanotubes can enhance cell tracking, sense the environment, deliver transfection agents, and integrate scaffolding with the host body in tissue engineering materials. One way to make a material stronger is to add these nanotubes to it. The formation of biocomposites involving these nanotubes and specific biodegradable or bioresorbable polymers carries the unfortunate possibility of trapping within human cells. A neural network could be created through the formation of cell-bridges, which are made easier by the presence of nanotubes. The cytotoxicity of carbon nanotubes has been proven in a large number of scientific articles. For example, after 6 hours of incubation, the nanotubes showed a significant increase in cytotoxicity when exposed to alveolar macrophages. Inadequate biodegradability further limits the use of these materials in implant applications [48].

Carbon nanotubes and other nanoparticles are so small that they can be breathed and deposited deep into the lungs, where they can access areas that larger particles cannot. Due to their size, they can pass through cell and tissue membranes, which gives them the potential to translocate from their point of exposure to the organ they are targeting in the body [49]. Three established criteria quantify a particle's toxicity:

- The particle's surface area in relation to its mass particles with a large surface area are better able to interact with cell membranes to absorb and transport toxic substances;
- Particle retention time; harm is more likely when particles stay in contact with the cell membrane for an extended period of time.
- Included in this factor are the element's properties, the particles' chemical reactivity or inherent toxicity, and their mobility (via clearance or migration to nearby tissue).

Nanoclays are often made from layered silicate, a polymer known for its ordered layered structure. Nanopolymer composites typically exhibit three distinct structural types. The exfoliation and dispersion properties of the clay structure inside the polymer matrix are improved by newly created nanoparticles that are supported by nanoclay [50]. The regulated morphology spanning micro to nano cellular structures and a high cell density made possible by porous ceramic materials or scaffolds composed of PLA/nanoclay biocomposites has recently attracted a lot of attention from researchers looking for ways to aid in tissue regeneration.

Recent research suggests that biocomposites shouldn't include more than 5% nanoclay in the polymer, as doing so could significantly diminish mechanical properties like flexural strength. The addition of nanoclay to the biocomposites may alter their chemical structures, which could lead to their softening [51]. Because of its structural and chemical similarities to the mineral phase of human bone, nano-hydroxyapatite has found use in orthopedic and dental implant applications. A biocomposite was created, enhancing polyamide 66 (PA66) with nano-hydroxyapatite (n-HA). Biodegradability can be controlled, according to the study. A significant improvement in the biocomposites' thermal stability compared to pure PA 66 was observed when hydrogen bonds were established between the amide group of PA 66 and the hydroxyl group (-OH) of n-HA. Even though the crystallinity percentage is smaller, this causes the T to be larger [52].

12.7 Biocomposites Derived from Silk

It was found that biodegradable polymers can be synthesized in an environmentally sustainable fashion by mixing silk sericin with various resins. The superior ability of sericin-containing polyurethane foams to absorb and desorb moisture was discovered. Superior mechanical and thermal properties of sericin-infused polyurethane have been synthesized. Sericin is very compatible with polyvinyl alcohol (PVA) and other water-soluble polymers. The physical properties and fine structure of sericin and PVA composite films were studied. Another invention detailed a hydrogel membrane that was cross-linked with PVA and sericin using dimethylurea. The membrane was suitable for usage as a functional film due to its high moisture content, longevity, and remarkable strength. When wet, silk fibroin film allows dissolved oxygen to pass through easily, but when dry, it is too delicate to be utilized on its own. Films and threads made of the biodegradable and biocompatible polymer chitosan are a good alternative. Researchers have proposed silk fibroin and chitosan blends as possible biomedical composites due to the fact that the crystallinity and mechanical characteristics of silk fibroin are significantly improved with increasing concentrations of chitosan. Silk fibroin and alginate form a biocomposite in the form of a composite sponge. The biomedical and biotechnological fields stand to gain from silk fibroin's versatility in treatment, its lack of toxicity, and its compatibility with both living things and the environment. Dry silk fibroin sponges don't have enough mechanical properties to be used as wound dressings. To make them better, you can mix silk fibroin films with polymers, both natural and synthetic, like polysaccharide-sodium alginate. The effects of biocomposites made of silk and poly(butylene succinate) (PBS) were studied. The tensile strength, fracture toughness, impact resistance, and thermal stability of biocomposites are significantly affected by their production procedures. Furthermore, scanning electron microscopy (SEM) imaging showed that the silk threads adhered strongly to the PBS matrix [53].

Tussah silk, *Bombyx mori*, and twisted *B. mori* were tested for their tensile strengths. The tensile strength and extensibility of tulsah silk were higher than those of other types of silk. Still, there was hardly a difference in the stiffness of the samples. This could be because of variations in silkworm care, cocoon making, and spinning circumstances. When comparing *B. mori* silk fiber to Tussah silk fiber, the Weibull analysis showed that the former had better experimental measurement reproducibility. This might be because the degumming process changes the microstructure of

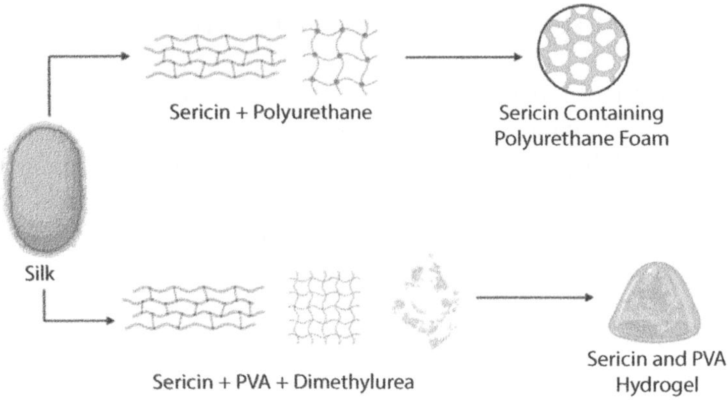

Figure 12.2 Biocomposites derived from silk.

the fiber. Reinforcing biodegradable polymer using silk fiber drastically changes its mechanical properties. Researchers found that compared to an unaltered sample, adding silk fiber to poly(lactic acid) (PLA) increased its ductility by 53% and elastic modulus by 40%. The amount of silk fiber in the composites affected how biodegradable the silk/PLA biocomposites were. The ability to regulate the resorbability of biocomposites utilized by humans is crucial for the utilization of this novel material in bone plate development [54]. Figure 12.2 shows the procedure for obtaining biocomposites from silk.

12.8 Immunomodulatory Properties: *In Vitro* and *In Vivo* Evidence

The immunomodulatory features of chitosan-hydroxyapatite biocomposites can be better understood through interactions with biological systems, such as studying inflammatory indicators in a targeted manner and determining their biocompatibility [55].

12.8.1 Immunomodulatory Assessment *In Vitro* (HAC and HACF Scaffolds)

In a controlled laboratory environment, the immunomodulatory and anti-inflammatory effects of hydrogels derived from marine biopolymers were evaluated. The mRNA expression of tumor interleukin-6 (IL-6), necrosis factor-alpha (TNF-α), and nuclear factor kappa-light-chain-enhancer of activated B cells (NF-κB) was evaluated in order to ascertain the levels of important inflammatory indicators. Fibroblast cells from mice L929 were used to analyze these markers in a cytotoxic setting [56]. Pro-inflammatory cytokines, such as TNF-α, IL-6, and IL-1β, activate inflammatory cells in the course of acute responses. NF-κB plays an important role in the signaling pathways that cause these cytokines to be produced. To determine if the biomaterial causes an inflammatory reaction, it is necessary to measure the expression levels of certain markers when they come into contact with the material [57]. When the mRNA expressions of IL-6, TNF-α, and NF-κB were compared in control cells with those in HAC and HACF hydrogels, no significant alterations were seen. So, HAC and HACF are completely safe for human consumption and have no negative effects on health. Because of their immunomodulatory and anti-inflammatory properties, natural polysaccharides, including chitosan, alginate, and fucoidan in HACF are believed by the authors to be responsible for the positive reaction of the synthesized hydrogels. Based on these results, it seems that hydrogels can be very biocompatible and non-inflammatory. Sulphated fucoidan's negatively charged ions improve its interaction with biological systems, which is why HACF outperformed HAC to a lesser extent, as stated by the researchers. Importantly for implantable or tissue engineering materials, our *in vitro* results show that HACF and HAC biocomposites do not seem to make fibroblast cells noticeably inflammatory *via* the examined cytokines [58]. The following is usually what is done in standard *in vitro* experiments:

- **Negative Control (Untreated Cells):** Standard gene expression levels were determined by cultivating cells that were not treated.
- **Positive Control (Pro-inflammatory Stimulus):** To ensure the sensitivity of the assay and the cells' reactivity to inflammatory stimuli, they were treated with known inflammatory inducers, such as lipopolysaccharide (LPS). As an example, macrophages and fibroblasts have been activated by LPS, resulting in an upregulation of cytokine production such as IL-6 and TNF-α. These checks and balances make sure that any changes in cytokine expression are due to the biomaterial itself and not to random chance [59].

12.9 Evaluation in a Stem Cell Model of Dermatitis (Chitosan-Xanthan-Hydroxyapatite Composite)

Researchers in this study analyzed four distinct treatment groups: G1, which received no treatment at all, G2, which received normal hydrocortisone ointment, G3, which received the biomaterial without conditioned medium, and G4, which received the biomaterial with conditioned medium. For 5 days after induction, treatments were administered either topically or as a single dressing (biomaterial) [60]. Skin sample analysis distinguished the groups according to tissue structure and inflammatory markers. Both the untreated G1 and the hydrocortisone-treated G2 groups exhibited a notable inflammatory influx, epidermal rupture, and hyperplasia. Many of the mast cells in G1 were degranulated, and their number was the largest. Mast cells play an important role in dermatitis and allergic reactions by secreting histamine and other mediators that exacerbate inflammation, itching, and angiogenesis. The G4 group showed a decrease in plasma histamine levels, which could be attributed to lower mast cell activity. In this group, biomaterial was administered together with conditioned medium. Importantly, G3 (CM-free biomaterial) and G4 (CM-containing biomaterial) showed better tissue regeneration, fewer mast cells, and reduced skin thickening than G1 and G2 [61]. Multiple inflammatory mediators' expression levels were also examined in the study. The endogenous anti-inflammatory mediator annexin A1 (AnxA1) is well known for its ability to decrease histamine release and mast cell degranulation. A family of enzymes known as intracellular Janus kinases (JAKs) transmits signals from surface cytokine receptors to the nucleus. Because the JAK/STAT signaling pathway is activated by Janus kinase (JAK)-1 and JAK-3, itching in dermatitis is frequently linked to inflammatory cytokines such as IL-31. The inflammatory cascade is significantly influenced by the cytokines interleukin IL-6, (IL)-1β, and tumor necrosis factor (TNF)-α. In dermatitis lesions, mast cells are the primary source of the pro-angiogenic material vascular endothelial growth factor (VEFG) [62]. Specifically, in G4, the number of cells expressing AnxA1 and JAK-1 and JAK-3 was decreased following biomaterial treatment, as compared to untreated G1 and G2. Compared to groups G1 and G2, treatments in G3 and G4 reduced levels of VEFG, IL-1β, IL-6, and TNF-α. The findings show that in a dermatitis model, the G3 biomaterial (a combination of chitosan, xanthan, and hydroxyapatite) can greatly decrease inflammation and enhance tissue regeneration, especially when mixed with G4, the conditioned medium. An immediate indication of the anti-inflammatory and immunomodulatory action is the decrease in mast cells, plasma histamine, pro-inflammatory cytokines (TNF-α, IL-1β, IL-6), and proteins associated with the pathways of inflammation and itching (JAK1/3) and angiogenesis (VEFG). G3 and G4 are more capable of tissue regeneration when inflammation is reduced. It is well established that inflammatory cells, such as fibroblasts and macrophages, can accelerate the healing process of wounds; this phenomenon is partially explained by biomaterials derived from chitosan [63].

The hydroxyapatite, xanthan gum, and chitosan scaffold showed functional integration with host tissue in an animal model of dermatitis. This integration was most noticeable in the domains of tissue regeneration and regulation of angiogenic and inflammatory processes. A variety of functional endpoints were used to assess the level of integration with host tissue and its therapeutic potential. Results from immunohistochemistry and enzyme-linked immunosorbent assay (ELISA) showed reduced levels of TNF-α, IL-1β, and IL-6, indicating greater biocompatibility, and histological investigation corroborated tissue regeneration by revealing improved skin structure. One way to indirectly evaluate angiogenesis is by looking at decreased VEFG expression, which suggests a regulated vascular response [64]. Mesenchymal stem cells (MSCs) and other cell types are cultured in serum-free media for 24 to 48 hours to create conditioned medium. The next step is to collect the supernatant, filter it to eliminate any cell debris, and concentrate it if needed. Some studies freeze-dry the conditioned media to make it more stable before adding it to scaffolds, which is usually done by soaking, rehydrating, or encapsulating. The concentration of bioactive components and the intended therapeutic effect usually dictate the dosage of CM. For instance, rehydrating the CM into the scaffold produced a secretome-enriched alginate/extracellular matrix hydrogel patch with a high concentration of bioactive ingredients [65]. To assess the therapeutic efficiency and sustained release profile of a secretome-enriched alginate/extracellular matrix hydrogel patch, for instance, researchers assessed the release of bioactive factors such as VEFG, IGFBPs, HGF, IL-8, and IL-6 [66].

12.10 Conclusion

Biocomposites are becoming useful for biomedical purposes because they can combine strength with biological function. A biocomposite's biocompatibility is how well it can do what it was meant to do without hurting the host. Immunogenicity, on the other hand, is how probable the item is to make the immune system react. A well-designed biocomposite should show higher biocompatibility and lower immunogenicity. Both the matrix and the reinforcing phases must be properly chosen and built for this to operate. These phases can be natural or man-made polymers, ceramics, or bioactive substances. People often utilize surface modification, sterilization methods, and add anti-inflammatory compounds to make objects more compatible and less prone to inducing immunological reactions. Some natural compounds can trigger immune responses despite being bioactive, and certain degradation products of polymers can induce inflammation or toxicity. Therefore, extensive *in vitro* and *in vivo* testing is essential to evaluate the long-term biological interactions of biocomposites. In conclusion, biocomposites provide significant advantages in tissue engineering, implants, and drug delivery systems; yet, their effectiveness relies on a delicate balance between mechanical performance, biocompatibility, and immunogenicity. Future research ought to continue the advancement of intelligent, adaptive biocomposites that alleviate unwanted reactions while improving integration with host tissues.

AI disclosure statement: The author acknowledges the use of AI-based tools for minor language editing or grammatical correction. However, all intellectual contributions, analyses, and conclusions presented in this chapter are entirely the author's own.

References

1. Arif, Z.U., Khalid, M.Y., Noroozi, R., Sadeghianmaryan, A., Jalalvand, M., Hossain, M., Recent advances in 3D-printed polylactide and polycaprolactone-based biomaterials for tissue engineering applications. *Int. J. Biol. Macromol.*, 218, 930–68, 2022 Oct 1.
2. Syduzzaman, M., Rumi, S.S., Fahmi, F.F., Akter, M., Dina, R.B., Mapping the recent advancements in bast fiber reinforced biocomposites: A review on fiber modifications, mechanical properties, and their applications. *Results Mater.*, 20, 100448, 2023 Dec 1.
3. Francis, A.P., Augustus, A.R., Chandramohan, S., Bhat, S.A., Priya, V.V., Rajagopalan, R., A review on biomaterials-based scaffold: An emerging tool for bone tissue engineering. *Mater. Today Commun.*, 34, 105124, 2023 Mar 1.
4. Swetha, M., Sahithi, K., Moorthi, A., Srinivasan, N., Ramasamy, K., Selvamurugan, N., Biocomposites containing natural polymers and hydroxyapatite for bone tissue engineering. *Int. J. Biol. Macromol.*, 47, 1, 1–4, 2010 Jul 1.

5. Oliveira, J.M., Rodrigues, M.T., Silva, S.S., Malafaya, P.B., Gomes, M.E., Viegas, C.A., Dias, I.R., Azevedo, J.T., Mano, J.F., Reis, R.L., Novel hydroxyapatite/chitosan bilayered scaffold for osteochondral tissue-engineering applications: Scaffold design and its performance when seeded with goat bone marrow stromal cells. *Biomaterials*, 27, 36, 6123–37, 2006 Dec 1.
6. Rajput, S., Malviya, R., Sridhar, S.B., Osteoporosis Risk Assessment and Individualized Feature Analysis Using Interpretable XAI and RAI Techniques, in: *Explainable and Responsible Artificial Intelligence in Healthcare*, pp. 89–113, Scrivener Publishing, Beverly, Massachusetts, 2025 Mar 13.
7. Annabi, N., Tamayol, A., Uquillas, J.A., Akbari, M., Bertassoni, L.E., Cha, C., Camci-Unal, G., Dokmeci, M.R., Peppas, N.A., Khademhosseini, A., 25th anniversary article: Rational design and applications of hydrogels in regenerative medicine. *Adv. Mater.*, 26, 1, 85–124, 2014 Jan.
8. Islam, M.M., Shahruzzaman, M., Biswas, S., Sakib, M.N., Rashid, T.U., Chitosan based bioactive materials in tissue engineering applications-A review. *Bioact. Mater.*, 5, 1, 164–83, 2020 Mar 1.
9. Ielo, I., Calabrese, G., De Luca, G., Conoci, S., Recent advances in hydroxyapatite-based biocomposites for bone tissue regeneration in orthopedics. *Int. J. Mol. Sci.*, 23, 17, 9721, 2022 Aug 27.
10. Venkatesan, J., Bhatnagar, I., Kim, S.K., Chitosan-alginate biocomposite containing fucoidan for bone tissue engineering. *Mar. Drugs*, 12, 1, 300–16, 2014 Jan 16.
11. Desai, N., Rana, D., Salave, S., Gupta, R., Patel, P., Karunakaran, B., Sharma, A., Giri, J., Benival, D., Kommineni, N., Chitosan: a potential biopolymer in drug delivery and biomedical applications. *Pharmaceutics*, 15, 4, 1313, 2023 Apr 21.
12. Liang, W., Zhou, C., Bai, J., Zhang, H., Long, H., Jiang, B., Wang, J., Huang, X., Zhang, H., Zhao, J., Prospective applications of bioactive materials in orthopedic therapies: A review. *Heliyon*, 10, 16, 2024 Aug 30.
13. de Sousa Victor, R., Marcelo da Cunha Santos, A., Viana de Sousa, B., de Araújo Neves, G., Navarro de Lima Santana, L., Rodrigues Menezes, R., A review on Chitosan's uses as biomaterial: Tissue engineering, drug delivery systems and cancer treatment. *Materials*, 13, 21, 4995, 2020 Nov 6.
14. Seves, A., Romanò, M., Maifreni, T., Sora, S., Ciferri, O., The microbial degradation of silk: a laboratory investigation. *Int. Biodeterior. Biodegrad.*, 42, 4, 203–11, 1998 Nov 1.
15. M. Nomura, Y. Iwasa, H. Araya, Moisture absorbing and desorbing polyurethane foam and its production. Japan Patent, 1995.
16. H. Hatakeyama, Biodegradable sericin-containing polyurethane and its production. Japan Patent, 1996.
17. Ishikawa, H., Nagura, M., Tsuchiya, Y., Fine structure and physical properties of blend film composed of silk sericin and poly (vinyl alcohol). *Sen'i Gakkaishi*, 43, 6, 283–7, 1987 Jun 10.

18. K. Nakamura and Y. Koga, Sericin-containing polymeric hydrous gel and method for producing the same. Japan patent, 2001.
19. Park, S.J., Lee, K.Y., Ha, W.S., Park, S.Y., Structural changes and their effect on mechanical properties of silk fibroin/chitosan blends. *J. Appl. Polym. Sci.*, 74, 11, 2571–5, 1999 Dec 9.
20. P.J. S., Singh, K., Kokkranikal, J., Bharadwaj, R., Rai, S., Antony, J., Service quality and customer satisfaction in hospitality, leisure, sport and tourism: An assessment of research in web of science. *J. Qual. Assur. Hosp. Tour.*, 24, 1, 24–50. 2023 Jan 2.
21. Tushir, R., Das, S., Chauhan, A., Gurjar, S., Bhanwala, L., Patnaik, S., Jain, A., Exponential Global Health Hazards of Pharmaceutical Pollution and Its Preventive Measures: A Comprehensive Review. *J. Environ. Health Sustainable Dev.*, 2025 Mar 30.
22. Kweon, H., Ha, H.C., Um, I.C., Park, Y.H., Physical properties of silk fibroin/ chitosan blend films. *J. Appl. Polym. Sci.*, 80, 7, 928–34, 2001 May 16.
23. De Winter, J.C., Zadpoor, A.A., Dodou, D., The expansion of Google Scholar versus Web of Science: a longitudinal study. *Scientometrics*, 98, 2, 1547–65, 2014 Feb.
24. Lee, S.M., Cho, D., Park, W.H., Lee, S.G., Han, S.O., Drzal, L.T., Novel silk/ poly (butylene succinate) biocomposites: the effect of short fibre content on their mechanical and thermal properties. *Compos. Sci. Technol.*, 65, 3-4, 647–57, 2005 Mar 1.
25. Chan, A.L. and Chow, T.T., Investigation on energy performance and energy payback period of application of balcony for residential apartment in Hong Kong. *Energy Build.*, 42, 12, 2400–5, 2010 Dec 1.
26. Rana, A., Mittal, A., Vashist, C., Rajput, S., Sridhar, S.B., Malviya, R., Exploration of 4D Printing and its Applications in the Biomedical Sciences. *Curr. Pharm. Des.*, 2025.
27. Burgers, L.E., Raza, K., Van Der Helm-Van, A.H., Window of opportunity in rheumatoid arthritis–definitions and supporting evidence: from old to new perspectives. *RMD Open*, 5, 1, e000870, 2019 Apr 1.
28. Bordes, P., Pollet, E., Avérous, L., Nano-biocomposites: biodegradable polyester/nanoclay systems. *Prog. Polym. Sci.*, 34, 2, 125–55, 2009 Feb 1.
29. Zhang, X., Li, Y., Lv, G., Zuo, Y., Mu, Y., Thermal and crystallization studies of nano-hydroxyapatite reinforced polyamide 66 biocomposites. *Polym. Degrad. Stab.*, 91, 5, 1202–7, 2006 May 1.
30. Frumento, D. and &xxx021A;ălu, #0218., Immunomodulatory Potential and Biocompatibility of Chitosan–Hydroxyapatite Biocomposites for Tissue Engineering. *J. Compos. Sci.*, 9, 6, 305, 2025 Jun 17.
31. Dong, C. and Lv, Y., Application of collagen scaffold in tissue engineering: recent advances and new perspectives. *Polymers*, 8, 2, 42, 2016 Feb 4.
32. Rahman, M.S., Rana, M.M., Spitzhorn, L.S., Akhtar, N., Hasan, M.Z., Choudhury, N., Fehm, T., Czernuszka, J.T., Adjaye, J., Asaduzzaman, S.M., Fabrication of biocompatible porous scaffolds based on hydroxyapatite/

collagen/chitosan composite for restoration of defected maxillofacial mandible bone. *Prog. Biomater.*, 8, 3, 137–54, 2019 Sep.
33. Zhang, T., Jin, L., Fang, Y., Lin, F., Sun, W., Xiong, Z., Fabrication of biomimetic scaffolds with oriented porous morphology for cardiac tissue engineering. *J. Biomater. Tissue Eng.*, 4, 12, 1030–9, 2014 Dec 1.
34. Meng, Z., He, J., Li, J., Su, Y., Li, D., Melt-based, solvent-free additive manufacturing of biodegradable polymeric scaffolds with designer microstructures for tailored mechanical/biological properties and clinical applications. *Virtual Phys. Prototyping*, 15, 4, 417–44, 2020 Oct 1.
35. Szymczyk-Ziółkowska, P., Łabowska, M.B., Detyna, J., Michalak, I., Gruber, P., A review of fabrication polymer scaffolds for biomedical applications using additive manufacturing techniques. *Biocybern. Biomed. Eng.*, 40, 2, 624–38, 2020 Apr 1.
36. Kayser, L.V. and Lipomi, D.J., Stretchable conductive polymers and composites based on PEDOT and PEDOT: PSS. *Adv. Mater.*, 31, 10, 1806133, 2019 Mar.
37. Ahmad Ruzaidi, D.A., Mahat, M.M., Mohamed Sofian, Z., Nor Hashim, N.A., Osman, H., Nawawi, M.A., Ramli, R., Jantan, K.A., Aizamddin, M.F., Azman, H.H., Robin Chang, Y.H., Synthesis and characterization of porous, electro-conductive chitosan–gelatin–agar-based PEDOT: PSS scaffolds for potential use in tissue engineering. *Polymers*, 13, 17, 2901, 2021 Aug 28.
38. Feng, P., Wu, P., Gao, C., Yang, Y., Guo, W., Yang, W., Shuai, C., A multimaterial scaffold with tunable properties: toward bone tissue repair. *Adv. Sci.*, 5, 6, 1700817, 2018 Jun.
39. Zhang, T., Wang, Y., Kong, L., Xue, Y., Tang, M., Threshold dose of three types of quantum dots (QDs) induces oxidative stress triggers DNA damage and apoptosis in mouse fibroblast L929 cells. *Int. J. Environ. Res. Public Health*, 12, 10, 13435–54, 2015 Oct.
40. Rajput, S., Malviya, R., Srivastava, S., Ahmad, I., Rab, S.O., Uniyal, P., Cardiovascular disease and thrombosis: Intersections with the immune system, inflammation, and the coagulation system. *Ann. Pharm. Fr.*, 83, 2, 228–250, 2025 Mar 1, Elsevier Masson.
41. Lock, A., Cornish, J., Musson, D.S., The role of *in vitro* immune response assessment for biomaterials. *J. Funct. Biomater.*, 10, 3, 31, 2019 Jul 12.
42. Zhang, Y., Li, X., Chihara, T., Dong, H., Kagami, H., Effect of TNF-α and IL-6 on compact bone-derived cells. *Tissue Eng. Regener. Med.*, 18, 3, 441–51, 2021 Jun.
43. Zhang, T., Shi, X., Huang, Y., Gong, Y., He, Y., Xiao, D., Wang, S., Zhao, C., Oxidized fucoidan-based nanocomposite hydrogel for cryptotanshinone delivery and prevention of postoperative abdominal adhesions. *J. Controlled Release*, 382, 113733, 2025 Jun 10.
44. Kaur, H., Gogoi, B., Sharma, I., Das, D.K., Azad, M.A., Pramanik, D.D., Pramanik, A., Hydrogels as a potential biomaterial for multimodal therapeutic applications. *Mol. Pharmaceutics*, 21, 10, 4827–48, 2024 Sep 18.

45. Pires, P.C., Mascarenhas-Melo, F., Pedrosa, K., Lopes, D., Lopes, J., Macário-Soares, A., Peixoto, D., Giram, P.S., Veiga, F., Paiva-Santos, A.C., Polymer-based biomaterials for pharmaceutical and biomedical applications: A focus on topical drug administration. *Eur. Polym. J.*, 187, 111868, 2023 Apr 3.
46. Li, X., Chen, M., He, X., Cong, J., Zhao, W., Fu, Y., Lu, C., Wu, C., Pan, X., Quan, G., Biomineralized *in situ* catalytic nanoreactor integrated microneedle patch for on demand immunomodulator supply to combat psoriasis. *Theranostics*, 14, 17, 6571, 2024 Oct 7.
47. Kim, Y., Zharkinbekov, Z., Raziyeva, K., Tabyldiyeva, L., Berikova, K., Zhumagul, D., Temirkhanova, K., Saparov, A., Chitosan-based biomaterials for tissue regeneration. *Pharmaceutics*, 15, 3, 807, 2023 Mar 1.
48. Parham, M.M., Krishnan, B., Huttenbach, Y.T., A comparison of mast cells in skin biopsies of cutaneous mastocytosis with other inflammatory dermatoses: A study of 33 cases. *J. Cutan. Pathol.*, 51, 11, 876–80, 2024 Nov.
49. Ozpinar, E.W., Frey, A.L., Cruse, G., Freytes, D.O., Mast cell–biomaterial interactions and tissue repair. *Tissue Eng. Part B: Rev.*, 27, 6, 590–603, 2021 Dec 1.
50. Parisi, J.D., Corrêa, M.P., Gil, C.D., Lack of endogenous annexin A1 increases mast cell activation and exacerbates experimental atopic dermatitis. *Cells*, 8, 1, 51, 2019 Jan 15.
51. Pupjalis, D., Goetsch, J., Kottas, D.J., Gerke, V., Rescher, U., Annexin A1 released from apoptotic cells acts through formyl peptide receptors to dampen inflammatory monocyte activation *via* JAK/STAT/SOCS signalling. *EMBO Mol. Med.*, 3, 2, 102–14, 2011 Feb 2.
52. Ogulur, I., Mitamura, Y., Yazici, D., Pat, Y., Ardicli, S., Li, M., D'Avino, P., Beha, C., Babayev, H., Zhao, B., Zeyneloglu, C., Type 2 immunity in allergic diseases. *Cell. Mol. Immunol.*, 17, 1–32, 2025 Feb.
53. Reay, S.L., Marina Ferreira, A., Hilkens, C.M., Novakovic, K., The Paradoxical Immunomodulatory Effects of Chitosan in Biomedicine. *Polymers*, 17, 1, 19, 2024 Dec 25.
54. Lee, C.G., Da Silva, C.A., Dela Cruz, C.S., Ahangari, F., Ma, B., Kang, M.J., He, C.H., Takyar, S., Elias, J.A., Role of chitin and chitinase/chitinase-like proteins in inflammation, tissue remodeling, and injury. *Annu. Rev. Physiol.*, 73, 1, 479–501, 2011 Mar 17.
55. Al Mahmud, M.Z., Exploring the versatile applications of biocomposites in the medical field. *Bioprinting*, 36, e00319, 2023 Dec 1.
56. Kumar, S., Malviya, R., Uniyal, P., Srivastava, S., Sridhar, S.B., Rab, S.O., Singh, B., Plant Fiber Composites for biomedical applications: advances and prospective. *Curr. Pharm. Biotechnol.*, 26, 13, 2111–26, 2025 Oct.
57. Haraguchi, K., Biocomposites, in: *Encyclopedia of Polymeric Nanomaterials*, pp. 1–8, Springer, Berlin, Heidelberg, 2014.
58. Anderson, J.M. and Voskerician, G., The challenge of biocompatibility evaluation of biocomposites, in: *Biomedical composites*, pp. 325–353, Woodhead Publishing, Cambridge, United Kingdom, 2010 Jan 1.

59. Dorozhkin, S.V., Biocomposites and hybrid biomaterials based on calcium orthophosphates. *Biomatter*, 1, 1, 3–56, 2011 Jul 1.
60. Dorozhkin, S.V., Calcium orthophosphate-containing biocomposites and hybrid biomaterials for biomedical applications. *J. Funct. Biomater.*, 6, 3, 708–832, 2015 Aug 7.
61. Ramakrishna, S., Huang, Z.M., Kumar, G.V., Mayer, J., Batchelor, A.W., An introduction to biocomposites, World Scientific, 2004 Jun 3.
62. Giardino, R., Fini, M., Aldini, N.N., Parrilli, A., Testing the *in vivo* biocompatibility of biocomposites, in: *Biomedical Composites*, pp. 385–410, Woodhead Publishing, Cambridge, United Kingdom, 2010 Jan 1.
63. Sanjaykumar, S.G., Malviya, R., Srivastava, S., Ahmad, I., Uniyal, P., Singh, B., Nisar, N., Chitosan-peptide composites for tissue engineering applications: advances in treatment strategies. *Curr. Protein Pept. Sci.*, 26, 3, 185–200, 2025 Mar.
64. Rajput, S., Sharma, P.K., Malviya, R., Kaushik, N., Curcumin-Enclosed Nanoparticles for Cancer Therapy. *Curr. Nutr. Food Sci.*, 20, 5, 615–35, 2024 Jun 1.
65. Norrrahim, M.N., Nurazzi, N.M., Shazleen, S.S., Najmuddin, S.U., Yasim-Anuar, T.A., Naveen, J., Ilyas, R.A., Biocompatibility, biodegradability, and environmental safety of PLA/cellulose composites, in: *Polylactic Acid-Based Nanocellulose and Cellulose Composites*, pp. 251–264, CRC Press, Florida, USA, 2022 Apr 20.
66. Revati, R., Majid, M.A., Ridzuan, M.J., Mamat, N., Cheng, E.M., Alshahrani, H.A., *In vitro* biodegradation, cytotoxicity, and biocompatibility of polylactic acid/napier cellulose nanofiber scaffold composites. *Int. J. Biol. Macromol.*, 223, 479–89, 2022 Dec 31.

Part 3

REGULATORY CONSIDERATIONS AND FUTURE PERSPECTIVES OF 3D AND 4D PRINTING

13

Ethical and Regulatory Considerations in 3D and 4D Printing for Healthcare

Komal Gupta[1,2*], Niranjan Kaushik[1], Amit Singh[3] and Vikram Sharma[2]

[1]Department of Pharmacy, School of Medical as well as Allied Sciences, Galgotias University, Greater Noida, Uttar Pradesh, India
[2]Department of Pharmacy, Galgotias College of Pharmacy, Greater Noida, Uttar Pradesh, India
[3]Department of Pharmacy, Sharda University, Greater Noida, Uttar Pradesh, India

Abstract

Tissues as well as organ bioprinting, Stratified medicine, advanced prosthetics and dynamic medical devices that act to environmental indications are all enabled through quick advancement of 3D and 4D printing technologies that are entirely transforming the field of healthcare. These advancements hold the potential to enhance patient care; however, they additionally pose difficult legal and ethical challenges that have to be carefully assessed. The application of healthcare with 3D and 4D printing presents ethical concerns about affordability, equity, and accessibility. As these technologies evolve, there is increasing concern regarding the possibility of socioeconomic inequities causing unequal access to stratified therapies. Furthermore, considering the experimental nature of many bioprinted goods and dynamic 4D-printed devices that adapt to environmental stimuli, patient safety always needs to come foremost. Since the development of patient-specific medical devices frequently depends on the exchange of confidential digital files that need to be protected from abuse or illegal access, data security and privacy concerns are particularly critical. Another level of complexity is added by the legal consequences of bioprinting biological tissues and organs that pose concerns about the constraints of individual improvement and the relevance of human life. Addressing laws and regulations, the fast pace at how rapidly technology is developing raises substantial challenges to established frameworks. Traditional models for assessing safeguards and effectiveness of healthcare equipment frequently lack the necessary equipment to examine the dynamic behavior of 4D-printed devices

**Corresponding author*: shrutiaibs@gmail.com

Rishabha Malviya and Mukesh Roy (eds.) Biocomposites in 3D and 4D Printing: Materials, Techniques, and Applications in Regenerative Medicine, (351–392) © 2026 Scrivener Publishing LLC

and the distinct characteristics of 3D-printed structures. Regulatory agencies must deal with obstacles of material standardization, reproducibility, and stability and biocompatibility of printed items over the long term. Additionally, using bioprinting raises new regulatory challenges such as whether to categorize bioprinted tissues as medicines, biologics, or medical devices, each of which has its specific approval process. International distinctions in regulatory norms render it more challenging for these technologies to be developed and distributed globally, which restricts research and accessibility to patients. This chapter investigates the relation between ethics as well as regulations for healthcare 3D and 4D printing, highlighting the necessity of a multidisciplinary strategy to overcome these barriers. This emphasizes how crucial it is to have high ethical standards and flexible legal frameworks in order to optimize the beneficial social implications of these revolutionary innovations while lowering any possible risks and inequalities.

Keywords: Bioprinting, personalized, biocompatibility, prosthetics, multidisciplinary

13.1 Introduction

13.1.1 The Role of 4D and 3D Printing Technologies in Healthcare

Smart materials, sometimes called sensitive or adaptable materials, are made to continuously respond to environmental inputs, changing their properties and applications to meet a particular set of requirements [1–6]. Stimuli such as mechanical pressure, light, humidity, electromagnetic radiation, heat, ionic strength, and pH frequently cause these compounds to react [7]. Additionally, they are capable of simulating functions involving compose memory, self-evolution, self-treatment, that self-assemble, recognizing, and actuating [8–11].

By means of the advancement as well as investigation of systems that are structural, researchers were compelled for developing a variety of approaches to enhancing behavior and functionality with regard to certain objectives (Figure 13.1) [12]. Smart materials have surpassed traditional manufacturing because of the natural control and mechanical limitations of this sort of equipment, becoming becoming more intricate in terms of topologies, micro- and nano structures mechanical characteristics, physical-chemical elements [13].

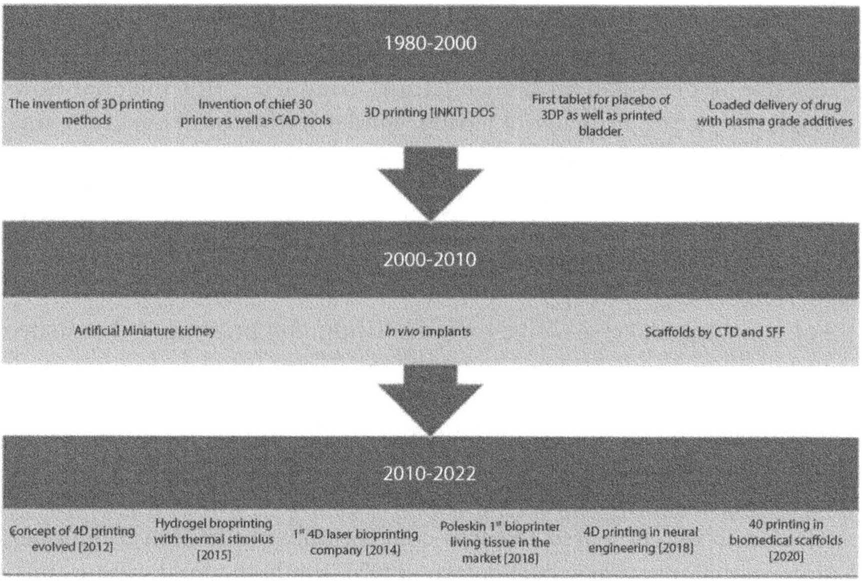

Figure 13.1 The timeline provides a brief overview of the development of 4D printing technology.

13.1.2 History of 3D and 4D Printing

Rapid prototyping, also referred to fabrication with added substances, is a new method that shown great promise throughout industrial sector [14–16]. This technology has advanced significantly, which has developed into a useful instrument for a variety of disciplines, including science, engineering, design, production, and research [15]. The three-dimensional model was created as a result of several disciplines coming together. A collection of procedures known as three-dimensional printing combine regulated components to create three-dimensional objects [17]. Usually, this procedure is carried out layer by layer. 3D printing, often known as additive manufacturing, is quickly emerging as most popular and cutting-edge technologies of the twenty-first century [18]. Several manufacturing processes, such as Selective laser sintering, binder jetting, material extrusion, and material jetting, are referred to as additive manufacturing. These processes include the selective placement of materials through layer by layer method [19].

Layer-by-layer material deposition is characteristic that distinguishes additive manufacturing systems. In addition to the mechanical and pneumatic deposition of semisolid materials, material extrusion (A) is typically used to deposit thermoplastic materials. Both material jetting (B) and binder jetting (C) compose use of renowned inkjet heads; however just a adhesive is deposited in binder jetting, complete print material flows through the nozzle in material jetting. Binder jetting has benefit for supporting powder bed, which eliminates requirement of sacrificial material as well as support structures. When a laser selectively fuses a powder bed, this process observed through selective laser sintering (D). Finally, required component produced by specifically polymerising liquid resin vat using stereolithography (E).

For consumer-level additive manufacturing, 3D printing is dominated by two main methods: stereolithography (SLA) and fused deposition modeling (FDM). These procedures build things *via* gradually adding components, layer by layer. FDM extrudes semi-liquid plastic in the desired pattern, whereas SLA uses UV light to selectively cure resin. This technology's explosive expansion has made major advancements possible. In particular, when compared to some conventional production techniques, 3-dimensional printing—particularly FDM method—hold significantly decreased weight of the good, it is the building process durations, and manufacturing prices while minimizing waste. As a result, the average customer can now afford 3D printing [20, 21] and time consumption in 3D printing when compared to traditional manufacturing methods.

Recent advancements in three-dimensional (3D) printable advanced substances have enabled a more advanced kind of "dimensional printing" called "four-dimensional (4D) printing." The idea of 4D Printing enables graphic things and customize the structural morphology with reaction towards external inputs and recognizes time as a fourth dimension [22, 23]. Over the past few years, excipients production, also known as 3D printing, which possesses become an adaptable method along valuable substitute for traditional production with regard to layer-by-layer production of several chemicals. This has led to establishment of numerous healthcare technologies, flexible electronics, soft robotics, scaffolds, actuators, and wearable devices.

The 3D and 4D printing technologies are comparable. The printed object responds to variables like temperature along with humidity, among others. For altering their appearance in accordance with specifications, adding another level of metamorphosis over time. It is a brand-new, modern technology that is quickly spreading across several fields, including computer science, engineering, medicine, chemistry, material science, and basic sciences. Based on the principles of stereolithography, this technology uses ultraviolet radiation (UV) to cure a material layer by layer while it is being printed [2, 25]. It is an efficient and accurate method to manufacture soft structures that may be customized. It makes use of particular substances with self-deformation properties that scientists utilize for research [3, 4].

13.1.3 Importance of Ethical and Regulatory Frameworks

Patient perspectives and acceptability are also influenced by ethical issues associated with customized 3D-printed treatments. Patients appreciate open and honest communication about how their personal information, particularly medical imaging data used for 3D printing, is used. Building patient trust and ensuring ethical practice require informed consent and privacy protection [5, 25], adhering to patient autonomy and advancing moral healthcare practices require integrating patients in shared decision-making processes, including conversations about the advantages, disadvantages, and available options of customized 3D-printed solutions. Furthermore, the possible overuse of 3D-printed therapies may raise ethical questions for patients. Unnecessary use of technology for 3D printing may result from appeal of artistic endeavour and customization [26, 27].

Furthermore, it is necessary to address the issue of fair access to 3D and 4D printing technology. If innovations are exclusively available to specific populations or geographical areas, there is a chance that they will worsen already-existing healthcare inequities [28]. This emphasizes the necessity of legal frameworks that support just and equal access to state-of-the-art medical technology.

Ensuring patient safety and addressing ethical issues are crucial tasks for regulation. To ensure that 3D- and 4D-printed medical devices fulfill strict safety and effectiveness requirements, regulatory agencies must set regulations for their approval and usage. Regulators must modify existing regulations that would not adequately account for the distinctive features of printing in 3D and 4D products due to flexible nature among such technologies.

Furthermore, the substances utilized for printing both 3D and 4D must be taken into account during the regulatory process because they may have an impact on functioning and biocompatibility [29]. Thorough review and ongoing post-implementation monitoring are necessary to guarantee that materials are safe and effective for patient usage [30].

13.2 Scope and Objectives of the Chapter

The technologies of three-dimensional and four-dimensional printing include developing at rapid pace for healthcare industry, offering revolutionary possibilities for tissue engineering, implants, prosthetics, and personalized treatment. But these developments also provide important legal and ethical problems that need to be appropriately considered. The ethical issues surrounding patient autonomy, informed consent, and privacy are addressed in this chapter, with a focus on bioprinting and regenerative medicine. In order to make sure that these innovative technologies don't contribute to already-existing healthcare inequities, it also examines at accessibility and equity issues.

The chapter offers a comprehensive review of worldwide frameworks regulating the authorization and application of bioprinted tissues and 3D and Medical devices made with 4D printingfrom an ethical view point. It examines rules established by significant regulatory agencies like the FDA and EMA, pointing out any shortcomings and difficulties in guaranteeing efficacy, safety, and compliance. Furthermore, as material biocompatibility, durability, and long-term performance must be carefully evaluated to avoid potential hazards, patient safety and quality control are crucial factors. The advancement of medical 3D and 4D printing applications further complicated *via* legal and intellectual property issues. The chapter examines ownership rights, liability concerns, and the complex nature of patenting, especially when printed medical devices malfunction or have unfavorable impacts. In order to achieve a balance between safety measures and technological advancement, this chapter concludes with legislative ideas and ethical criteria for responsible innovation.

13.3 Detail Understanding for Printing *Via* 3D and 4D Printing in Healthcare

13.3.1 Printing in 3D

3D printing, also known for speedy prototyping or excipients manufacture, is essential to the development of pharmaceuticals because it enables the creation of scaffolds that can either replace or repair conventional distribution methods (Figure 13.2). 3D printing is a bottom-up approach. A computer program uses the cross-sectional information obtained from cutting an electromagnetic radiation (MR) or digital image of the object

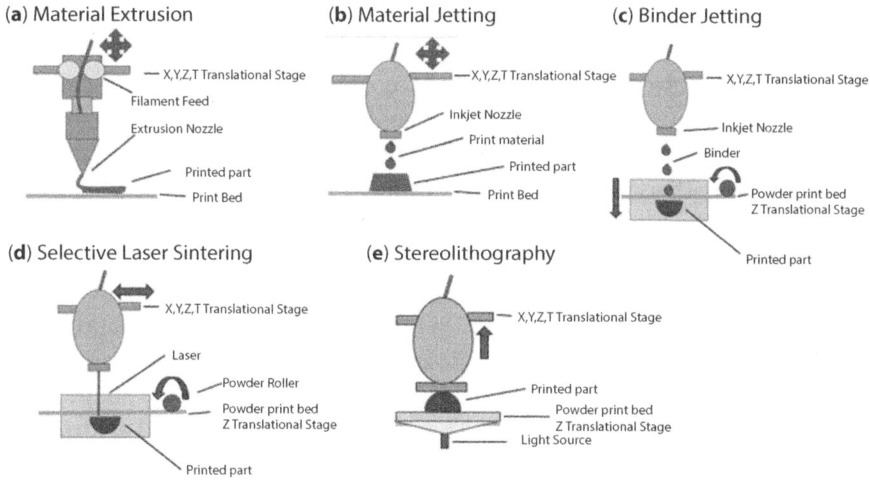

Figure 13.2 Standard processes for additive manufacturing.

to guide manufacture. This allows for layer-by-layer fabrication of more intricate structures, which is particularly helpful for implants made for a single person [31]. SLS, or selective laser sintering, method for producing excipients; FDM or fused deposition modeling is popular fast prototyping technique; as well as 3D printing with light assistance (also known as stereolithography, or SLA) main subcategory of 3D printing which uses brittle polymeric materials to produce products. Bioprinting, which uses cell-containing polymeric hydrogels, is another kind. These tactics are described in the list of concepts that follows.

13.3.2 Conventional Techniques for 3D Printing

Nozzle-based technologies, which print inks one layer at a moment within spatial control to create a specific design, are used in most conventional 3D printing techniques. The methods include light-assisted, near-field electrospinning (NFES), extrusion-based, and inkjet-based methods [32, 33].

13.3.2.1 FDM

Crump created and patented FDM in 1989 is most widely used rapid development techniques [34]. Among the industries that have used these technologies are automotive, aviation, verification regarding design and wellness to name some [35]. In order to generate a three-dimensional shape, FDM first melts a polymer that is thermoplastic which was introduced for device (in a type of particles or a thread) inside reaction vessel. The polymer is then removed onto the platform through nozzle as well as imprinted layers by layer. During the production process, computer software moves nozzle's position into plane of x & y to produce optimal design. The nozzle moves a preset distance towards z-axis for printing next layer after previous layer gets complete. The same process is continued once desired shape has been created [36]. The final version's level of fine detail depends upon numerous parameters, constituting nozzle size, printing quality; angle, as well as distance between fibers in subsequent layers [37]. FDM's primary benefit is their potential for numerous varieties of molds from various components. This process entails manipulating nozzles that hold multiple thermoplastics so that they extrude one after the other, creating a finished shape with a range of characteristics. The FDM's simplicity, cost, and speed are further advantages [38, 39].

The lack of pharmaceutical-grade, biocompatible components that can be employed as thermoplastics is a disadvantage of the technology. Finding components with right melting viscosity—that ought to sufficient for deposition and low in order to extrude—is also challenge [40].

13.3.2.2 SLA, or Light-Assisted 3D Printing

SLA, or light-assisted 3D printing, thousands of micromirrors regulated through computerized Micromirror configuration, thus try crosslinking A polymerizable photo pigments into preferred form [41]. Many intricate biological structures can be rapidly printed using this method, but only if photocrosslinkable dyes are utilized [42]. Its rapid printing speed (200–1,600 mm/s), exceptional cell survival, and great clarity (based on micromirror studies) are its key features. Furthermore, medications may be included with the ink before printing, resulting in matrix inclusion and regulated drug release. Primary drawback constitutes a limited inks range that nearly always rely on UV crosslinking, which may increase the risk of cancer development because of the generation of reactive oxygen moieties [43].

13.3.2.3 SLS

Deckard developed and patented the SLS excipient production process in 1989 [44]. With this technique, A thin coating of granular substance (metallic substances, ceramics, as well as plastic polymeric materials) is dispersed throughout form of powder bed and melted using a laser as the power source. To construct the two-dimensional (2D) form, the computer tells beam how to burn material to fuse altogether. When a layer is formed, generated layer shifts down to single thickness level, as well as piston sinters the next layer by spreading the newly powder layer upon the topmost station part. The process continued till final component gets produced [45].

After the fabrication is complete, the leftover material is cleaned with a brush or compressed air. One advantage of SLS is its capacity to create large, complex scaffolds. Another advantage is that SLS doesn't actually need extra support networks during the manufacturing process because the object being heated is embedded to hard powder particles as well as no covered layer gets required. Similar to FDM, SLS utilizes a manufacturing process without the use of solvents, meaning that no solvent residue is present in the final product. The primary drawback of SLS is the fact that its surface is simply untidy, also requires to be polished as round particles are fused to create the product, which finally results in a rough texture [46].

13.3.2.4 Extrusion-Based 3D Printing

Layer-by-layer ink layering is made possible by extrusion-based 3D printing, which creates a continuing filament by forcing ink through a small tip using pressure or other pressure changes [47]. It is possible to print thermoplastic materials into the required shape by employing FDM and a temperature-controlled extruder. Extrusion-based 3D printing, as opposed to inkjet-based printing, keeps a constant flow, which could allow inks to be printed smoothly onto such a surface. Before printing, a medication can be added to the ink to enable continuous release of the medicine by microbial oxidation. However, larger constructions are less practical due to extruder printing pace, which somewhat slow (10–50 m/s) [48].

13.3.2.5 Inkjet 3D Printing

In the 1970s, The Corporation of Hewlett-Packard developed inkjet, first excipient printing technology manufacturing, because merely 2 dimensional printing method. In 1992, three-dimensional printing device was built by coupling this technique with a cylinder and an elevated station that can move along the z-axis [49]. Piezoelectric inkjet printers and heat are being used more and more for tissue engineering applications. For thermal inkjet printing, an ink cartridge contains a prepolymer mixture known as ink, which may also contain cells. Once the cartridge is placed on printer head which is computer-controlled, the droplets of ink are expelled from it by tiny bubbles of air created through providing heat to the printing face (Table 13.1) [50].

Table 13.1 A brief overview of three-dimensional printing methods: workings, advantages, and disadvantages.

3D printer	Mode of action	Advantages	Disadvantages
FDM (Fused Deposition Modelling)	Through a heated nozzle, drug-loaded filament is extruded. The printer head is moved along the x-y axis to release the molten extrudate onto a construction plate as it solidifies at room temperature. Layer-by-layer 3D printing of a medication is made possible by build plate being successively lowered towards vertical z-axis.	• Able to create compositions that have a continuous rapid releases • Can increase the solubility of medications that aren't very soluble by creating amorphous solid dispersions • capacity to produce multi-drug combinations by multi-nozzle printing; • Affordable • Compact, easy to use	• Perhaps inappropriate for drugs that are sensitive to heat • Formulating the first filament feedstock can be difficult. • Required low about of drug loading. • scaling up is difficult
Stereolithography	A photo polymerizable resin is exposed through high energy light like UV to which induces polymerisation and solidify the materials. Very time possible the resin is considered to be solidify through defined depth, and platform was transferred to little bit down in vertical direction and the layer with is constructed is resin recoated. The procedure is constantly repeated to build a 3D printing medicine.	• Mostly researched to manufacturing of medical devices and rapid-release pharmaceutical medicines. • High precision as well as resolution (better than other 3D printing technology), that create possibility for producing intricate shapes; • Solubility can be modified • Suitable for creating polypills with several layers.	• Could not be appropriate for medications that are photosensitive; • Possible problems with material toxicity

(*Continued*)

Table 13.1 A brief overview of three-dimensional printing methods: workings, advantages, and disadvantages. (*Continued*)

3D printer	Mode of action	Advantages	Disadvantages
Binder jet printing	A nozzle that holds a binder liquid deposits the liquid on level powder surface by transfering through an x-y axis. The powder particles are bound along by liquid, which results in layer solidification. After that, fabrication build plate is lowered towards vertical z-axis. A 3D-printed medication is formed through creating a thin layer of powder on top and repeating the procedure one after the other.	• Used to create the first 3D-printed medication ever approved by the US Food and Drug Administration (FDA); • Able to create immediate and sustained-release formulations; • Capable of producing zero order release (a drug released at a constant rate) and delayed release formulations; • High resolution allows the formation of complex geometries	• High cost; • Absence of transportable tools.
Semi-solid extrusion	In order to extrude drug-loaded semi-solid substance (like paste or gel), a syringe-based tool head is used. The extrudate is released into build plate by moving the printer head along the x-y-z axis, where it solidifies at room temperature.	• Capable for producing numerous preparations, like oral films, polypills, as well as immediate-release and controlled-release dosage forms; • Suitable for the creation of chewable and appetising formulations	• Low throughput; • Only appropriate for medications that can be produced as semi-solids; • Low resolution in comparison to other 3D printing methods

(*Continued*)

Table 13.1 A brief overview of three-dimensional printing methods: workings, advantages, and disadvantages. (*Continued*)

3D printer	Mode of action	Advantages	Disadvantages
Selective laser sintering	Powdered particles bound together by selective partial or complete melting caused by laser that aimed design a specific design on powder bed. A new powdered layer is applied to sintered material by roller after layer has been sintered. In order to produce 3D-printed medication, this procedure continues layer by layer.	• Have an ability to formulate various preparation, such as controlled-release as well as immediate-release dosage and medical devices; • Have an ability to form highly porous dosage forms (dissolving quickly); • capable to produce complex geometries through a high-resolution process; • Suitable for producing polypills.	• It might not be appropriate for medications that are photosensitive orthermosensitive; • Post-processing is necessary; • Precise control over powder flow characteristics is necessary.
Direct powder extrusion	Using an extrusion-based method, a powder hopper is filled with mixture of preparations loaded with medicines. The hopper produces molten extrudate that hardens at room temperature onto an assembly plate after feeding into warmed single screw extrusion in the print head. To create a 3D-printed medication layer by layer, the build plate is lowered successively along the vertical z-axis.	Producing compositions that have a prolonged and rapid absorption • Capable of production of amorphous solid dispersions, which can make medications more soluble when they are poorly soluble; • Scalable (as proved by Triastek, which obtained FDA experimental new drug application clearance for a formulation made with a comparable technique).	• Fairly new 3D printing technique in pharmaceuticals; • May not be appropriate for medications that are heat-sensitive.

13.3.3 3D Printing with Laser Technology

Laser aided 3D printing is an additional printing process that makes use of pulsed beam light, contributing level, as well as receiver layer. This method works by applying ink on a ribbon that has a very small layer of energy-absorbing material on it. The ribbon and the receptor are oriented parallel to one another. When pulsating laser source focused upon radiation absorbing material, a vapor bubble is produced. The pressure from this bubble distorts the ink and creates droplets. These hydrogel droplets laden with cells are directed toward the sensor, where they are collected and crosslinked [51]. This technology has the advantage of neither clogging nor physically stressing the cells because it does not require a nozzle and uses a nonintrusive printing technique. They all increase the vitality of cells. However, this technical approach appears to be more expensive than some other printers [52].

13.3.4 Novel Techniques for Printing with 3D

New 3D printing techniques aim get beyond the geometrical limitations by combining advanced biomaterials with conventional 3D printing. For example, multi-material three-dimensional printing, which usually makes use of an extruder 3D printer, may produce intricate structures that resemble genuine tissues by employing the mini-tissue technique. However, the method gets hampered through clutter as well as slow printing speeds [53]. Other here developed technique for 3D printing precise architectural elements utilizing a variety of materials and minimal assistance. Reduced ink mixing and a large variety of high-precision constructions are the primary advantages of this approach; nevertheless, expansion, automation, resolution, and ink selection are the drawbacks [54]. The necessary design is extruded in printing fluid meant for sacrifice by micro-extrusion printer utilized to integrated 3D printing. This deadly bath is removed through cross-linking relevant printed construct which allows the final build to be subject to fewer geometrical constraints. The sluggish speed of printing as well as prerequisite to obtain a suitable sacrificial print solution is major disadvantages of this technique [55].

13.3.5 4D Printing

Structures made *via* additive manufacturing comprised of skilled (sophisticated) materials can change itself into certain form or fulfill planned goal based upon response available upon environment; thus methods also known in form of "4D printing" [56, 57]. 4D fabrication makes use of the same sophisticated manufacturing techniques and equipment that were previously discussed in the three-dimensional printing section. The primary difference among the two types of printing is type of material incorporated. For 3D printing, categorized as a 4D-printed object, one must possess a minimum of one smart activity, which includes "composite materials" as well as "self-actuation" [58]. 4D printers provide a number of advantages over 3D printers.

Both temporal and geographical regulation for manufactured commodities is made possible by fourth dimension's addition, time frame, to the 3 dimensional frameworks. One of the primary issues with 3D printing is resolved by 4D printing, which creates dynamic and animated buildings. Typically, intelligent components are defined as "materials that exhibit changes in their chemical or physical characteristics in a controlled and beneficial way after being exposed to an external stimulus, such as humidity, temperature, light, a magnetic field, or pH." The previously mentioned input types may alter the form, shade, functionality, and other physical or chemical properties of a 4D-printed product.

Owing to material's intelligence, the 4D-printed items' fully programmable state and functionality eliminate the need for extra post-processing tools or processes, speed up manufacturing, as well as to certain cases, also helps for the process of application [56]. Such as, contouring digital scaffolding that exhibits compaction before *in vivo* purpose can be applied in non-invasive therapies as well as assemble themselves to desired complex form because of dynamics reaction after implant [59].

13.3.5.1 Components Affecting 4D Printing

The type of extra production method, the adaptability of the material, the stimuli, the interaction between the stimuli or sections, and the mathematical model of the material transformation are the five primary variables that affect the operation of 4D printing.

13.3.5.2 4D Printing Applications in the Delivery of Medications

The application of four-dimensional printing to create tissues and organs is described in a number of biomedical research studies. Examples of 4D bioprinting include the cell-filled structures production, self-folding tubes, and biosensors. Although several studies did not explicitly mention the use of 4D printing, the designs did include some 4D printing components. Systems for administering drugs made with 4D printable are relatively novel concept. Ever since the 4D printing concept was first established in bioprinting technology [60], the portion discusses what might is categorized to four-dimensional printing in delivery of drug industry as well as the application programs.

13.3.5.2.1 Drug Delivery Methods Using 4D-Printed Hydrogels

Hydrogels ought to be specified as the first use. Whether they composed of natural and synthetic polymers, they have distinct capacity to absorb or retain huge amounts of liquid. Because hydrogels are biodegradable and biocompatible, they have also been found to be used as medication delivery systems. In relation to the fourth dimension of hydrogels, some hydrogels are highly suitable for 4D delivery methods due to their stimulus sensitivity, which may help with the extended release of medications. Numerous studies have demonstrated the four-dimensional manufacturing of hydrogels using a variety of methods and bio-ink formulas. Few, though, have actually used them as 4D-printed drug delivery devices [61].

13.3.5.2.2 Oral Formulations for 4D Printing

Melocchi and colleagues created the remarkable shape-memory extending gastroretentive delivery system (SMX GRDDS) [62]. For FDM, a Kloner3D® 240 Twins using a 0.5 millimeter nozzle was utilized. The initial drug used in the printing process was allopurinol (ALP), an antagonist of xanthine oxidase. The printing material was a blend of poly(vinyl alcohol) (PVA) and glycerol (GLY). A circular, helix-shaped structure measuring 15 millimeters in wide and seventeen millimeters in length was produced at a rate of 23 mm/s using the PVA05GLY-ALP combo. The group gave a demonstration of how to use printing molds to print different forms. A capsule meant for oral ingestion contained the SMX was GRDDS in a temporary form. The capsule immediately began to degrade when the structure that was printed was subjected to a concentration of 0.1 mol/L HCl at 37°C, which mimics the acidic conditions in GIT. Over course of 2 hours, the technique showed a constant drug release regardless for the printed version [62].

13.3.5.2.3 Development of Microneedles by 4D Printing

Innovative 4D printing was used by Han *et al.* [63] to produce a tiny needle matrix that enhanced tissue adhesion. Phenylbis (2,4,6 trimethylbenzoyl), Sudanese I as a light absorber, and phosphate oxide as an photoinitiator were all 3D printed by the scientists utilizing a custom designed projected microstereolithography. Han *et al.* [63] printed a horizontally curved prong on the microneedles to encourage tissue attachment for the microneedle matrix. This prong could expand into an angled shape during post-printing disintegration. Compared to a needle without barbs, the barbs enhanced skin adherence eighteen times more.

It is believed that adhesion is essential whenever a cutaneous drug-eluting device is gradually releasing its medication. The fluorescent rhodium B molecule, a model medication, was used to illustrate transdermal drug transport [63].

13.3.5.2.4 Drug-Eluting Implants Made with 4D Printing

There are numerous uses for implants that respond to stimuli used to provide medication. Using state-of-the-art 4D printing techniques, several groups have created devices with the potential to administer drugs devices [64, 65]. Few have effectively inserted a (model) drug into an input-responsive 3D-printed implant. Salimi *et al.* [66] presented novel thermoplastic polyurethane (TPU) that was used as a framework to enable medication-eluting devices. A pressurized extrusion bioprinter was employed by the team (CELLINK, Sweden) and ultraviolet light-emitting diode curing equipment to make a bar-shaped implant made from TPU as well as poly(ethylene glycol) (PEG). The bar-shaped implant was filled with paracetamol (16%) as a prototype drug. In a mouse fibroblast cell line, the implant demonstrated acceptable biocompatibility but no appreciable harmful effect.

The material did deliver paracetamol in a controlled manner and was both flexible and stiff. Salimi *et al.* [66] noted implant deformation both before and after drug release, despite the fact that the implant had several positive features. This suggests that phase transition components of TPU need for modification.

13.3.5.2.5 Magnetically Activated Drug Delivery System

Magneto-restrictive components that are activated by an external magnetic field could be used to create a 4D-printed material. This could aid in the targeted administration of medication for the best dosage and the fewest possible negative effects. For example, Gazzaniga *et al.* [67] developed production employing hydrogel bilayer and a conventional lithographic technique. By changing its structure in response to a specific pH, Drug release was facilitated by the pH-responsive characteristics of one of the layers. However, another layer contained metallic particles that ensured site-specific drug administration and enabled magnetic directionality of the formulation. The tailored delivery of anti-cancer drugs is one application for this particular discovery. Drug release can be stimulated by a low oxygen pressure, a certain pH in the tissue of tumor, and cancerous tissue's microenvironment [68].

13.3.5.2.6 pH-Controlled Medication Delivery Method

A medication delivery device with pH-directed controlled release using intelligent materials that respond specifically to pH changes is feasible. Mathur *et al.* [69] developed a mucoadhesive drug delivery system which changes form as well as sticks onto gut wall when penetrates small intestinal tract (pH 6.5) using a pH-sensitive formulation. These self-folding formulations demonstrated longer residence duration, less drug interaction with intestinal fluid, and increased drug absorption through the mucosal epithelium [69].

13.4 Ethical Considerations in 3D and 4D Printing

13.4.1 Safety as an Ethical Consideration in 3D Printing

One of the primary ethical issues with the use of 3D bioprinting is safety. 3D bioprinting has special difficulties in establishing acceptable safety standards, in contrast to traditional biomedical breakthroughs that are subjected to stringent clinical studies and pharmacovigilance. Although complete safety is unachievable, risk thresholds are largely determined by ethical considerations [70, 71].

Generally speaking, biomedical safety is controlled by evaluating known risks and putting preventative measures in place. Nonetheless, new technologies like nanotechnology and genetic engineering frequently adhere to the "preventative principle," which calls for taking precautions even when there is uncertainty. Although this approach applies to environmental concerns, it is usually not required for medical applications, since risk mitigation and identification provide safety [72, 73].

When it comes to 3D bioprinting, two main safety issues surface:

1. Biological Material Safety: Making sure that printable transplants are free of contaminants and infections that could endanger the health of the general public as well as specific patients.

2. Transplant Functionality: Assessing the successful integration of bioprinted organs and tissues into the recipient's body, given that these constructs have no prior history of functioning in a living creature [74].

Although xenotransplantation and traditional organ transplantation offer valuable insights, 3D bioprinting poses unprecedented risks that require thorough testing to guarantee public and individual safety [75].

13.4.2 Consent as an Ethical Aspect of 3D Printing

One of the main ethical rules guiding the use of medical devices that are bioprinted and 3D-printed is informed consent. It protects patient autonomy by guaranteeing that people have authority over choices that impact their biological state [74]. This is different from conventional medical paternalism, which gave doctors complete control over treatment choices. Historical ethical transgressions, increased medical expertise, and the complexity of healthcare have all contributed to the trend toward patient autonomy [75]. Physicians are required to adequately inform patients about planned medical interventions, including possible alternatives, in order to obtain their informed consent. According to their own preferences, patients are free to accept or reject treatment. Doctors, as specialists, are in charge of suggesting suitable treatment alternatives, thus this does not imply that patients can order medical treatments [76, 77].

When a doctor rejects a patient's request for ethical or scientific reasons, conflicts may result. In these situations, a doctor could decide not to perform the desired intervention. On the other hand, under the context for informed consent, patients maintain the right for declining treatment after a medicinal relationship has been developed. The informed consent procedure has some limitations in emergency situations. Practitioners have a moral duty to carry out needed medical procedures based on their knowledge if obtaining authorization is not possible because of time restrictions [78]. Nonetheless, since advance directives represent individual liberty and moral medical practice, they must nonetheless be followed when they are accessible [79].

Informed permission is essential for making ethical decisions in the realm of 3D bioprinting, especially when it comes to experimental techniques, new biomaterials, and possible long-term hazards [80, 81]. Upholding ethical principles in this developing sector requires preserving openness, honoring patient preferences, and striking a balance between patient autonomy and medical knowledge [82].

The significance of patient education and genuine consent
Making sure that informed permission is sincere is crucial issue in moral use of bioprinting and 3D printing. When a patient gives their true consent, they do so after being given full, objective, and intelligible information from a doctor. However, a number of things, such as patient misconceptions, media impact, and subjective doctor communication, can compromise the validity of consent [83].

Complications in Obtaining Genuine Consent
• Limitations in Physician Communication: Although it is the duty of medical personnel to convey accurate data, there is no impartial method to assess how well they do so. Since patients mostly rely on spoken explanations and the chance to ask questions, written consent forms are insufficient on their own. A patient's capacity to understand medical ideas, the doctor's choice of important details, and the potential for inadvertent bias in communication all affect the accuracy of information [84]. These difficulties raise the possibility of manipulation, even inadvertently.

Media Impact and Misinformation: The media regularly sensationalizes new medical technology, such as 3D bioprinting, promising groundbreaking breakthroughs without sufficient scientific backing [85]. Based on inflated or deceptive information, patients may develop preconceived ideas about novel treatments, resulting in biases that run counter to medical knowledge. Because of media-driven expectations, patients may choose to forego traditional treatment choices in favor of experimental bioprinting applications. Informed consent becomes more difficult due to the doctor-patient interaction being disrupted [86].

13.4.3 Patient Education's Contribution to Ethical Decision Making

Since bioprinting technologies are developing so quickly, it is not realistic to assume that all doctors will be up to date on the most recent developments. Additionally, patients frequently carry out independent research, compiling fragmented data from non-scientific sources, social media, and the internet. This leads to a misunderstanding, necessitating organized patient education to guarantee well-informed choices.

Approaches for patient awareness should be created to assist patients in addressing these problems:

1. Examine medical information critically, separating media exaggerations from scientific facts.

2. Recognize the advantages, disadvantages, and unknowns of new 3D bioprinting applications.
3. Have meaningful conversations with doctors that are grounded in factual information rather than myths propagated by the media.

When it comes to creating educational programs, medical associations, patient representation groups, scientific publications, and regulatory agencies are essential. These programs ought to be created with the goal of closing the knowledge gap so that patient can make decisions based on reliable medical data rather than rumors or false information [87]. Generally speaking, obtaining true informed permission in 3D bioprinting necessitates a thorough patient education program in addition to physician communication. Moral decision-making can be enhanced by encouraging consciousness and critical thought, guaranteeing that patients make decisions that respect their autonomy as well as medical facts.

13.4.4 Ethical Aspects of 3D Printing: Enhancement

When 3D bioprinting is used for purposes other than medical treatment, such as enhancement—the alteration of youthful human organs and tissues in order to enhance biological functioning rather than cure disease—the ethical complexity of the technology rises dramatically. In contrast to therapeutic procedures, augmentation calls into question the conventional role of doctors in giving consent to the procedure, necessitating a more thorough ethical analysis [88].

The Moral Challenge of Enhancement

- Right to Medical Responsibility vs. Right to Self-Determination: People are free to improve their bodies as a sign of their individuality, so long as it doesn't violate the rights of others. Physicians must carefully weigh the hazards and moral ramifications of changing a healthy human body when augmentation entails surgical procedures.
- Physician Responsibilities and Risk Awareness: Because enhancement treatments don't cure illnesses, doctors are more responsible for making sure patients are aware of all the possible hazards [89]. This comprises:
  1. Providing thorough and precise details regarding potential issues.

2. Evaluating the patient's comprehension of the gravity of the intervention, since operations that could upset a normally healthy body are usually not in line with rational decision-making.

- Development of Novel Physiological Materials: Another ethical consideration is the possibility of bioprinting completely original biological materials that are not present in nature. Risk assessment is made much more difficult by the uncertainty surrounding the long-term effects of introducing artificial human body organs and tissues.

In conclusion, there exist serious ethical questions about risk management, doctor accountability, and patient autonomy that are brought up by enhancement through 3D bioprinting. Medical practitioners have to carefully negotiate individual preferences with ethical safety issues, making sure that improvements do not result in unanticipated harm, even if people may want to improve their bodily functions [90].

13.4.5 Ethical Issues with 3D Printing: Data Processing from Cell Donors

There are serious ethical questions about donor data confidentiality when cells are collected and used for bioprinting. Maintaining privacy and control over biological data is crucial in order to avoid misuse, discrimination, or unforeseen personal repercussions, as these data include sensitive personal information [91].

Data Sensitivity and Informational Self-Determination

- Idea for "informational self-determination" holds that people need to have immediate authority over their personal information, especially if it exists about biological or health information.
- Sensitive information—such as inherited, biometric, and medical data—can affect a person's social standing, job, coverage, and personal existence, making its preservation vital, even while basic personal information (such as name and address) is largely used for identification [92].

Maintaining Research Needs and Data Privacy in Balance

- Research on bioprinting uses genetic and biological information to create novel transplant materials, treatments, and diagnostic instruments.
- A contradiction arises between facilitating scientific breakthroughs that benefit society and safeguarding the privacy of personal data. A balanced strategy is required to support medical research without jeopardizing donor autonomy, even if ethical standards place a higher priority on individual rights than on societal interests.

13.4.6 Legal and Ethical Safeguards

- Anonymous Biological Material: Early research phases can lower privacy hazards by using non-identifiable donor samples.
- Strict Protection of Information in Clinical Research: Strong confidentiality protocols must be put in place since clinical applications need identifiable biological material to be compatible with transplants.
- Informed Consent: To ensure openness and moral adherence, donors must expressly consent to usage and replication of its biological component.

With conclusion, ethical frameworks must strike a balance between the demands of scientific advancement and human privacy, even while data protection is crucial in bioprinting research. Ethical issues about the processing of cell donors' data can be successfully addressed by putting in place stringent confidentiality regulations, informed consent procedures, and anonymization whenever feasible.

Access to Bioprinting and Ethical Issues in 3D Printing
Equity in Access to Bioprinted Organs
Given the acute organ shortage, access to biological resources for transplanting is a crucial ethical concern. The goal of current priority systems is to guarantee equitable distribution according to objective standards such as medical urgency. By creating customized grafts from patient's own cells, bioprinting addresses scarcity concerns if it proves to be a feasible substitute [93].

Public Healthcare Systems and Financial Barriers

However, if 3D bioprinting is still too expensive for national healthcare systems, access is not assured. Cost-efficiency studies that contrast the expenses of bioprinting with those of traditional transplantation are necessary to determine whether it is eligible for public funding. Important things to think about are:

1. Safety and Functionality: Before being used in public healthcare, bioprinted organs must show sustained dependability.
2. Prioritization Frameworks: A transparent and equitable patient selection process that reflects existing transplant waiting lists will be required if financial constraints prohibit universal access [94].

Bioprinting is a promising answer to the organ shortage problem, but in order to guarantee fair and moral access within healthcare systems, its cost-effectiveness and distribution must be carefully considered.

13.4.7 Clinical Trials

Strict legal and ethical guidelines must be adhered to in clinical trials utilizing biological materials, 3D-printed implants, and gadgets [95]. Only in the event there are no similarly viable alternatives and the potential hazards to the individual are reasonable in comparison to the anticipated benefits may these trials move forward. The research protocol also needs to be approved by ethical and scientific authorities. Patients must give their express consent, be fully informed, and have the freedom to change their minds at any moment. For children and people with cognitive impairments, special safeguards are in place to guarantee that their involvement is in their greatest interests and that a legal representative gives their consent [96, 97]. Selecting experimental techniques when there isn't a recognized substitute is a major obstacle in these experiments [98].

Withdrawing from a bioprinted tissue implant may necessitate further surgery, which carries its own hazards, in contrast to drug trials, where stopping a substance usually results in a cessation of symptoms. In these circumstances, the doctor has a greater duty to consider the moral ramifications and make sure that the individual's autonomy is upheld while causing the least amount of harm possible [99].

13.5 Regulatory and Ethical Issues with 4D Printing in Healthcare

13.5.1 4D-Printed Medical Devices' Ethical Concerns

Prior to delving into the ethical concerns of 4D-printed medical equipment, it is crucial to examine the groundbreaking and motivational research that has already been done for subject of 3D printing techniques [27, 100], or 3D printing for the medical field [71]. Experts generally concur, based on a selection of publications, that 3D-printed medical equipment may raise ethical concerns about the following issues:

1. Safety: It is believed that centralized manufacturing is giving way to decentralized manufacturing as 3D printing redefines product design and production. Although the hazards of decentralized quality assurance should be carefully considered and create ethical problems, this can encourage point-of-care solutions and personalization in the medical area [101].
2. Environmental effects: Because materials or even medical devices can be downloaded or printed through cloud, the aforementioned decentralized production should reduce transportation-related environmental effects. Despite the apparent benefits, there are still certain points to be encountered. These constitute use of new substances like distributing threads or laminating powders, whose hazards to human health and the environment should be methodically addressed while also relating to safety. In decentralized production and supply chains, it is also more difficult to track and reclaim used products *via* reconditioning, reuse, or recycling, which could have detrimental effects on the environment [102]. Medical equipment manufactured using 3D (and 4D) technology should have life cycle evaluations done and compared to the current state of affairs.
3. Economic struggles: The biomedicine sector's reformulation *via* customized 3D-printed solutions affect labor need because automation would reduce the requirement for workers in conventional manufacturing chains. However, other roles will be required, such as additive manufacturing designers, technological builders, materials researchers, and

supervisors and technicians of printing "farms." The ultimate objective would be to support decent work globally by creating linked design & printing hubs and achieving high-quality, customized, and reasonably priced medical equipment through this paradigm shift that encompasses information and best practices sharing.

4. Liability and intellectual property: Historically, the makers of medical devices were in charge of their efficacy and safety as well as any potential harm that patients might sustain from defective equipment. The chain of responsibility for 3D-printed medical devices should likely be modified, particularly if decentralized printing of online-shared designs becomes commonplace. Regulators should carefully examine the accountability for developers (and users) sharing medical product designs (as well as design modifications) *via* on the internet collaboration surroundings, as well as the procedures put in place to monitor design modifications, record concepts and their evolutions, and maintain quality control in delocalized production. Within the same time, it is important to investigate novel approaches to identifying and rewarding proprietary information in the newly developed interactive design and fabrication platforms [103].
5. Privacy and patient data: Users and designers who collaborate on developing and distributing 3D printable models, changes to preexisting designs, and accessories with printing with 3D hardware and software are just a few examples of how 3D printing has developed alongside the makers' movement. These collaborative platforms have seen the emergence of numerous medical applications for 3D printing, and the outcomes are shared *via* platforms such as GitHub, GrabCAD, and Thingiverse, to name a few. Among the often shared "objects" in these digital infrastructures are 3D printable representations for various human body parts, such as faces, bones or digitized heads, and other features. Most of the models produced using particular applications for segmented and 3D geometry generation in conjunction with healthcare imaging and scanning technology. The privacy of patients or users may be significantly impacted by this information interchange, which should be taken into account when putting *ad hoc* procedures for data management into place. In any event, if patient data is going to be

utilized to support studies and developments in the field of 3D-printed medical devices, informed authorization should be acquired.

6. Equity and universal access to health care: 3D (as well as 4D, which consists) technology developers should concentrate on using these cutting-edge tools to ensure that medical solutions are as readily available as possible, even resulting in customized solutions that are more cost-effective for users and health care systems while also being technically superior to the state-of-the-art. This might be achievable using knowledge-based additive manufacturing techniques.
7. Human enhancement: In addition to being utilized for body repair and rebuilding, 3D printing technology, as well as orthoses and prostheses, can also be used for human improvement and even for pointless body alteration procedures (such as subdermal horn implants). It is important to carefully examine these more recent activities and the resources allocated to them because they raise ethical concerns. Discussions concerning the development of human capacities and the availability of technology that increase human skills are crucial.
8. Abstinence of standards and regulations: In general, 3D printing technologies and medical equipment appear to be developing far more quickly than the biomedical industry's standards and regulations. The drastic changes in healthcare which 3D (and 4D) printing methods bring about is still beyond the scope of current rules, especially the more recent ones such as the EU Healthcare Device Regulation 2017/745. These include future do-it-yourself medical gadgets, which are difficult to monitor and control and may end up being dangerous, in addition to open-source medical equipment [104].

13.5.2 Ethical and Regulatory Highlights

13.5.2.1 Regulatory Compliance

To guarantee safety and effectiveness for the structures and equipment being created, regulatory bodies must thoroughly examine and approve any new technologies for medicine, including 4D printing. In many countries, agencies like the United States Drug and Food Administration, or

FDA, and the European Medicines Agency (EMA) regulate medical items, including those created *via* 4D printing.

Since the licensing procedure for 4D-printed biomedical equipment frequently consists of several steps, such as preclinical research, clinical studies, and post-market surveillance, navigating the complicated regulatory environment can be difficult. Furthermore, considering how young 4D printing technology is, there might be exceptions to current rules or a need for new ones that are especially designed to handle the special characteristics of 4D-printed products, like their versatility and shape-changing capabilities. To fully comprehend the specifications and approval procedures for 4D-printed biomedical equipment, manufacturers and researchers must collaborate closely with regulatory bodies. This entails making certain that products exhibit security and effectiveness with respect to experimental as well as clinical trials, in a way in which devices utilized are biocompatible, and that fabrication procedures adhere to standards of quality.

13.5.2.2 Ethical Compliance

To guarantee the proper development and application of this technology, a number of ethical problems are raised by the application of 4D printing technology in the field of biological engineering. Patient confidentiality, informed consent, or fair access for 4D-printed equipment is a few important ethical factors. Since the creation of customized 4D-printed equipment frequently necessitates access to private medical information, such as a patient's unique anatomy or physiological parameters, patient privacy is an important consideration. Techniques for protecting patient confidentiality and ensuring that sensitive data is handled securely throughout the design and production processes must be developed by researchers and manufacturers.

Another crucial ethical factor is informed permission, especially when it comes to clinical trials incorporating 4D-printed items. The possible advantages and disadvantages of using 4D-printed equipment, in addition to any alternatives, should be thoroughly explained to patients taking part in these trials.

Concerns about equitable availability of 4D-printed biomedical equipment also exist because the high expenses of their development and production may prevent some patients from receiving them. In order to ensure that everyone, regardless of socioeconomic background, may benefit from this novel approach to healthcare, researchers and politicians must investigate ways to lower the overall price of 4D printing technologies.

Apart from the aforementioned broad constraints of four-dimensional printing in biomedical engineering, it is crucial to acknowledge for every particular application in this domain may provide a distinct set of difficulties. Diverse projects have very diverse needs and goals, hence customized methods and approaches to problem-solving are required. There are unique problems associated with every combination of materials, methods, and applications that need for careful research. The intended usefulness, compliance with biological structures, complexity of structure, and scalability are a few examples of factors that may present unique challenges that require careful consideration. As a result, each application must be handled separately, with in-depth investigation and analysis carried out to pinpoint and resolve the unique difficulties related to that project. We can successfully negotiate 4D printing's constraints and realize the complete promise to broad field of biomedical engineering by doing this [70].

13.5.3 Ethics Concerns with 3D Printers

The last decade has seen additive manufacturing become increasingly popular, particularly since home 3D printers became accessible. Two ethical concerns are brought up by the possibility that many products may be produced domestically in the future. First, there are safety concerns. Enforcing our existing safety guidelines on this new production model is challenging because they rely on centralized manufacturing assumptions. I make the case, using existing US legislation as an example that although consumers are unable to accurately assess all pertinent risks, they nevertheless want protection; any rules will most likely be limited to ideas and not actual objects. Second, there are concerns about intellectual property. Existing intellectual property laws do not provide protection against this for many products, but it is now feasible to scan them and print replicas when used in conjunction with a 3D scanner. Innovative things are protected by patents; everything else is appropriately unprotected. Creativity is the foundation of intellectual property, yet when new technologies emerge, the definition of creativity shifts [71].

13.5.3.1 Legal and Ethical Concerns with Bioprinting and 3D Printing

An integrative patient-specific bioprinted product requires computational simulation using the patient's anatomical data and attributes.

Regulatory and ethical issues

There are many ethical and legal issues with the use of cells from individuals and animals in CTE. The use of embryonic stem cells from humans (hESCs), which are produced from the inner volume of embryonic cells in the blastocyst stage, is at the center of the ethical controversy. Although these cells are special because they are pluripotent, meaning they can differentiate into many kinds of cells, their use presents serious moral and ethical issues, especially when it comes to the embryo's status and the consequences of destroying it [105]. The UK has led the way in creating legislative frameworks for controlling hESC research, mirroring the global drive to address these ethical issues through legislation and policy [106].

The ethical environment around CTE has grown even more complex with the introduction of biotechnological breakthroughs like iPSCs. By transforming mature cells to an embryonic state, iPSCs provide a substitute to hESCs that might get around some of the moral dilemmas surrounding the use of embryos. They do, however, present fresh difficulties with regard to the long-term consequences and safety of using these cells in therapeutic settings. Important considerations in this field include overcoming genetic manipulation issues and ensuring legal acquisition of donor cells for iPSCs [107].

The clinical use of cell-based therapy in CTE is not exempt from regulatory issues that arise in the lab. These treatments must pass rigorous scientific & clinical testing to prove their safety and effectiveness before they are made available to patients. Complex regulatory regimes that differ between nations and regions oversee this process, making standardization and international cooperation difficult. Further aggravating the regulatory environment is the requirement that good manufacturing practices, or GMP, be followed when scaling up production processes for clinical applications [105]. In order to make it easier to translate research into clinical practice, efforts must be made to unify these regulations [108, 109].

Aside from these technical and scientific difficulties, ethical issues also include privacy and informed permission. Obtaining cells—whether from individuals, embryos, or other sources like adipose tissue or umbilical cord blood—needs thorough consent procedure which upholds rights as well as donor autonomy. Donor privacy needs to be protected, particularly if biological information is involved with iPSCs. To properly handle these issues and ensure the privacy and security of all parties involved are upheld, rules and regulations are required [106, 110]. In addition, the discipline of CTE needs to address accessibility and justice concerns.

In order to prevent inequalities in access to state-of-the-art medical treatments, it is crucial to think about how these technologies may be made available fairly as they develop. In the framework of global medical care, where resource discrepancies can make injustices worse, this question is especially crucial. The moral and legal issues surrounding CTE are complex and ever-changing. They cover the sourcing and application of cells, especially hESCs, as well as the complexities of various regulatory systems and the wider ramifications of quickly developing biotechnologies. A comprehensive strategy that incorporates regulatory vigilance, ethical inspection, and a dedication to equity and justice is needed to address these issues.

To guarantee the responsible and advantageous development of CTE, it is crucial that the field continuously adapts and takes into account both current frameworks and new ethical considerations [105, 109].

13.6 Prospects for the Future

13.6.1 3D Bioprinting

Reproductive substances

The ability to recreate generative components, especially lines of embryonic cells as well as perhaps early stages of embryogenesis, is a futuristic view of 3D bioprinting. Generally speaking, the utilization of reproductive material requires the donors' express, informed agreement. Both donors must give their approval when using embryonic material. Given that embryos are made of biological products that contains the genetic identity of the donor as well as that genes' shared some hereditary characteristics utilizing dual genitors, protecting donor data is also a concern [97]. As a result, all of the previously specified assurances about safeguarding private information must also be taken into account here. One potential solution to issues with the accessibility of reproductive material, specifically ova, is bioprinting.

These days, many couples' access to assisted reproduction is severely hampered by the limited supply of eggs. In an attempt to promote private ova donation between friends and family, numerous nations have dropped the donor anonymity restriction (which was put in place to preserve privacy). However, the issue contributes to the global illegal market for ova (and consequently, women) trafficking. There may be a solution there if available ova are used to produce early embryos that will be used to extract immortal stem cells as the starting material for bioprinting.

Even if the technique could solve these issues, from an ethical perspective, the genitors' specific consent for bioprinting will still be required because any intentional processing of the unique recurrent substance by the clinic with the intention of using reproduced embryos by third parties raises the challenge of confidential information protection because this is recognized material. This implies that even extra original reproductive material, or extra embryos, is still directly controlled by the donors and is not regarded as "leftover" that the clinic can freely dispose of. Future biological printing using developing immortal stem cell lines may provide more significant ethical issues because these blastocyst cells have the capacity to develop into full human organisms after completing the entire process of reproduction. This fact is comparable to that has been in-depth explored regarding the potential use of human reproductive cloning in the future. In fact, bioprinted immortal lines of stem cells will produce organisms that are genetically identical to the clones since they will have the same genetic profile. A fundamental argument against the production of clones of humans is because act of deliberately producing similar twins diminishes the dignity of human since they would have to live with the weight of proving their individuality in contrast to their "designing." The same reasoning applies to bioprinting since it merely modifies the process of creating identical humans, not the goal itself. However, this is the sole ethical justification for opposing the bioprinting of human stem cell lines. However, since there is no goal of producing identical humans, bioprinting stem cells for the sole purpose of making transplant material does not raise the same ethical issues as "therapeutic cloning" [111].

13.6.2 Development in 4DP

A lot of work has been done in the last several years to create efficient 4DP systems. The 4DP process is made simpler and more effective by the development of six distinct software solutions to support its various stages [112]. These can be categorized as host/firmware, printing management software, modeling, slicing, simulation, and monitoring, in that order. Shape-changing systems are structurally controlled by a few commercial computer programs.

MIT's computational science and Intelligent Systems Lab's Foundry, Mosaic Manufacturing's CANVAS software, Autodesk's Project Cyborg, and Monolithic multi-material voxel software are examples of current technologies for self-assemblage modeling and printing parameter optimization. Additionally, by giving the structures nodes, routes, edges, polygons, vertices, and folds, programming tools like Origamizer or E-Origami

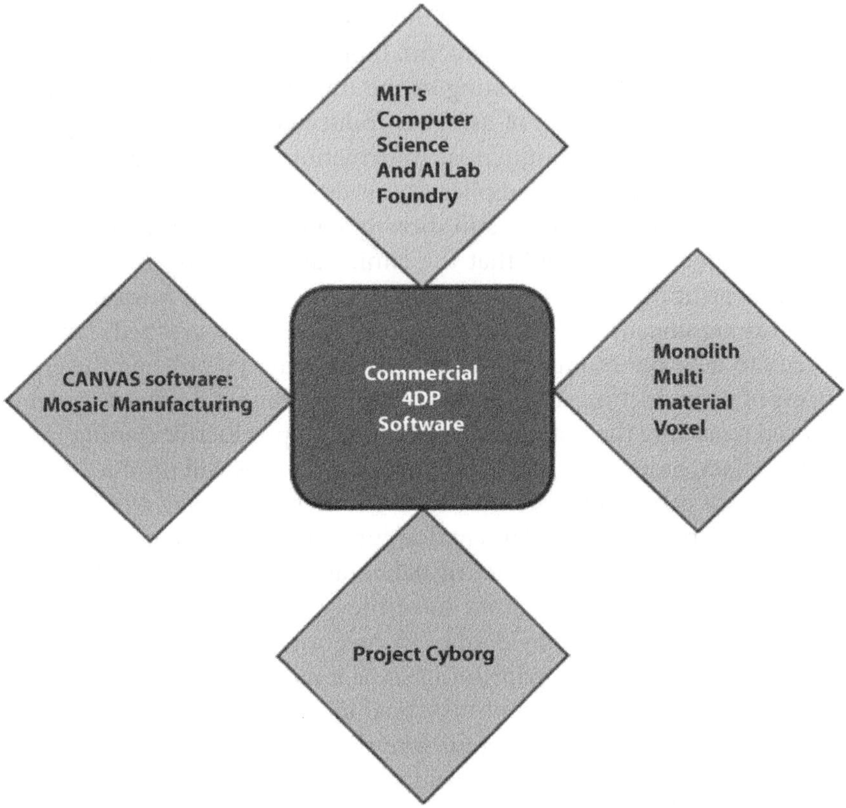

Figure 13.3 A few 4DP programs that are sold commercially are Project Cyborg, Canvas, Monolith, and Foundry multi-material voxels. These programs are utilized to regulate the shape-changing device in an organized way and to make 4DP simpler, more efficient, and less expensive.

Systems may help create intricate origami shapes [113]. As represented within Figure 13.3, production of 4D-printed goods requires collective utilization of this program.

13.6.3 Development against 3D Printing

4DP technologies application can result in products of which characteristics as well as sizes are subject to alter gradually in reaction to modifications in the immediate surroundings, including variations in light and temperature. Here are a few benefits of 4DP:

- Features of the printed item: 4DP allows consumers for printing items using intelligent product that possess numerous uses in engineering, materials research, and healthcare.
- Printed object resolution: Using 4D P technology allows for the printing of high-resolution objects, which is challenging to accomplish with 3D printing.
- Shape alteration in reaction to stimuli: A 4D-printed object's shape varies as time passes in response to many factors like light, pH, temperature, or so forth.
- Novel Method: Throughout the product's design and development phases, this technology innovates it.
- Self-Assembly: Because intelligent materials are used in this process, items with self-assembling qualities are probably going to be produced.
- Cost-effective: Compared to 3D printing, printed goods are less expensive to manufacture.

The literature divides 4DP technologies into four main printing methods: inkjet head 3DP, powder bed, fused deposition modeling, and stereolithography. Superior thermal and mechanical characteristics of flexible materials are mainstay for these technologies. However, this technology's future value rests in redefining manufacturing-related industries that can address the shortcomings of current manufacturing techniques. In contrast to 4DP, which only uses materials that react to variations in humidity, temperature, pressure, etc. over time, 3DP can produce fixed non-temporal output (dimension) using numerous substances, including metals, powders, thermoplastic polymers that are UV-curable resins, and many more [114].

When compared to the 3DP, it 4DP is more adaptable and dynamic. In contrast to 3D, 4DP makes use of dynamic, stimuli-responsive smart materials. In addition, 4DP has a broader range of applications than 3DP.

13.7 4DP Utilization in Healthcare

There are numerous healthcare applications for the 4DP. It is employed in the creation of artificial tissue, grafts, and medicinal devices like smart stents. Additionally, it is employed in the printing of organs like the liver, heart, etc. The smart devices made with this type of technology are more body-adaptable and biocompatible. The most current method for creating stimuli-responsive stents of similar size has been shown to be 4D bioprinting. For stents, several 4D bioprinting methods and materials have been created. Stents would conform to their proper size and shape by self-deformation after implantation [115, 116]. As a result, fewer surgical invasions may occur. Tissue engineering, particularly skin bioprinting, is a promising treatment option for deep burns, wounds from surgery, and skin fragility diseases.

Printed tissue is believed to offer a quicker recovery, less discomfort, and possibly a more palatable outcome than grafts of skin from the patient's unaffected organs. The biological printing of organs including the liver, kidney, heart, and others depends heavily on the time-dependent component of 4DP. Drug delivery is another potential use for 4DP technology. Localized drug release within the body can be achieved by achieving a large transition temperature range and lowering the threshold threshold for thermoresponsive substances closer to the normal physiological temperature. Because of their greater area and low weight, porous polymers are used as drug carriers for medication delivery [117, 118].

AI disclosure statement: The author acknowledges the use of AI-based tools for minor language editing or grammatical correction. However, all intellectual contributions, analyses, and conclusions presented in this chapter are entirely the author's own.

References

1. Bastawrous, S., Wu, L., Strzelecki, B., Levin, D.B., Li, J.S., Coburn, J., Ripley, B., Establishing quality and safety in hospital-based 3D printing programs: patient-first approach. *Radiographics*, 41, 4, 1208–29, 2021 Jul.
2. Ji, Y., Luan, C., Yao, X., Fu, J., He, Y., Recent progress in 3D printing of smart structures: classification, challenges, and trends. *Adv. Intell. Syst.*, 3, 12, 2000271, 2021 Dec.
3. Khoo, Z.X., Teoh, J.E., Liu, Y., Chua, C.K., Yang, S., An, J., Leong, K.F., Yeong, W.Y., 3D printing of smart materials: A review on recent progresses in 4D printing. *Virtual Phys. Prototyping*, 10, 3, 103–22, 2015 Jul 3.
4. Breger, J.C., Yoon, C., Xiao, R., Kwag, H.R., Wang, M.O., Fisher, J.P., Nguyen, T.D., Gracias, D.H., Self-folding thermo-magnetically responsive soft micro-grippers. *ACS Appl. Mater. Interfaces*, 7, 5, 3398–405, 2015 Feb 11.
5. Sydney Gladman, A., Matsumoto, E.A., Nuzzo, R.G., Mahadevan, L., Lewis, J.A., Biomimetic 4D printing. *Nat. Mater.*, 15, 4, 413–8, 2016 Apr.

6. Wang, X., Guo, Q., Cai, X., Zhou, S., Kobe, B., Yang, J., Initiator-integrated 3D printing enables the formation of complex metallic architectures. *ACS Appl. Mater. Interfaces*, 6, 4, 2583–7, 2014 Feb 26.
7. Kim, Y., Yuk, H., Zhao, R., Chester, S.A., Zhao, X., Printing ferromagnetic domains for untethered fast-transforming soft materials. *Nature*, 558, 7709, 274–9, 2018 Jun 14.
8. Sharma, P., Jain, V., Tailang, M., Selection and role of polymers for designing of a drug carrier, in: *Drug carriers*, IntechOpen Limited, 2022 Mar 19.
9. Liu, X., Yuk, H., Lin, S., Parada, G.A., Tang, T.C., Tham, E., de la Fuente-Nunez, C., Lu, T.K., Zhao, X., 3D printing of living responsive materials and devices. *Adv. Mater.*, 30, 4, 1704821, 2018 Jan.
10. Rana, A., Mittal, A., Vashist, C., Rajput, S., Sridhar, S.B., Malviya, R., Exploration of 4D Printing and its Applications in the Biomedical Sciences. *Curr. Pharm. Des.*, 2025.
11. He, F., Ou, Y., Liu, J., Huang, Q., Tang, B., Xin, F., Zhang, J., Jiang, M., Chen, S., Yu, Z., 3D Printed biocatalytic living materials with dual-network reinforced bioinks. *Small*, 18, 6, 2104820, 2022 Feb.
12. Balasubramanian, S., Aubin-Tam, M.E., Meyer, A.S., 3D printing for the fabrication of biofilm-based functional living materials. *ACS Synth. Biol.*, 8, 7, 1564–7, 2019 Jul 19.
13. Sharma, P. and Jain, V., An overview of current advances and pharmaceutical uses of 3D and 4D printing. *Explor. Med.*, 4, 4, 560–75, 2023 Aug 31.
14. Tofail, S.A., Koumoulos, E.P., Bandyopadhyay, A., Bose, S., O'Donoghue, L., Charitidis, C., Additive manufacturing: scientific and technological challenges, market uptake and opportunities. *Mater. Today*, 21, 1, 22–37, 2018 Jan 1.
15. Basgul, C., Yu, T., MacDonald, D.W., Siskey, R., Marcolongo, M., Kurtz, S.M., Structure–property relationships for 3D-printed PEEK intervertebral lumbar cages produced using fused filament fabrication. *J. Mater. Res.*, 33, 14, 2040–51, 2018 Jul.
16. McGilvray, K.C., Waldorff, E.I., Easley, J., Seim, H.B., Zhang, N., Linovitz, R.J., Ryaby, J.T., Puttlitz, C.M., Evaluation of a polyetheretherketone (PEEK) titanium composite interbody spacer in an ovine lumbar interbody fusion model: biomechanical, microcomputed tomographic, and histologic analyses. *Spine J.*, 17, 12, 1907–16, 2017 Dec 1.
17. Kjar, A. and Huang, Y., Application of micro-scale 3D printing in pharmaceutics. *Pharmaceutics*, 11, 8, 390, 2019 Aug 3.
18. Bernard, A. and Fischer, A., New trends in rapid product development. *CIRP Ann.*, 51, 2, 635–52, 2002 Jan 1.
19. Pathak, K., Saikia, R., Das, A., Das, D., Islam, M.A., Pramanik, P., Parasar, A., Borthakur, P.P., Sarmah, P., Saikia, M., Borthakur, B., 3D printing in biomedicine: Advancing personalized care through additive manufacturing. *Explor. Med.*, 4, 6, 1135–67, 2023 Dec 29.

20. Birtchnell, T. and Urry, J., 3D, SF and the future. *Futures*, 50, 25–34, 2013 Jun 1.
21. Low, Z.X., Chua, Y.T., Ray, B.M., Mattia, D., Metcalfe, I.S., Patterson, D.A., Perspective on 3D printing of separation membranes and comparison to related unconventional fabrication techniques. *J. Membr. Sci.*, 523, 596–613, 2017 Feb 1.
22. Holzmann, P., Breitenecker, R.J., Soomro, A.A., Schwarz, E.J., User entrepreneur business models in 3D printing. *J. Manuf. Technol. Manage.*, 28, 1, 75–94, 2017 Feb 6.
23. Zhou, W., Qiao, Z., Nazarzadeh Zare, E., Huang, J., Zheng, X., Sun, X., Shao, M., Wang, H., Wang, X., Chen, D., Zheng, J., 4D-printed dynamic materials in biomedical applications: chemistry, challenges, and their future perspectives in the clinical sector. *J. Med. Chem.*, 63, 15, 8003–24, 2020 Apr 7.
24. Roy, A., Hossain, M.S., Bhowmick, A., Nizhum, N., Kumar, S., Prospects of 4d printing in pharmaceuticals. *Pharmacologyonline*, 3, 292–302, 2020.
25. Seok, J., Yoon, S., Ryu, C.H., Kim, S.K., Ryu, J., Jung, Y.S., A personalized 3D-printed model for obtaining informed consent process for thyroid surgery: a randomized clinical study using a deep learning approach with mesh-type 3D modeling. *J. Pers. Med.*, 11, 6, 574, 2021 Jun 18.
26. Rajput, S., Malviya, R., Sridhar, S.B., Osteoporosis Risk Assessment and Individualized Feature Analysis Using Interpretable XAI and RAI Techniques, in: *Explainable and Responsible Artificial Intelligence in Healthcare*, pp. 89–113, Springer Nature Singapore Pte Ltd., Singapore, 2025 Mar 13.
27. Dodds, S., 3D printing raises ethical issues in medicine. *ABC Science*, 2015 Feb.
28. Fox, M. and Peregrin, T., 3-D printing: Revolutionizing preoperative planning, resident training, and the future of surgical care. *Bull. Am. Coll. Surg.*, 101, 7, 9–18, 2016 Jul.
29. Hessel, V., Stoudemire, J., Miyamoto, H., Fisk, I.D., *In-Space Manufacturing and Resources*, Wiley, Hoboken, NJ, USA, 2022.
30. Rather, H.A., Vya, J., Singh, S., Biomaterials for 3D Printing of Patient-Specific Organ Models. *Biosens. Nanotheranostics*, 2, 1, 1–2, 2023 May 22.
31. Peltola, S.M., Melchels, F.P., Grijpma, D.W., Kellomäki, M., A review of rapid prototyping techniques for tissue engineering purposes. *Ann. Med.*, 40, 4, 268–80, 2008 Jan 1.
32. Papaioannou, T.G., Manolesou, D., Dimakakos, E., Tsoucalas, G., Vavuranakis, M., Tousoulis, D., 3D bioprinting methods and techniques: applications on artificial blood vessel fabrication. *Acta Cardiologica Sin.*, 35, 3, 284, 2019 May.
33. Jeong, H.J., Nam, H., Jang, J., Lee, S.J., 3D bioprinting strategies for the regeneration of functional tubular tissues and organs. *Bioengineering*, 7, 2, 32, 2020 Mar 31.

34. S.S. Crump, inventor; Stratasys Inc, assignee, Apparatus and method for creating three-dimensional objects. United States patent US 5,121,329, 1992 Jun 9.
35. Casavola, C., Cazzato, A., Moramarco, V., Pappalettere, C., Orthotropic mechanical properties of fused deposition modelling parts described by classical laminate theory. *Mater. Des.*, 90, 453–8, 2016 Jan 15.
36. Xu, N., Ye, X., Wei, D., Zhong, J., Chen, Y., Xu, G., He, D., 3D artificial bones for bone repair prepared by computed tomography-guided fused deposition modeling for bone repair. *ACS Appl. Mater. Interfaces*, 6, 17, 14952–63, 2014 Sep 10.
37. Yuan, B., Zhou, S.Y., Chen, X.S., Rapid prototyping technology and its application in bone tissue engineering. *J. Zhejiang Univ.-Sci. B*, 18, 303–15, 2017 Apr.
38. Wang, X., Jiang, M., Zhou, Z., Gou, J., Hui, D., 3D printing of polymer matrix composites: A review and prospective. *Composites, Part B*, 110, 442–58, 2017 Feb 1.
39. Thavornyutikarn, B., Chantarapanich, N., Sitthiseripratip, K., Thouas, G.A., Chen, Q., Bone tissue engineering scaffolding: computer-aided scaffolding techniques. *Prog. Biomater.*, 3, 61–102, 2014 Dec.
40. Chia, H.N. and Wu, B.M., Recent advances in 3D printing of biomaterials. *J. Biol. Eng.*, 9, 1–4, 2015 Dec.
41. Hribar, K.C., Soman, P., Warner, J., Chung, P., Chen, S., Light-assisted direct-write of 3D functional biomaterials. *Lab Chip*, 14, 2, 268–75, 2014.
42. Tumbleston, J.R., Shirvanyants, D., Ermoshkin, N., Janusziewicz, R., Johnson, A.R., Kelly, D., Chen, K., Pinschmidt, R., Rolland, J.P., Ermoshkin, A., Samulski, E.T., Continuous liquid interface production of 3D objects. *Science*, 347, 6228, 1349–52, 2015 Mar 20.
43. Wang, Z., Kumar, H., Tian, Z., Jin, X., Holzman, J.F., Menard, F., Kim, K., Visible light photoinitiation of cell-adhesive gelatin methacryloyl hydrogels for stereolithography 3D bioprinting. *ACS Appl. Mater. Interfaces*, 10, 32, 26859–69, 2018 Jul 19.
44. C.R. Deckard, inventor; University of Texas System, assignee, Method and apparatus for producing parts by selective sintering. United States patent US 4,863,538, 1989 Sep 5.
45. Rajput, S., Malviya, R., Uniyal, P., Advances in the treatment of kidney disorders using mesenchymal stem cells. *Curr. Pharm. Des.*, 30, 11, 825–40, 2024 Mar 1.
46. Bai, J., Goodridge, R.D., Yuan, S., Zhou, K., Chua, C.K., Wei, J., Thermal influence of CNT on the polyamide 12 nanocomposite for selective laser sintering. *Molecules*, 20, 10, 19041–50, 2015 Oct 20.
47. Li, J., Chen, M., Fan, X., Zhou, H., Recent advances in bioprinting techniques: approaches, applications and future prospects. *J. Transl. Med.*, 14, 1–5, 2016 Dec.

48. Zein, I., Hutmacher, D.W., Tan, K.C., Teoh, S.H., Fused deposition modeling of novel scaffold architectures for tissue engineering applications. *Biomaterials*, 23, 4, 1169–85, 2002 Feb 15.
49. Huang, Y., Zhang, X.F., Gao, G., Yonezawa, T., Cui, X., 3D bioprinting and the current applications in tissue engineering. *Biotechnol. J.*, 12, 8, 1600734, 2017 Aug.
50. Cui, X., Boland, T., DD'Lima, D., K. Lotz, M., Thermal inkjet printing in tissue engineering and regenerative medicine. *Recent Pat. Drug Deliv. Formul.*, 6, 2, 149–55, 2012 Aug 1.
51. Gruene, M., Unger, C., Koch, L., Deiwick, A., Chichkov, B., Dispensing pico to nanolitre of a natural hydrogel by laser-assisted bioprinting. *Biomed. Eng. Online*, 10, 1–1, 2011 Dec.
52. Mandrycky, C., Wang, Z., Kim, K., Kim, D.H., 3D bioprinting for engineering complex tissues. *Biotechnol. Adv.*, 34, 4, 422–34, 2016 Jul 1.
53. Kang, H.W., Lee, S.J., Ko, I.K., Kengla, C., Yoo, J.J., Atala, A., A 3D bioprinting system to produce human-scale tissue constructs with structural integrity. *Nat. Biotechnol.*, 34, 3, 312–9, 2016 Mar.
54. Grigoryan, B., Sazer, D.W., Avila, A., Albritton, J.L., Padhye, A., Ta, A.H., Greenfield, P.T., Gibbons, D.L., Miller, J.S., Development, characterization, and applications of multi-material stereolithography bioprinting. *Sci. Rep.*, 11, 1, 3171, 2021 Feb 4.
55. Wu, W., DeConinck, A., Lewis, J.A., Omnidirectional printing of 3D microvascular networks. *Adv. Mater.*, 23, 24, H178–83, 2011 Jun 24.
56. Tibbits, S., 4D printing: multi-material shape change. *Archit. Des.*, 84, 1, 116–21, 2014 Jan.
57. Choi, J., Kwon, O.C., Jo, W., Lee, H.J., Moon, M.W., 4D printing technology: a review. *3D Print. Addit. Manuf.*, 2, 4, 159–67, 2015 Dec 1.
58. Li, X., Shang, J., Wang, Z., Intelligent materials: a review of applications in 4D printing. *Assem. Autom.*, 37, 2, 170–85, 2017 Apr 3.
59. Miao, S., Zhu, W., Castro, N.J., Leng, J., Zhang, L.G., Four-dimensional printing hierarchy scaffolds with highly biocompatible smart polymers for tissue engineering applications. *Tissue Eng. Part C Methods*, 22, 10, 952–63, 2016 Oct 1.
60. Wan, Z., Zhang, P., Liu, Y., Lv, L., Zhou, Y., Four-dimensional bioprinting: Current developments and applications in bone tissue engineering. *Acta Biomater.*, 101, 26–42, 2020 Jan 1.
61. Mukhopadhyay, P., Sarkar, K., Bhattacharya, S., Bhattacharyya, A., Mishra, R., Kundu, P.P., pH sensitive N-succinyl chitosan grafted polyacrylamide hydrogel for oral insulin delivery. *Carbohydr. Polym.*, 112, 627–37, 2014 Nov 4.
62. Melocchi, A., Uboldi, M., Inverardi, N., Briatico-Vangosa, F., Baldi, F., Pandini, S., Scalet, G., Auricchio, F., Cerea, M., Foppoli, A., Maroni, A., Expandable drug delivery system for gastric retention based on shape memory polymers:

Development *via* 4D printing and extrusion. *Int. J. Pharm.*, 571, 118700, 2019 Nov 25.

63. Han, D., Morde, R.S., Mariani, S., La Mattina, A.A., Vignali, E., Yang, C., Barillaro, G., Lee, H., 4D printing of a bioinspired microneedle array with backward-facing barbs for enhanced tissue adhesion. *Adv. Funct. Mater.*, 30, 11, 1909197, 2020 Mar.
64. Song, Z., Ren, L., Zhao, C., Liu, H., Yu, Z., Liu, Q., Ren, L., Biomimetic non-uniform, dual-stimuli self-morphing enabled by gradient four-dimensional printing. *ACS Appl. Mater. Interfaces*, 12, 5, 6351–61, 2020 Jan 10.
65. Zhang, F., Wang, L., Zheng, Z., Liu, Y., Leng, J., Magnetic programming of 4D printed shape memory composite structures. *Compos. Part A Appl. Sci. Manuf.*, 2019.
66. Salimi, S., Wu, Y., Barreiros, M.E., Natfji, A.A., Khaled, S., Wildman, R., Hart, L.R., Greco, F., Clark, E.A., Roberts, C.J., Hayes, W., A 3D printed drug delivery implant formed from a dynamic supramolecular polyurethane formulation. *Polym. Chem.*, 11, 20, 3453–64, 2020.
67. Gazzaniga, A., Foppoli, A., Cerea, M., Palugan, L., Cirilli, M., Moutaharrik, S., Melocchi, A., Maroni, A., Towards 4D printing in pharmaceutics. *Int. J. Pharm.: X*, 5, 100171, 2023 Dec 1.
68. Pingale, P., Dawre, S., Dhapte-Pawar, V., Dhas, N., Rajput, A., Advances in 4D printing: from stimulation to simulation. *Drug Delivery Transl. Res.*, 13, 1, 164–88, 2023 Jan.
69. Mathur, V., Agarwal, P., Srinivasan, V., Panwar, A., Vasanthan, K.S., Facet of 4D printing in biomedicine. *J. Mater. Res.*, 38, 1, 2–18, 2023 Jan 14.
70. Ramezani, M. and Mohd Ripin, Z., 4D printing in biomedical engineering: Advancements, challenges, and future directions. *J. Funct. Biomater.*, 14, 7, 347, 2023 Jun 29.
71. Neely, E.L., The risks of revolution: Ethical dilemmas in 3D printing from a US perspective. *Sci. Eng. Ethics*, 22, 1285–97, 2016 Oct.
72. Patuzzo, S., Goracci, G., Gasperini, L., Ciliberti, R., 3D bioprinting technology: scientific aspects and ethical issues. *Sci. Eng. Ethics*, 24, 335–48, 2018 Apr.
73. World Health Organization, The importance of pharmacovigilance, in: *The Importance of Pharmacovigilance*, World Health Organization, Geneva, 2002.
74. Andorno, R., The precautionary principle: A new legal standard for a technological age. *J. Int. Biotechnol. Law*, 2004.
75. Caplan, A.L., Is xenografting morally wrong? *Transplant. Proc.*, 24, 2, 722–727, 1992.
76. McNeiil, P.M., Experimentation on human beings. A companion to bioethics, 1998.
77. Beauchamp, T.L. and Childress, J.F., *Principles of biomedical ethics*, Edicoes Loyola, Oxford University Press, 1994.
78. Patuzzo, R., Maurichi, A., Camerini, T., Gallino, G., Ruggeri, R., Baffa, G., Mattavelli, I., Tinti, M.C., Crippa, F., Moglia, D., Tolomio, E., Accuracy and

prognostic value of sentinel lymph node biopsy in head and neck melanomas. *J. Surg. Res.*, 187, 2, 518–24, 2014 Apr 1.
79. Beauchamp, T.L., The "four principles" approach, in: *Meaning and medicine: a reader in the philosophy of health care*, vol. 22, pp. 147–55, Routledge, New York, 1999 Jul.
80. Cantor, N.L., A patient's decision to decline life-saving medical treatment: Bodily integrity versus the preservation of life. *Rutgers Law Rev.*, 26, 228, 1972.
81. Moskop, J.C. and Iserson, K.V., Triage in medicine, part II: Underlying values and principles. *Ann. Emergency Med.*, 49, 3, 282–7, 2007 Mar 1.
82. Den Hartogh, G., *What Kind of Death? The Ethics of Determining One's Own Death*, Routledge, 2022 Oct 31.
83. Beauchamp, T.L. and Childress, J.F., Morality and ethical theory, in: *Applied Ethics: Critical Concepts in Philosophy*, pp. 75–95, Routledge, London, 2002.
84. i Tubau, J.M., El respecte a l'autonomía, segons Beauchamp i Childress, en l'obra, Principles of Biomedical Ethics. *Bioètica debat: tribuna oberta de l'Institut Borja de Bioètica*, 17, 62, 7–10, 2011.
85. Tan, C. and Neo, R., Illustrating the contribution of Confucian philosophy through a reinterpretation of Beauchamp and Childress' principle of respect for autonomy. *Monash Bioeth. Rev.*, 7, 1–5, 2025 Jun.
86. Bosma, P.J., Chowdhury, J.R., Bakker, C., Gantla, S., de Boer, A., Oostra, B.A., Lindhout, D., Tytgat, G.N., Jansen, P.L., Elferink, R.P., Chowdhury, N.R., The genetic basis of the reduced expression of bilirubin UDP-glucuronosyltransferase 1 in Gilbert's syndrome. *N. Engl. J. Med.*, 333, 18, 1171–5, 1995 Nov 2.
87. Gilbert, J.K., Science communication: towards a proper emphasis on the social aspects of Science and Technology. *Alexandria: Revista de Educacao em Ciencia e Tecnologia*, 1, 1, 3–25, 2008.
88. Maas, T.Y., Pauwelussen, A., Turnhout, E., Co-producing the science–policy interface: towards common but differentiated responsibilities. *Humanit. Soc. Sci. Commun.*, 9, 1, 1–1, 2022 Mar 23.
89. Ozbolat, I.T., Peng, W., Ozbolat, V., Application areas of 3D bioprinting. *Drug Discovery Today*, 21, 8, 1257–71, 2016 Aug 1.
90. Petróczi, A. and Aidman, E., Psychological drivers in doping: The life-cycle model of performance enhancement. *Subst. Abuse Treat. Prev. Policy*, 3, 1–2, 2008 Dec.
91. Rouvroy, A. and Poullet, Y., The right to informational self-determination and the value of self-development: Reassessing the importance of privacy for democracy, in: *Reinventing data protection?*, pp. 45–76, Springer Netherlands, Dordrecht, 2009.
92. Fagan, M.B., Is there collective scientific knowledge? Arguments from explanation. *Philos. Q.*, 61, 243, 247–69, 2011 Apr 1.
93. Atella, V., Schafheutle, E., Noyce, P., Hassell, K., Affordability of medicines and patients' cost-reducing behaviour: empirical evidence based on SUR

estimates from Italy and the UK. *Appl. Health Econ. Health Policy*, 4, 23–35, 2005 Mar.

94. Vidalis, T., 3D printing and bioprinting: ethical and legal issues, in: *3D Printing: Applications in Medicine and Surgery*, pp. 51–68, Elsevier, 2020 Jan 1.
95. Ciliberti, R., Bonsignore, A., Bonzano, C., Ventura, F., Licata, M., Taking care of life: the new Italian law on post-mortem donation for study purposes, training and scientific research. *Ann. Anat.-Anat. Anz.*, 236, 151712, 2021 Jul 1.
96. Gropper AL. The curious disappearance of choice of law as an issue in Chapter 15 cases. *Brook. J. Corp. Fin. & Com. L.*, 9, 57, 2014.
97. Knaapen, M., Ploem, M.C., Kruijt, M., Oudijk, M.A., van der Graaf, R., Bet, P.M., Bakx, R., van Heurn, L.E., Gorter, R.R., van der Lee, J.H., Low-risk trials for children and pregnant women threatened by unnecessary strict regulations. Does the coming EU Clinical Trial Regulation offer a solution? *Eur. J. Pediatr.*, 179, 1205–11, 2020 Aug.
98. Nieminen, S., Assisted Reproductive Technologies and Medically Assisted Reproduction in the Context of the European Convention on Human Rights: Legal and Social Perspectives.
99. Arnull, A., Private Applicants and the Action for Annulment under Article 173 of the EC Treaty. *Common Mark. Law Rev.*, 32, 7, 1995.
100. Maric, J., Rodhain, F., Barlette, Y., 3D Printing Trends-Discussing Societal Environmental and Ethical Implications. *Manag. Des. Technol. Organisationnelles*, 6, 1, 127–38, 2017.
101. Wild, S., 3D printing and the murky ethics of replicating bones, Undark Magazine, 2020 Oct.
102. Vermeulen, N., Haddow, G., Seymour, T., Faulkner-Jones, A., Shu, W., 3D bioprint me: a socioethical view of bioprinting human organs and tissues. *J. Med. Ethics*, 43, 9, 618–24, 2017 Sep 1.
103. Kirillova, A., Bushev, S., Abubakirov, A., Sukikh, G., Bioethical and legal issues in 3D bioprinting. *Int. J. Bioprinting*, 6, 3, 272, 2020 Apr 28.
104. Lantada, A.D., Ethical issues of 4D printed medical devices. *IEEE Pulse*, 14, 2, 23–8, 2023 May 25.
105. Malviya, R., Rajput, S., Vaidya, M., *Artificial Intelligence for Bone Disorder: Diagnosis and Treatment*, John Wiley & Sons, 2024 Feb 21.
106. Fritz, J.R., Pelaez, D., Cheung, H.S., Current challenges in cartilage tissue engineering: A review of current cellular-based therapies. *Curr. Rheumatol. Rev.*, 5, 1, 8–14, 2009 Feb 1.
107. King, N.M. and Perrin, J., Ethical issues in stem cell research and therapy. *Stem Cell Res. Ther.*, 5, 1–6, 2014 Dec.
108. Johnstone, B., Alini, M., Cucchiarini, M., Dodge, G.R., Eglin, D., Guilak, F., Madry, H., Mata, A., Mauck, R.L., Semino, C.E., Stoddart, M.J., Tissue engineering for articular cartilage repair—the state of the art. *Eur. Cell Mater.*, 25, 248, e67, 2013 May 2.

109. Lammi, M.J., Piltti, J., Prittinen, J., Qu, C., Challenges in fabrication of tissue-engineered cartilage with correct cellular colonization and extracellular matrix assembly. *Int. J. Mol. Sci.*, 19, 9, 2700, 2018 Sep 11.
110. Jeyaraman, M., Jeyaraman, N., Nallakumarasamy, A., Ramasubramanian, S., Yadav, S., Critical challenges and frontiers in cartilage tissue engineering. *Cureus*, 16, 1, 2024 Jan 28.
111. Joshi, S., Rawat, K., Rajamohan, V., Mathew, A.T., Koziol, K., Thakur, V.K., Ass, B., 4D printing of materials for the future: Opportunities and challenges. *Appl. Mater. Today*, 18, 100490, 2020 Mar 1.
112. Sundaram, S., Kim, D.S., Baldo, M.A., Hayward, R.C., Matusik, W., 3D-printed self-folding electronics. *ACS Appl. Mater. Interfaces*, 9, 37, 32290–8, 2017 Sep 20.
113. Chung, S., Song, S.E., Cho, Y.T., Effective software solutions for 4D printing: A review and proposal. *Int. J. Precis. Eng. Manuf. Green Technol.*, 4, 359–71, 2017 Jul.
114. Javaid, M. and Haleem, A., 4D printing applications in medical field: a brief review. *Clin. Epidemiol. Glob. Health*, 7, 3, 317–21, 2019 Sep 1.
115. Dababneh, A.B. and Ozbolat, I.T., Bioprinting technology: a current state-of-the-art review. *J. Manuf. Sci. Eng.*, 136, 6, 061016, 2014 Dec 1.
116. Kumar Gupta, D., Ali, M.H., Ali, A., Jain, P., Anwer, M.K., Iqbal, Z., Mirza, M.A., 3D printing technology in healthcare: applications, regulatory understanding, IP repository and clinical trial status. *J. Drug Targeting*, 30, 2, 131–50, 2022 Feb 7.
117. Vali, A., Aristova, M., Vakil, P., Abdalla, R., Prabhakaran, S., Markl, M., Ansari, S.A., Schnell, S., Semi-automated analysis of 4D flow MRI to assess the hemodynamic impact of intracranial atherosclerotic disease. *Magn. Reson. Med.*, 82, 2, 749–62, 2019 Aug.
118. Shen, B., Erol, O., Fang, L., Kang, S.H., Programming the time into 3D printing: current advances and future directions in 4D printing. *Multifunct. Mater.*, 3, 1, 012001, 2020 Jan 23.

Index

Printed and bound by CPI Group (UK) Ltd, Croydon, CR0 4YY

29/09/2026

14971772-0002